# CONTEMPORARY MATHEMATICS

## Titles in this Series

## Titles in this Series

**Volume**

21 **Topological methods in nonlinear functional analysis,** S. P. Singh, S. Thomeier, and B. Watson, Editors

22 **Factorizations of $b^n \pm 1$, $b = 2, 3, 5, 6, 7, 10, 11, 12$ up to high powers,** John Brillhart, D. H. Lehmer, J. L. Selfridge, Bryant Tuckerman, and S. S. Wagstaff, Jr.

23 **Chapter 9 of Ramanujan's second notebook—Infinite series identities, transformations, and evaluations,** Bruce C. Berndt and Padmini T. Joshi

24 **Central extensions, Galois groups, and ideal class groups of number fields,** A. Fröhlich

25 **Value distribution theory and its applications,** Chung-Chun Yang, Editor

26 **Conference in modern analysis and probability,** Richard Beals, Anatole Beck, Alexandra Bellow and Arshag Hajian, Editors

27 **Microlocal analysis,** M. Salah Baouendi, Richard Beals and Linda Preiss Rothschild, Editors

28 **Fluids and plasmas: geometry and dynamics,** Jerrold E. Marsden, Editor

29 **Automated theorem proving,** W. W. Bledsoe and Donald Loveland, Editors

30 **Mathematical applications of category theory,** J. W. Gray, Editor

31 **Axiomatic set theory,** James E. Baumgartner, Donald A. Martin and Saharon Shelah, Editors

32 **Proceedings of the conference on Banach algebras and several complex variables,** F. Greenleaf and D. Gulick, Editors

33 **Contributions to group theory,** Kenneth I. Appel, John G. Ratcliffe and Paul E. Schupp, Editors

34 **Combinatorics and algebra,** Curtis Greene, Editor

35 **Four-manifold theory,** Cameron Gordon and Robion Kirby, Editors

CONTEMPORARY MATHEMATICS

Volume 34

# Combinatorics and Algebra

Curtis Greene, Editor

AMERICAN MATHEMATICAL SOCIETY

Providence • Rhode Island

**PROCEEDINGS OF THE AMS-IMS-SIAM JOINT SUMMER RESEARCH CONFERENCE IN THE MATHEMATICAL SCIENCES ON COMBINATORICS AND ALGEBRA**

HELD AT THE UNIVERSITY OF COLORADO, BOULDER
JUNE 5–11, 1983

These proceedings were prepared by the American Mathematical Society with partial support from the National Science Foundation Grant MCS 8218075.

1980 *Mathematics Subject Classification*. Primary 05A15, 05A17, 05A19, 20C30, 22E46.

---

**Library of Congress Cataloging in Publication Data**

Main entry under title:

Combinatorics and algebra.

(Contemporary mathematics, ISSN 0271-4132; v. 34)

"Proceedings of the AMS-NSF Joint Summer Research Conference on Combinatorics and Algebra, held at the University of Colorado, Boulder, during June 5–11, 1983"—Introd.

Bibliography: p.

1. Algebra—Congresses. 2. Combinatorial analysis—Congresses. 3. Representations of groups—Congresses. 4. Representations of algebras—Congresses. I. Greene, Curtis. II. AMS-NSF Joint Summer Research Conference on Combinatorics and Algebra (1983; University of Colorado) III. American Mathematical Society. IV. National Science Foundation (U.S.) V. Series; Contemporary mathematics (American Mathematical Society); v. 34.

QA150.C647 1984 513 84-18608
ISBN 0-8218-5029-6 (alk. paper)

---

TABLE OF CONTENTS

TABLE OF CONTENTS

# INTRODUCTION

This volume contains the Proceedings of the AMS-NSF Joint Summer Research Conference on COMBINATORICS AND ALGEBRA held at the University of Colorado, Boulder, during June 5-11, 1983. The conference organizing committee consisted of Adriano Garsia, Curtis Greene, Gian-Carlo Rota, and me.

Although combinatorial techniques have pervaded the study of algebra throughout its history, it is only in recent years that any kind of systematic attempt has been made to understand the connections between algebra and combinatorics. Both subjects have been greatly enriched as a result of this undertaking. The present Research Conference drew together specialists in both algebra and combinatorics, and provided an invaluable opportunity for them to collaborate.

The topic most discussed at the conference was representation theory, in particular the representation theory of the symmetric group and complex general linear group. The close connections with combinatorics, especially the theory of Young tableaux, was evident from the pioneering work of G. Frobenius, I. Schur, A. Young, H. Weyl, and D. E. Littlewood. In response to popular demand Phil Hanlon gave an introductory survey of this subject, whose inclusion in this volume should make many of the remaining papers more accessible to a reader with little background in representation theory.

Ten of the papers in this volume (excluding Hanlon's survey) impinge on representation theory in various ways. Some are directly concerned with the groups, Lie algebras, etc., themselves, while others deal with purely combinatorial topics which arose from representation theory and suggest the possibility of a deeper connection between the combinatorics and the algebra.

The remaining papers are concerned with a wide variety of topics. There are valuable surveys on the classical subject of hyperplane arrangements and its recently discovered connections with lattice theory and differential forms, and on the surprising connections between algebra, topology, and the counting of faces of convex polytopes and related complexes. There also appears an instructive example of the interplay between combinatorial and algebraic properties of finite lattices, and an interesting illustration of combinatorial reasoning to prove a fundamental algebraic identity.

In addition to the talks represented by papers in this volume, there were three other stimlulating lectures by I. G. Macdonald, David Anick, and Ranee Gupta. A highly successful problem session was held during the conference, and a list of the problems presented appears at the end of this volume.

As chairman of the Organizing Committee for the Research Conference I wish to thank the American Mathematical Society for its support, and especially Mrs. Carole Kohanski for her unending efforts to insure that the conference was a success. I am also grateful to Curtis Greene for serving as editor of these Proceedings.

Richard P. Stanley
Cambridge, MA
July, 1984

## LIST OF PARTICIPANTS

| | |
|---|---|
| AISSEN, Michael | Rutgers University |
| ANICK, David | M.I.T. |
| BANNAI, Eiichi | Ohio State University |
| BAYER, Margaret | Cornell University |
| BENNETT, M.K. | University of Massachusetts |
| BILLERA, Louis J. | Cornell University |
| BJÖRNER, Anders | University of Stockholm |
| BOGART, Ken | Dartmouth College |
| CALDERBANK, Robert | Bell Labs |
| CHAO, C.Y. | University of Pittsburgh |
| CHEN, Young-Ming | Western Illinois University |
| COLLINS, Karen L. | M.I.T. and Bell Labs |
| COPPERSMITH, Don | IBM Research |
| CRAPO, Henry | University of Quebec |
| DECKHART, Robert W. | Miami University |
| DUFFUS, Dwight | Emory University |
| EALY, Clifton (Jr.) | Northern Michigan University |
| EDELMAN, Paul H. | University of Pennsylvania |
| FERGUSON, Pamela | University of Miami |
| GARSIA, Adriano | U.C.S.D. |
| GESSEL, Ira | M.I.T. |
| GREENE, Curtis | Haverford College |
| GRIGGS, Jerry | University of South Carolina |
| GUPTA, Ranee | I.A.S., Brown University |
| HANLON, Philip | M.I.T. |
| HILLER, Howard | Columbia University |
| HSU, D. Frank | Fordham University |
| KAHN, Jeff | Rutgers University |
| KLEINSCHMIDT, Peter | University of Washington |
| KUNG, Joseph | Northern Texas State University |
| KUSTIN, Andrew | University of South Carolina |
| LEE, Carl | University of Kentucky |

## LIST OF PARTICIPANTS

| | |
|---|---|
| LIEBLER, Robert | Colorado State University |
| MACDONALD, Ian G. | Queen Mary College, Univ. of London |
| MC FARLAND, Robert L. | |
| MERRIS, Russell | California State University, Hayward |
| MEYEROWITZ, Aaron | Colorado State University |
| MISRA, Kailash | University of Virginia |
| NIEDERHAUSEN, Heinrich | University of Toronto |
| PARSHALL, Brian | University of Virginia |
| PROCTOR, Robert | U.C.L.A. |
| REMMEL, Jeffrey | U.C.S.D. |
| REGEV, Amitai | Brandeis University |
| RIVAL, Ivan | University of Calgary |
| ROTA, Gian-Carlo | M.I.T. |
| SAGAN, Bruce | University of Michigan |
| SCHMIDT, Frank | Southern Illinois University |
| SIMION, Rodica | Southern Illinois University |
| SMITH, Kenneth | Colorado State University |
| SOLOMON, Louis | University of Wisconsin |
| STANLEY, Richard P. | M.I.T. |
| STRUIK, Ruth R. | University of Colorado |
| SULANKE, Robert | Boise State University |
| SUNLEY, Judith | National Science Foundation |
| SUTHERLAND, David | North Texas State University |
| SVRTAN, Dragutin | U.C. Berkeley |
| TERAO, Hiroaki | University of Wisconsin |
| TOWBER, Jacob | Yale |
| VINCE, Andrew | University of Florida |
| WACHS, Michelle | University of Miami |
| WALKER, James | Caltech |
| WANG, Stuart | Oakland University |
| WEBB, Ursula | University of Illinois |
| WHITE, Dennis | University of Minnesota |
| WHITE, Neil | University of Florida |
| WORLEY, Dale R. | M.I.T. |
| YUZVINSKY, Sergey | University of Oregon |

Contemporary Mathematics
Volume 34, 1984

# AN INTRODUCTION TO THE COMPLEX REPRESENTATIONS OF THE SYMMETRIC GROUP AND GENERAL LINEAR LIE ALGEBRA

Phil Hanlon

This article is an introduction to some of the topics discussed at this conference. It assumes a familiarity with linear algebra but no formal training in representation theory. It is not intended as a survey paper in representation theory, to the contrary it emphasizes results pertinent to the talks given at the conference. In particular the approach throughout is very combinatorial. It is hoped that this article will help place in perspective the more technical results to follow.

## Part I. Characters of the Symmetric Groups.

### SECTION 1: REPRESENTATIONS OF FINITE GROUPS

Let G be a finite group with identity e, and let V be a finite dimensional complex vector space. By $G\ell(V)$ we mean the group of nonsingular linear transformations of V.

DEFINITION 1: A representation $\phi$ of G on V is a group homomorphism of G into V. We say that V is a G-module.

Note that $\phi(e)$ is the identity transformation and that $\phi(g^{-1}) = \phi(g)^{-1}$ for all $g \varepsilon G$.

One simple example of a representation is the case where V is one dimensional and $\phi(g)$ is the identity map for all $g \varepsilon G$. This is called the trivial representation of G. Note that trace $(\phi(g)) = 1$ for all $g \varepsilon G$.

We obtain another example of a representation by letting V be the complex vector space of dimension $|G|$ having basis $\{x_{g_1}, x_{g_2}, \ldots\}$ indexed by the group elements. Define $\phi$: $G \to G\ell(V)$ by

$$\phi(h)(x_g) = x_{gh^{-1}}$$

Then $\phi(h_1)\phi(h_2)(x_g) = \phi(h_1)x_{gh_2^{-1}} = x_{gh_2^{-1}h_1^{-1}} = \phi(h_1h_2)(x_g)$

so $\phi$ is a group homomorphism. We call $\phi$ the regular representation. Note that


0271-4132/84 $1.00 + $.25 per page

$$\text{trace}(\phi(h)) = \begin{cases} |G| & \text{if } h = e \\ 0 & \text{if } h \neq e. \end{cases}$$

Let $\phi: G \to G\ell(V)$ and $\psi: G \to G\ell(W)$ be representations of G. We say $\phi$ and $\psi$ are similar if there exists a nonsingular linear transformation S from V onto W such that $S\phi(g) = \psi(g)S$ for all $g \in G$. We consider similar representations to be the same.

A subspace W of V is G-invariant if $\phi(g)W = W$ for all $g \in G$. In this case, W is a G submodule of V. We say V is irreducible if the only G-invariant subspaces of V are $\{0\}$ and V.

A fundamental theorem of representation theory is the following result known as Maschke's theorem.

THEOREM 1: ([3], pg. 21) Let V be a finite dimensional G-module. Then we can write $V = V_1 \oplus \ldots \oplus V_\ell$ where each $V_i$ is an irreducible G-submodule of V.

The decomposition of V as a direct sum of irreducibles is not unique. However if W is an irreducible G-module, the number of times W appears in the direct sum is unique. This number is called the multiplicity of W in V and is denoted $m_W(V)$. Thus we can write

$$V \cong \bigoplus_W m_W(V)W$$

in a unique way where the sum is over the irreducible G-modules W and where $m_W(V)W$ denotes the direct sum of $m_W(V)$ copies of W. To decompose V as a G-module is to write V in the above form (i.e. to determine the multiplicities $m_W(V)$).

One needs very little information about the representation $\phi$ to decompose V.

DEFINITION 2: The character $\chi_V$ of $\phi$ is the function from G to given by

$$\chi_V(g) = tr(\phi(g)).$$

We say $\chi_V$ is an irreducible character of G if V is an irreducible representation. There are several observations that can be made at this point.

(1) Since $\phi(e)$ is the identity, $\chi_V(e) = \dim(V)$.

(2) If $V = U_1 \oplus U_2$ where $U_1$ and $U_2$ are G submodules then $\chi_V = \chi_{U_1} + \chi_{U_2}$. In particular

$$\chi_V = \sum_W m_W(V)\chi_W.$$

(3) If g,h are in G then $\phi(ghg^{-1}) = \phi(g)\phi(h)\phi(g)^{-1}$. Consequently $\phi(ghg^{-1})$ and $\phi(h)$ are similar matrices hence have the same eigenvalues. So

$$\chi_V(ghg^{-1}) = \chi_V(h).$$

(4) Consider the two examples of representations given earlier.

(A) If V is the trivial representation we have $\chi_V(g) = 1$ for all V. In this case V is clearly irreducible so $\chi_V$ is an irreducible character.

(B) If V is the regular representation we have $\chi_V(e) = |G|$ and $\chi_V(g)=0$ for $g\neq e$.

A <u>class function</u> of G is a function from G to $\mathbb{C}$ which is constant on conjugacy classes. By (3), each character of G is a class function. Let $CF_G$ denote the vector space of class functions of G. Clearly the dimension of $CF_G$ is the number of conjugacy classes of G. Define the non-degenerate, symmetric bilinear form $<,>$ on $CF_G$ by

$$\langle U,V\rangle = \frac{1}{|G|} \sum_{g\in G} U(g)\, \overline{V(g)}.$$

<u>THEOREM 2</u>: Let $\Phi$ denote the set of irreducible characters of G. Then

(A) ([3], pg.49) $|\Phi|$ is the number of conjugacy classes of G.

(B) ([3], pg.39) If $\chi,\phi\in\Phi$ then $\langle\chi,\phi\rangle = \delta_{\chi\phi}$.

(C) $\Phi$ is an orthonormal basis for $CF_G$.

(D) If V is a G-module and W is an irreducible G-module then $m_W(V) = \langle\chi_V,\chi_W\rangle$.

It is customary to list the values of the irreducible characters of G in an array whose rows are indexed by characters and whose columns are indexed by the conjugacy classes of G. This array is called the <u>character table</u> of G. We always index the first row of this table by the trivial character and the first column by the conjugacy class containing only the identity e of G. So the first row of our character table is all 1's and the first column consists of non-negative integers, these being the dimensions of the irreducible G-modules. It is an interesting fact that columns of this matrix are orthogonal ([3],pg.51).

To end this section we decompose the regular representation of G. Let V be the regular representation and let $\rho$ be its character. Let W be any irreducible representation with character $\chi$. Then

$$m_W(V) = \langle\rho,\chi\rangle = \frac{1}{|G|} \sum_{\sigma\in G} \rho(\sigma)\, \overline{\chi(\sigma)}$$

$$= \frac{1}{|G|} \langle\rho(e),\chi(e)\rangle \qquad (\text{since } \rho(g) = 0 \text{ for } g\neq e)$$

$$= \chi(e) = \dim(W).$$

So in the regular representation each irreducible G-module W occurs as many times as its dimension.

SECTION 2: THE CHARACTERS OF $S_f$

We now turn to the case where G is $S_f$. It is well-known that two permutations are conjugate in $S_f$ if and only if they have the same cycle structure. Hence the conjugacy classes of $S_f$ are indexed by the set $\pi(f)$ of partitions of the integer f. We write partitions in the form $1^{j_1}2^{j_2}3^{j_3}\ldots$ which means that the partition is composed of $j_1$ ones, $j_2$ twos, etc. We will also identify partitions $\alpha$ by their Ferrers' diagrams $F_\alpha$. We draw a row of i boxes for each part i in the partition and arrange these rows from longest on top to shortest on bottom. Thus the Ferrer's diagram of $1^2 2^2 4$ is

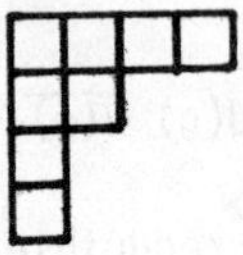

For a partition $\alpha$, let $\alpha_i$ denote the length of the $i^{th}$ row of its Ferrer's diagram.

By Theorem 2(A), there are as many irreducible characters of $S_f$ as there are partitions in $\pi(f)$. In fact there is a natural indexing $\alpha \to \chi^\alpha$ of irreducible characters by partitions which lets one compute the character values $\chi^\alpha(\beta)$ by a simple combinatorial rule.

DEFINITION 3: Let F be a Ferrer's diagram. A rim hook R of F is a rook-wise connected section of the southeast boundary of F with the property that $F\setminus R$ is again a Ferrer's diagram.

In the examples below, the shaded portions R of the first two Ferrer's diagrams are rim hooks. The shaded portions of the latter two are not.

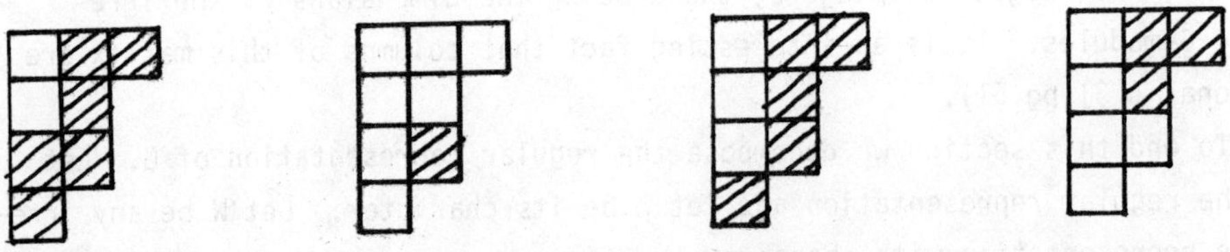

$\ell(R) = 6$ $h(R) = 4$ | $\ell(R) = 1$ $h(R) = 0$ | (not rook-wise connected) | ($F\setminus R$ not a Ferrer's diagram)

Let R be a rim hook of F. Define $\ell(R)$ to be the total number of boxes in R and define h(R) to be j-i where j is the last row containing a box of R and i is the first row containing a box of R.

The following remarkable theorem allows one to compute the value of the character $\chi^\alpha$ on the conjugacy class $\beta$ for $\alpha,\beta \in \pi(f)$.

<u>THEOREM 3:</u> ([2],pg.79). Let $\alpha$ and $\beta$ be in $\pi(f)$. Then

$$\chi^{\alpha}(\beta) = \sum_{R_1,\ldots,R_k} (-1)^{h(R_1)+\ldots+h(R_k)}$$

where the sum is over all $R_1,\ldots,R_k$ such that for each i, $\ell(R_i) = \beta_i$ and $R_i$ is a rim hook of $F\setminus(R_1 \cup R_2 \cup \ldots \cup R_{i-1})$.

As an example, let $\alpha$ = 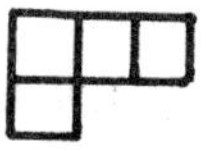 and let $\beta = 2^2$. There is a unique way to successively remove two rim hooks of length 2 from $\alpha$ this being 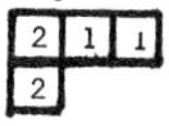 . Here we put 1's in the boxes occupied by $R_1$ and 2's in the boxes occupied by $R_2$. Thus $\chi$  $(2^2) = (-1)^{0+1} = -1$. There is no way to remove a rim hook of length 3 from  so $\chi$  $(31) = 0$. The reader is invited to work out that the remaining values of $\chi$  $(\beta)$ are

| $\beta$ | $1^4$ | $1^2 2$ | $2^2$ | 31 | 4 |
|---|---|---|---|---|---|
| $\chi$  $(\beta)$ | 3 | 1 | -1 | 0 | -1 |

Theorem 3 has some surprising corollaries.

<u>COROLLARY 3.1.</u> Let $\alpha$ and $\beta$ be in $\pi(f)$. Then

(A) $\chi^{\alpha}(\beta)$ is an integer

(B) $\chi^{f}(\beta) = 1$

(C) $\chi^{1^f}(\beta) = \mathrm{sgn}(\beta)$ .

(D) $\chi^{\alpha^t}(\beta) = \mathrm{sgn}(\beta)\chi^{\alpha}(\beta)$ where $\alpha^t$ is the partition obtained from $\alpha$ by reflecting across the main diagonal.

(E) $\chi^{\alpha}(1^f)$ is the number of standard Young tableau of shape $\alpha$ .

The proofs of these five statements are quite easy given the last theorem. Statement (A) is obvious. For (B) and (C) note that there is only one way to remove rim hooks of length $\beta_1,\ldots,\beta_k$ from  or from  . In the first case, each rim hook is a row (of height 0) and in the latter case each rim hook is a column (of heights $(\beta_1-1),\ldots,(\beta_k-1)$). To prove (D) note that the removal of the rim hooks $R_1,\ldots,R_k$ from $\alpha$ can be put in correspondence with the removal of the rim hooks $R_1^t,\ldots,R_k^t$ from $\alpha^t$. Since $h(R_i) + h(R_i^t) = \ell(R_i)=1 = \beta_i-1$ the result follows. To prove (E) note that $\chi^{\alpha}(1^f)$ is the number of ways to remove the f rim hooks $R_1,\ldots,R_f$ all of length 1. To each such removal, associate the standard Young tableau having an $f+1-i$ in the box corresponding to $R_i$. It is clear that this association sets up a one to one correspondence between terms in $\chi^{\alpha}(1^f)$ and standard Young tableau of shape $\alpha$.

The reader can check, using the rim-hook method, that the character table for $S_4$ is the following:

| $\alpha \backslash \beta$ | $1^4$ | $1^2 2$ | $2^2$ | $13$ | $4$ |
|---|---|---|---|---|---|
| $4$ | 1 | 1 | 1 | 1 | 1 |
| $31$ | 3 | 1 | -1 | 0 | -1 |
| $2^2$ | 2 | 0 | 2 | -1 | 0 |
| $21^2$ | 3 | -1 | -1 | 0 | 1 |
| $1^4$ | 1 | -1 | 1 | 1 | -1 |

The rim hook method shows how to determine the values of the character $\chi^\alpha$. However it does not prove that these functions $\chi^\alpha$ are characters. To do so, we must construct an $S_f$-module $S^\alpha$ having $\chi^\alpha$ as its character.

DEFINITION 4: Let $\alpha$ be in $\pi(f)$ and let $F_\alpha$ be the Ferrer's diagram of $\alpha$. An <u>$\alpha$-tableau</u> t is a filling of $F_\alpha$ with the numbers $1,2,\ldots,f$. Clearly there are $f!$ distinct $\alpha$-tableau. We say t is <u>standard</u> if the numbers in t increase along each row and column.

Given a box x in F, the <u>hook</u> of x is the set of all boxes which are either directly to the right or directly below x (including x itself). The <u>hook-length</u> $h_x$ is the size of the hook of x. For example, in the Ferrer's diagram below, each box x is filled with its hook-length $h_x$.

| | | |
|---|---|---|
| 4 | 3 | 1 |
| 2 | 1 | |

The following amazing theorem is due to Frame, Robinson and Thrall.

<u>THEOREM 4</u>: ([2], pg.77) The number of standard $\alpha$-tableau of is $f!/(\prod_{x \in F} h_x)$.

In the example above we have that f! divided by the product of the hook-lengths is $\frac{5!}{4\cdot 3\cdot 2\cdot 1\cdot 1} = 5$. The five standard $\alpha$-tableau are given below.

| | | | | |
|---|---|---|---|---|
| 123 | 124 | 125 | 134 | 135 |
| 45 | 35 | 34 | 25 | 24 |

An <u>$\alpha$-tabloid</u> is an equivalence class of $\alpha$-tableaux up to row equivalence. If t is an $\alpha$-tableau we write $\{t\}$ for the tabloid containing t. When we write an $\alpha$-tabloid we put lines between the rows to indicate that this is a tabloid, not a tableau. For example, $\begin{array}{c}\hline 12\\ \hline 34\\ \hline\end{array}$ indicates the tabloid containing $\begin{array}{c}12\\ 34\end{array}$ and so we have $\begin{array}{c}\hline 12\\ \hline 34\\ \hline\end{array} = \begin{array}{c}\hline 21\\ \hline 34\\ \hline\end{array} = \begin{array}{c}\hline 12\\ \hline 43\\ \hline\end{array} = \begin{array}{c}\hline 21\\ \hline 43\\ \hline\end{array}$.

Let $M^\alpha$ denote the complex vector space with basis the set of $\alpha$-tabloids. Clearly we have

$$\dim M^\alpha = \frac{f!}{\alpha_1!\alpha_2!\ldots\alpha_\ell!} .$$

It will turn out that the irreducible $S_f$-module $S^\alpha$ is a submodule of $M^\alpha$. We need to first define a representation of $S_f$ on $M^\alpha$.

For $\sigma\varepsilon S_f$ and t an $\alpha$-tableau, define $t\sigma$ to be the tableau obtained from t by permuting the entries according to $\sigma$. For example if $\sigma = (1,3,4)(2,5)$ then

$$\binom{125}{34}\sigma = \begin{matrix}352\\41\end{matrix} \qquad \text{and} \qquad \binom{521}{43}\sigma = \begin{matrix}253\\14\end{matrix}$$

Note that for the tableau $t_1$ and $t_2$ above we have $\{t_1\} = \{t_2\}$ and $\{t_1\sigma\}=\{t_2\sigma\}$. In fact, for all $\alpha$-tableaux $t_1,t_2$ and all $\sigma\varepsilon S_f$ we have $\{t_1\} = \{t_2\}$ if and only if $\{t_1\sigma\} = \{t_2\sigma\}$.

DEFINITION 5: Define $\phi$: $S_f \to G\ell(M^\alpha)$ by

$$\phi(\sigma)\{t\} = \{t\sigma\}.$$

This definition of $\phi$ makes $M^\alpha$ into an $S_f$-module.

To define $S^\alpha$ we need to consider $C_\alpha$, the group of all column permutations of the boxes of $F_\alpha$. $C_\alpha$ is isomorphic to a direct product of symmetric groups,

$$C_\alpha \cong S_{(\alpha^t)_1} \times S_{(\alpha^t)_2} \times \ldots \times S_{(\alpha^t)_k}.$$

For $\pi\varepsilon C_\alpha$ and t an $\alpha$-tableau let $t\pi$ be the $\alpha$-tableau obtained from t by permuting the entries according to $\pi^{-1}$. So i appears in box x of t if and only if i appears in box $\pi^{-1}(x)$ of $t\pi$.

DEFINITION 6: For t an $\alpha$-tableau, define $e_t \varepsilon M^\alpha$ by

$$e_t = \sum_{\pi\varepsilon C_\alpha} \text{sgn}(\pi)\{t\pi\}.$$

Let $S^\alpha$ be the subspace of $M^\alpha$ spanned by the $e_t$. It is not clear that $S^\alpha$ is $S_f$-invariant. However one can show that $\phi(\sigma)(e_t) = e_{t\sigma}$ (for $\sigma\varepsilon S_f$) hence $S^\alpha$ is an $S_f$-submodule of $M^\alpha$. We call $S^\alpha$ the <u>Specht module</u> associated with the shape $\alpha$.

<u>THEOREM 5</u>: Let $\alpha$ be in $\pi(f)$. Then

(A) ([2],pg.16) $S^\alpha$ is the irreducible $S_f$-module with character $\chi^\alpha$.

(B) ([2],pg.29) $S^\alpha$ has basis the set of $e_t$ where t is a standard $\alpha$-tableau. In particular

$$\dim S^\alpha = \frac{f!}{\prod_{x\varepsilon F_\alpha} h_x}$$

(C) ([2],pg.51) $S^\alpha$ occurs with multiplicity one in $M^\alpha$.

(D) ([2],pg.51) If $S^\alpha$ occurs with nonzero multiplicity in $M^\beta$ then $\beta$ is greater than or equal to $\alpha$ in the dominance ordering on partitions.

For a proof of Theorem 5 and a more detailed discussion of Specht modules, the reader is encouraged to look in James [2]. We end this section with the example $\alpha$ =  .

For $\alpha =$ 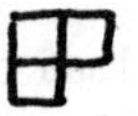 we have $f = 3$. The rim hook method gives that the character $\chi^\alpha$ has values $\chi^\alpha(1^3) = 2$, $\chi^\alpha(12) = 0$ and $\chi^\alpha(3) = -1$. According to Theorem 5(B), $S^\alpha$ has dimension 2 and basis $\{\beta_1, \beta_2\}$ where

$$\beta_1 = e_{\begin{smallmatrix}13\\2\end{smallmatrix}} = \begin{smallmatrix}\overline{13}\\ \overline{\underline{2}}\end{smallmatrix} - \begin{smallmatrix}\overline{23}\\ \overline{\underline{1}}\end{smallmatrix}$$

and $$\beta_2 = e_{\begin{smallmatrix}12\\3\end{smallmatrix}} = \begin{smallmatrix}\overline{12}\\ \overline{\underline{3}}\end{smallmatrix} - \begin{smallmatrix}\overline{32}\\ \overline{\underline{1}}\end{smallmatrix} .$$

We will compute the matrices of (1)(2)(3), (1,2)(3), and (1,2,3) with respect to this basis in order to evaluate the character of $S_3$ on $S^\alpha$.

Let tr denote the character of $S_3$ acting on $S^\alpha$. We have

1. (1)(2)(3): $\beta_1 \to \beta_1$ so $\phi((1)(2)(3)) = \begin{pmatrix}1&0\\0&1\end{pmatrix}$ and $\mathrm{tr}((1)(2)(3)) = 2$

$\beta_2 \to \beta_2$

2. (1,2)(3): $\beta_1 \to \begin{smallmatrix}\overline{23}\\ \overline{\underline{1}}\end{smallmatrix} - \begin{smallmatrix}\overline{13}\\ \overline{\underline{2}}\end{smallmatrix} = -\beta_1$

: $\beta_2 \to \begin{smallmatrix}\overline{21}\\ \overline{\underline{3}}\end{smallmatrix} - \begin{smallmatrix}\overline{31}\\ \overline{\underline{2}}\end{smallmatrix} = \beta_2 - \beta_1$ .

Thus $\phi((1,2)(3)) = \begin{pmatrix}-1&-1\\0&1\end{pmatrix}$ and $\mathrm{tr}((1,2)(3)) = 0$.

3. (1,2,3): $\beta_1 \to \begin{smallmatrix}\overline{21}\\ \overline{\underline{3}}\end{smallmatrix} - \begin{smallmatrix}\overline{31}\\ \overline{\underline{2}}\end{smallmatrix} = \beta_2 - \beta_1$

$\beta_2 \to \begin{smallmatrix}\overline{23}\\ \overline{\underline{1}}\end{smallmatrix} - \begin{smallmatrix}\overline{13}\\ \overline{\underline{2}}\end{smallmatrix} = -\beta_1$.

Thus $\phi((1,2,3)) = \begin{pmatrix}-1&-1\\1&0\end{pmatrix}$ and $\mathrm{tr}((1,2,3)) = -1$.

Hence tr equals $\chi^\alpha$ and so $S^\alpha$ is the irreducible $S_3$-module with character $\chi^\alpha$.

## SECTION 3: SYMMETRIC FUNCTIONS

Let $\chi$ be a character of $S_f$ (not necessarily irreducible) and let n be a positive integer. Define the characteristic of $\chi$, denoted $ch_\chi(x_1,\ldots,x_n)$ in the variables $x_1,\ldots,x_n$ by

$$ch_\chi(x_1,\ldots,x_n) = \frac{1}{f!} \sum_{\sigma \in S_f} \chi(\sigma)\left(\prod_{i=1}^{n} p_i(\underline{x})^{j_i(\sigma)}\right)$$

where $p_i(\underline{x})$ is the $i^{th}$ power sum symmetric function

$$p_i(\underline{x}) = (x_1^i + x_2^i + \ldots + x_n^i)$$

and $j_i(\sigma)$ is the number of i-cycles of $\sigma$.

We consider two examples.

1. Suppose $\rho$ is the character of the regular representation. So $\rho(e) = f!$ and $\rho(\sigma) = 0$ for $\sigma \neq e$. Then

$$ch_\rho(x_1,\dots,x_n) = (x_1+\dots+x_n)^f.$$

2. Suppose $f = 3$, $n = 2$ and $\chi = \chi$  .

Then $ch_\chi(x_1,x_2) = \frac{1}{6}(2(x_1+x_2)^3 - 2(x_1^3+x_2^3))$

$$= x_1^2x_2 + x_1x_2^2 .$$

Since each power sum symmetric function $p_i(\underline{x})$ is a symmetric function of $x_1,\dots,x_n$, the same is true of $ch_\chi(\underline{x})$.

DEFINITION 7: For $\alpha$ a partition of $f$ and $n$ a positive integer define $s_\alpha(x_1,\dots,x_n)$ to be the characteristic of the irreducible character $\chi^\alpha$.

A semistandard $\alpha$-tableau $t$ of type $1^n$ is a filling of $F_\alpha$ with numbers between 1 and $n$ so that the resulting tableau is weakly increasing along rows and strictly increasing along columns. For $\alpha = 21$, the semistandard $\alpha$-tableaux of type $1^2$ are $\begin{smallmatrix}11\\2\end{smallmatrix}$ and $\begin{smallmatrix}12\\2\end{smallmatrix}$. The reader can compare this with $s_\alpha(x_1,x_2)$ computed in the above example.

For $t$ a tableau of type $1^n$ define $\underline{x}^t$ to be $x_1^{\lambda_1(t)} x_2^{\lambda_2(t)} \dots x_n^{\lambda_n(t)}$ where $\lambda_i(t)$ is the number of times $i$ occurs in the tableau $t$. An elegant formula relating $s_\alpha(x_1,x_2)$ to semistandard tableau is the following.

THEOREM 6 : (see [2], pg 48). Let $\alpha$ be a partition of $f$ and let $n$ be a positive integer. Then

$$ch_{\chi^\alpha}(x_1,\dots,x_n) = s_\alpha(x_1,\dots,x_n) = \sum_t \underline{x}^t$$

where the sum on the right is over all semistandard tableaux $t$ of shape $\alpha$ and type $1^n$.

One can deduce three facts from this result which are not at all obvious.

(A) Clearly there are no column strict tableau of shape $\alpha$ if $n$ is less than $\ell(\alpha)$. Thus

$$ch_{\chi^\alpha}(x_1,\dots,x_n) = 0 \quad \text{if } n<\ell(\alpha).$$

(B) The coefficient of each monomial in $ch_{\chi^\alpha}(x_1,\dots,x_n)$ is non-negative.

(C) $\sum_t \underline{x}^t$ is a symmetric function of $x_1,\dots,x_n$.

There is yet one more commonly used definition of $s_\alpha(x_1,\dots,x_n)$. By observation (A) above we may assume $n$ is greater than or equal to $\ell(\alpha)$. Extend $\alpha$ to a partition with $n$ parts by saying that $\alpha_i = 0$ for $\ell(\alpha)<i\leq n$.

THEOREM 7: (see [ 2],pg. 74) We can express $s_\alpha(\underline{x})$ as a quotient of determinants by

$$s_\alpha(\underline{x}) = \frac{\det(x_j^{\alpha_i+(n-i)-1})}{\det(x_j^{(n-i-1)})} .$$

In our example above we had $\alpha = 21$. Then the quotient of determinants above is

$$\frac{\det\begin{bmatrix} x_1^3 & x_2^3 \\ x_1 & x_2 \end{bmatrix}}{\det\begin{bmatrix} x_1 & x_2 \\ 1 & 1 \end{bmatrix}} = \frac{x_1^3x_2-x_1x_2^3}{x_1-x_2} = x_1x_2(x_1+x_2) .$$

The reader can check that this agrees with our earlier computation of $s_\alpha(x_1,x_2)$.

The symmetric function $s_\alpha(x_1,\ldots,x_n)$ is called the Schur function of shape $\alpha$ in the variables $x_1,\ldots,x_n$. These will reappear in Part II of this article.

One further fact of interest is that for $n \geq f$ the Schur functions $S_\alpha(x_1,\ldots,x_n)$ with $\alpha$ a partition of f are linearly independent and span the vector space of symmetric functions in $x_1,\ldots,x_n$ which are homogeneous of degree f. In particular for any character $\chi$ of $S_f$, there is a unique way to write $ch_\chi(x_1,\ldots,x_n)$ as a linear combination of Schur functions. Also it is clear that if $\chi = \sum_\alpha m_\alpha(\chi)\chi^\alpha$ then

$$ch_\chi(x_1,\ldots,x_n) = \sum_\alpha m_\alpha(\chi)S_\alpha(x_1,\ldots,x_n) .$$

Hence we have a new way to decompose the character $\chi$, this being to write its characteristic as a sum of Schur functions.

For example, if $\rho$ is the regular representation of $S_f$ then $\rho = \sum_\alpha f_\alpha \chi^\alpha$ where $f_\alpha$ is the number of standard $\alpha$ Young tableaux of shape $\alpha$. Put in terms of characteristics this fact is

$$(x_1+\ldots+x_n)^f = \sum_{\alpha\in\pi(f)} f_\alpha S_\alpha(x_1,\ldots,x_n) .$$

PART II: The general linear Lie algebra.

## SECTION 4: REPRESENTATIONS OF $gl_n(\mathbb{C})$.

Throughout this section n is fixed. Let g be the vector space $gl_n(\mathbb{C})$ of $n\times n$ complex matrices and let G be the group $Gl(n,\mathbb{C})$ of $n\times n$ invertible complex matrices. For V a finite dimensional complex vector space End(V) denotes the space of linear transformations from V to itself.

For $a,b \in End(V)$ let $[a,b]$ denote the commutator $ab-ba$. Then $End(V)$ with this bracket multiplication is a Lie algebra. Since $G$ is just $End(\mathbb{C}^n)$ we have that $g$ with $[\ ,\ ]$ is also a Lie algebra. (In general, a complex Lie algebra is a vector space over $\mathbb{C}$ with a multiplication $[\ ,\ ]$ on which satisfies

(a) $[\ ,\ ]$ is bilinear

(b) $[a,a] = 0$ for all $a$

(c) The Jacobi identity:

$$[a,[b,c]] + [c,[a,b]] + [b,[c,a]] = 0 \ .$$

One can show that if $(\mathfrak{a},\cdot)$ is an associative algebra then the multiplication $[a,b] = a\cdot b-b\cdot a$ makes $\mathfrak{a}$ into a Lie algebra).

DEFINITION 7: A representation $\phi$ of $g$ on $V$ is a linear transformation from $g$ into $End(V)$ which satisfies $\phi([a,b]) = [\phi(a),\phi(b)]$ for all $a,b \in g$. In this case we say $V$ is a $g$-module.

Consider the following three examples of $g$-modules.

1. Let $V = \mathbb{C}^n$ so $End(V) = g$. Define $\phi$ to be the identity map. Then $\phi$ obviously is a representation of $g$, called the defining representation.

2. Let $V = g$. Define $\phi(a)(c)$ to be $\phi(a)(c) = [a,c] = ac-ca$. It is easy to see that $\phi$ is linear. To check the bracket condition, note that $[\phi(a),\phi(b)](c) = \phi(a)(\phi(b)(c)) - \phi(b)(\phi(a)(c))$
$= [a,[b,c]] - [b,[a,c]]$
$= [a,[b,c]] + [b,[c,a]] + [c,[a,b]] - [c,[a,b]]$
$= 0 + [[a,b],c] = \phi([a,b])(c)$.

This representation of $g$ on itself is called the adjoint representation.

3. Let $z$ be a complex number and let $V$ be the one dimensional complex vector space with basis element $e$. Define the linear map $\phi$ by $\phi(a)(e) = (z \text{ trace } (a))e$. To check the bracket condition, note that
$$[\phi(a),\phi(b)]e = z^2 \,(\text{trace}(a)\ \text{trace}(b)-\text{trace}(b)\text{trace}(a))e = 0$$
But $\text{trace}([a,b]) = 0$ hence $[\phi(a),\phi(b)] = \phi([a,b])$. We call $\phi$ $z$ times the trace and denote it $\phi = \tau_z$.

A $g$-module $V$ is irreducible if it contains no nontrivial $\phi(g)$-invariant subspaces. A result similar to Maschke's theorem for representations of finite groups holds again in this situation. (The theorem is due to Weyl).

THEOREM 6: ([1],pg.28) Let $V$ be a finite dimensional $g$-module. Then we can write $V$ as a direct sum of irreducible $g$-submodules.

As before, the above decomposition of $V$ is not unique. However, if $W$ is an irreducible $g$-module then the multiplicity of $W$ in $V$ is well-defined. Our goal in this section is to classify and construct the irreducible $g$-modules and to learn how to find the multiplicity of an irreducible module in a $g$-module $V$.

The root-space decomposition of g is the vector space decomposition $g = g_- \oplus H \oplus g_+$ where $g_-$ is the set of strictly lower triangular matrices, H is the set of diagonal matrices and $g_+$ is the set of strictly upper triangular matrices. We write $y_{ij}$ for the matrix with a one in the (i,j) position and zeros elsewhere. Then

$$g_+ = \langle y_{ij}: i<j \rangle$$

and
$$g_- = \langle y_{ij}: i>j \rangle .$$

Also note that $[h_1,h_2] = 0$ for all $h_1,h_2 \in H$. The subspace H of g is called the torus.

Let $\phi$ be a representation of g on V. For $h_1,h_2 \in H$ we have

$$\phi(h_1)\phi(h_2) - \phi(h_2)\phi(h_1) = [\phi(h_1),\phi(h_2)] = \phi([h_1,h_2]) = 0 .$$

Hence $\{\phi(h): h \in H\}$ is a set of commuting linear transformations in End(V). Hence we can choose a basis $\{v_\alpha\}$ for V with respect to which, each $\phi(h)$ is a diagonal matrix. We call $\{v_\alpha\}$ a basis of weight vectors for V. For each $\gamma$, define $\lambda^{(\gamma)}: H \to \mathbb{C}$ by

$$(*) \qquad \phi(h)(v_\alpha) = \lambda^{(\gamma)}(h)(v_\alpha).$$

We call $\lambda^{(\alpha)}$ the weight of $v_\alpha$.

Recall that the map $\phi$ is linear. From this and equation (*) one sees that each $\lambda^{(\gamma)}$ is linear. So if we let $\lambda^{(\gamma)}{}_i = \lambda^{(\gamma)}(y_{i,i})$ we have that $\lambda^{(\gamma)}(\mathrm{diag}(u_1,\ldots,u_n)) = (u_1,\ldots,u_n)\cdot(\lambda_1^{(\gamma)},\lambda_2^{(\gamma)},\ldots,\lambda_n^{(\gamma)})$. We identify $\lambda^{(\gamma)}$ with the vector $(\lambda_1{}^{(\gamma)}, \lambda_2{}^{(\gamma)},\ldots,\lambda_n{}^{(\gamma)})$ and so think of weights as vectors in $\mathbb{C}^n$.

DEFINITION 8: Let V be a finite dimensional g-module and let $\{v_\gamma\}$ be a basis of weight vectors of V. For each $\lambda \in \mathbb{C}^n$ let $\mu_\lambda(V)$, the multiplicity of $\lambda$ in V, be the number of $v_\gamma$ with weight $\lambda$.

The multiplicities $\mu_\lambda(V)$ are similar to character values for representations of finite groups in that they completely determine the representation. They also have interesting symmetry properties. For $(w_1,\ldots,w_n) = w$ in $\mathbb{C}^n$ and $\sigma \in S_n$ let $w\sigma$ be the vector in $\mathbb{C}^n$ obtained by permuting the coordinates of w according to $\sigma^{-1}$.

THEOREM 7: Let $V_1$ and $V_2$ be finite dimensional g-modules.

(A) $V_1$ and $V_2$ are isomorphic (as g-modules) if and only if $\mu_\lambda(V_1) = \mu_\lambda(V_2)$ for all $\lambda$ in $\mathbb{C}^n$.

(B) The multiplicity $\mu_\lambda(V_1)$ is a symmetric function of $\lambda$, i.e. $\mu_{\lambda\sigma}(V_1) = \mu_\lambda(V_1)$ for all $\lambda$ in $\mathbb{C}^n$ and all $\sigma \in S_n$.

Our strategy for classifying all the finite dimensional g-modules will be to build them up from particularly simple g-modules called polynomial representations. The two methods we use for building more complicated modules from simple ones will be first to take tensor products and second to take direct sums.

DEFINITION 9: Let U and V be finite dimensional g-modules with associated representations $\phi$: $g \to End(U)$ and $\psi: g \to End(V)$. The tensor product representation $\phi \otimes \psi$: $g \to End(V \otimes U)$ is defined by $(\phi \otimes \psi)(a)(v \otimes u) = (\phi(a)(v)) \otimes u + v \otimes (\psi(a)(u))$. **

We call $V \otimes U$ (with this tensor product representation) the tensor product module.

Let $\{v_\gamma\}$ be a basis of weight vectors for V with weights $\lambda^{(\gamma)}$, and let $\{u_\pi\}$ be a basis of weight vectors for U with weights $\beta^{(\pi)}$. Then $\{v_\gamma \otimes u_\pi\}$ is a basis of weight vectors for $V \otimes U$. From (**) one sees that the vector $v_\gamma \otimes u_\pi$ has weight $\lambda^{(\gamma)} + \beta^{(\pi)}$. Thus we have the next lemma.

Lemma: Let V and U be finite dimensional g-modules, and let $\lambda$ be in $\mathbb{C}^n$. Then

$$\mu_\lambda(V \otimes U) = \sum_{\beta \in \mathbb{C}^n} \mu_{\lambda-\beta}(V)\mu_\beta(U).$$

As a special case of this lemma, consider U equal to the one dimensional representation $\tau_z$ for $z \in \mathbb{C}$. Letting $\hat{z}$ denote $(z,z,\ldots,z)$ we have

$$\mu_\lambda(V \otimes \tau_z) = \mu_{\lambda-\hat{z}}(V).$$

So tensoring V by the representation $\tau_z$ has the effect of translating multiplicities by the vector $\hat{z} \in \mathbb{C}^n$. The following remarkable theorem holds.

THEOREM 8: Let V be an irreducible finite dimensional g-module. Then there exists a complex number z such that

$$V = V_0 \otimes \tau_z$$

and $\mu_\lambda(V_0) = 0$ unless every component of $\lambda$ is a non-negative integer.

DEFINITION 10: A finite dimensional g-module V is called a polynomial module (or polynomial representation) provided that $\mu_\lambda(V) = 0$ unless every component of $\lambda$ is a non-negative integer.

We can now state the general picture of a finite dimensional g-module V. By Theorem 6 we can write $V = \bigoplus_W m_W(V)W$ where the sum is over irreducible g-modules W. Each W can in turn be written as $W = W_0 \otimes \tau_z$ for z a complex number and W an irreducible polynomial representation of g. To fully understand the structure of V we need to study the irreducible polynomial representations of g.

SECTION 5: THE POLYNOMIAL REPRESENTATIONS OF $g\ell_n(\mathbb{C})$

The irreducible polynomial representations of g are indexed by partitions having no more than n rows. For the rest of this section let $\alpha$ be a partition with n or fewer rows (there is no restriction on what number $\alpha$ is a partition of).

DEFINITION 9: An $\alpha$-semitableau t is a filling of $F_\alpha$ from the set of numbers $\{1,2,\ldots,n\}$. In a semitableau, some numbers in $\{1,2,\ldots,n\}$ may be used more than once and some numbers may not be used at all. The semitableau t is semistandard if it is weakly increasing along rows and strictly increasing down columns. Lastly, an $\alpha$-semitabloid is an equivalence class of $\alpha$-semitableaux up to row equivalence.

Let $\tilde{M}^\alpha$ denote the complex vector space with basis the set of $\alpha$-semitabloids. Define a representation $\phi_\alpha$: $g\to\mathrm{End}(\tilde{M}^\alpha)$ as follows: for $\{t\}$ an $\alpha$ semitabloid,

$$\phi_\alpha(y_{ij})\{t\} = \sum_{\ell\varepsilon\{t\}} \{t\}[\ell\to\delta_{j\ell}i].$$

where the sum is over all numbers $\ell$ in $\{t\}$ and where $\{t\}[\ell\to\delta_{j\ell}i]$ is the semi-tabloid obtained from $\{t\}$ changing $\ell$ to i if $\ell=j$. For example,

$$\phi_{32}(y_{13})\left(\frac{\overline{332}}{\underline{34}}\right) = \frac{\overline{132}}{\underline{34}} + \frac{\overline{312}}{\underline{34}} + \frac{\overline{332}}{\underline{14}}$$

whereas

$$\phi_{32}(y_{13})\left(\frac{\overline{221}}{\underline{24}}\right) = 0 .$$

It is not difficult to show that $\phi_\alpha$ is a representation of g. Note that $\phi_\alpha(y_{ii})\{t\} = \sum_{\ell\varepsilon\{t\}} \{t\}[\ell\to\delta_{i\ell}i] = \sum_{\ell\varepsilon\{t\}} \delta_{i\ell}\{t\} = \lambda_i^t\{t\}$ where $\lambda_i^t$ is the number of times i occurs as an entry in t. So the $\alpha$-semitabloids are a basis of weight vectors for $\tilde{M}^\alpha$.

DEFINITION 10: For t an $\alpha$-semitabloid, define $e_t\varepsilon M$ by

$$e_t = \sum_{\pi\varepsilon C_\alpha} \mathrm{sgn}(\pi)\{t\pi\}.$$

Let $\tilde{S}^\alpha$ denote the g-submodule of $\tilde{M}^\alpha$ spanned by the $e_t$.

Note that the multiset of numbers occurring in $\{t\pi\}$ equals the multiset of numbers occurring in $\pi$. Thus $\phi_\alpha(y_{ii})\{t\pi\} = \lambda_i^t\{t\pi\}$ hence

$$\phi_\alpha y_{ii}e_t = \lambda_i^t e_t.$$

So each $e_t$ is a weight vector in $\tilde{S}^\alpha$. One can show that $\tilde{S}^\alpha$ has basis the set of $e_t$ such that t is a semistandard $\alpha$-tableau. Each $e_t$ is a weight vector of weight $\lambda^{(t)} = (\lambda_1^{(t)},\ldots,\lambda_n^{(t)})$where $\lambda_i^{(t)}$ is the number of times that i occurs as an entry of t. We compile all these facts in the following theorem.

THEOREM 9: ([2], section 26) Fix a positive integer n and let $g = g\ell_n(\mathbb{C})$.

(A) The irreducible polynomial representations are indexed by partitions $\alpha$ having no more than n rows. Corresponding to such an $\alpha$ is the irreducible g-module $\tilde{S}^\alpha$ constructed in this section.

(B) A basis of weight vectors for $\tilde{S}^\alpha$ is the set $\{e_t\}$ where t is a semi-standard $\alpha$-tableau. The corresponding weight $\lambda^{(t)}$ has $i^{th}$ component equal to the number of times that i occurs in t.

## SECTION 6: FORMAL CHARACTERS

A finite dimensional g-module V will be called integral provided that $\mu_\lambda(V) = 0$ unless $\lambda$ is in $\mathbb{Z}^n$. Thus every polynomial representation of g is integral and every integral representation is of the form $V_0 \times \tau_z$ where $V_0$ is polynomial and z is an integer. For the remainder of this section every representation of g is assumed to be integral.

DEFINITION: Let V be a finite dimensional representation of g. Define the formal character $f_V(x_1,\ldots,x_n)$ of V to be the generating function for the multiplicities $\mu_\lambda(V)$,

$$f_V(x_1,\ldots,x_n) = \sum_{\lambda\in\mathbb{Z}^n} \mu_\lambda(V)\, x_1^{\lambda_1} x_2^{\lambda_2} \cdots x_n^{\lambda_n} .$$

Before proceeding, we compute the formal characters of the three examples earlier in the section.

1. First let $V = \mathbb{C}^n$ and let $\phi$ be the defining representation. A basis of weight vectors for V is the set $\{e_i\}$ of unit coordinate vectors. Moreover, the vector $e_i$ has weight $e_i$ hence in this case $f_V(x_1,\ldots,x_n) = x_1 + x_2 + \ldots + x_n$.

2. Let $V = g$ and let $\phi$ be the adjoint representation. It is easy to check that the matrix $y_{ij}$ is a weight vector in this module of weight $e_i - e_j$. Hence

$$f_V(x_1,\ldots,x_n) = \sum_{i=1}^{n} \sum_{j=1}^{n} \frac{x_i}{x_j} .$$

3. Let V be the one dimensional vector space <e> and let $\phi$ be $\tau_m$ where $m\in\mathbb{Z}$. Then e is a weight vector with weight $m(1,1,\ldots,1)$. Hence

$$f_V(x_1,\ldots,x_n) = (x_1x_2\ldots x_n)^m .$$

The following facts about formal characters are direct consequences of earlier discussion. Let V and W be g-modules.

(A) V and W are isomorphic g-modules if and only if $f_V(\underline{x}) = f_W(\underline{x})$.

(B) $f_V(\underline{x})$ is a symmetric function of the $x_i$'s.

(C) $\dim(V) = f_V(1,1,\ldots,1)$.

(D) $f_{V \otimes W}(\underline{x}) = f_V(\underline{x}) f_W(\underline{x})$.

(E) $V$ is a polynomial representation of $g$ if and only if $f_V(\underline{x})$ is a polynomial in $x_1,\dots,x_n$.

In view of fact (A) it is worth learning more about how to compute formal characters.

1. Let $\alpha$ be a partition with no more than $n$ rows. We consider here how to compute the formal character of the irreducible representation $\tilde{S}^{\alpha}$. A basis of weight vectors with corresponding weights is given by Theorem 9(B) in Section 5. It follows that

$$f_{\tilde{S}^{\alpha}}(\underline{x}) = \sum_{t} x_1^{\lambda_1(t)} x_2^{\lambda_2(t)} \cdots x_n^{\lambda_n(t)}$$

where the sum is over all semistandard $\alpha$ tableaux of type $1^n$ and $\lambda_i^{(t)}$ is number of times that $i$ occurs in $t$. From Theorem 6 in Section 3 it follows that $f_{\tilde{S}^{\alpha}}(\underline{x})$ is just the Schur function $s_{\alpha}(x_1,\dots,x_n)$!

$$F_{\tilde{S}^{\alpha}}(x_1,\dots,x_n) = s_{\alpha}(x_1,\dots,x_n)$$

For (2) through (6) let $V$ and $W$ be finite dimensional g-modules with $\{V_\gamma\}$ and $\{W_\delta\}$ bases of weight vectors.

2. The space $V \oplus W$ is naturally a g-module with basis of weight vectors $\{V_\gamma\} \cup \{W_\delta\}$. Hence

$$f_{V \oplus W}(\underline{x}) = f_V(\underline{x}) + f_W(\underline{x})$$

3. Suppose $\phi$: $g \to \mathrm{End}(V)$ is a representation of $g$ on $V$. There is a representation $\phi^*$ of $g$ on $V^*$ given by

$$(\phi^*(a)(f))(v) = f(-\phi(a)(v)).$$

For each $V_\gamma$, let $f_\gamma$ be the linear functional defined by $f_\gamma(v_\beta) = \delta_{\gamma\beta}$. Then

$$(\phi^*(y_{ii})(f_\beta))(v_\gamma) = f(-\lambda_i^{(\beta)} v_\beta) = -\lambda_i^{(\beta)} (f_\gamma(v_\beta))$$

$$= \begin{cases} -\lambda_i^{(\gamma)} v_\gamma & \text{if } \gamma = \beta \\ 0 & \text{if } \gamma \neq \beta \end{cases}$$

Hence $\phi^*(y_{ii})(f_\gamma) = -\lambda_i^{(\gamma)} f_\gamma$ so $f_\gamma$ is a weight vector in $V^*$ with weight $-\lambda^{(\gamma)}$. Since $\dim V = \dim V^*$ it follows that $\{f_\gamma\}$ is a basis of weight vectors for $V^*$ so

$$f_{V^*}(x_1,\dots,x_n) = f_V\left(\frac{1}{x_1}, \frac{1}{x_2},\dots,\frac{1}{x_n}\right) .$$

4. The tensor algebra T(V) of V is the direct sum of the $p^{th}$ tensor powers of V for p = 0,1,2,... . It follows that

$$f_{T(V)}(\underline{x}) = \frac{1}{1-f_V(\underline{x})} .$$

5. The symmetric algebra Sym(V) has a basis of weight vectors consisting of all monomials $(\prod_\gamma v_\gamma^{i_\gamma})$ in the $V_\gamma$. Such a monomial has weight $\sum_\gamma i_\gamma \lambda^{(\gamma)}$. Thus

$$f_{Sym(V)}(\underline{x}) = \prod_\lambda \left(\frac{1}{1-x_\lambda^{\mu_\lambda(V)}}\right) .$$

6. Lastly, let T be the set of $\gamma$ which index the $V_\gamma$. Then the exterior algebra Ext(V) of V has a basis of weight vectors consisting of all $x_{\gamma_1} \wedge x_{\gamma_2} \ldots \wedge x_{\gamma_s}$ where $\{\gamma_1,\ldots,\gamma_s\}$ is a subset of T. The weight of this vector is $\sum_{i=1}^{s} \lambda^{(\gamma_i)}$. Hence

$$f_{Ext(V)}(\underline{x}) = \prod_\lambda (1+x_\lambda)^{\mu_\lambda(V)} .$$

As an example let $V = \mathbb{C}^n$ considered as a g-module in the usual way. The unit coordinate vectors $e_1,\ldots,e_n$ are a basis of weight vectors. The weight vector $e_i$ has weight $e_i$ and the formal character of V is $f_V(\underline{x}) = x_1+\ldots+x_n$. Thus the formal character of $T^fV$ is just

$$f_{T^fV}(\underline{x}) = (x_1+\ldots+x_n)^f .$$

By the example given at the end of Section 2, Part 1 we have

$$f_{T^fV}(\underline{x}) = \sum_{\alpha \vdash f} f_\alpha s_\alpha(x_1,\ldots,x_n).$$

This proves the famous result due to Schur that the g-module $T^fV$ decomposes as the direct sum of the irreducible g-modules $\tilde{S}^\alpha$ for $\alpha$ a partition of f. Moreover each $\tilde{S}^\alpha$ occurs exactly $f_\alpha$ times.

In this example we used properties of representations of the symmetric group to deduce a nontrivial fact about a representation of $g\ell_n(\mathbb{C})$. This sort of interplay is fruitful in the study of representations of $g\ell_n(\mathbb{C})$. The theory of symmetric functions establishes the connection between representations of the symmetric group and the general linear Lie algebra by virtue of the fact that a symmetric function of degree f in $x_1,\ldots,x_n$ can be viewed as the characteristic of a representation of $S_f$ or as the formal character of a representation of $g\ell_n(\mathbb{C})$.

SECTION 7: REPRESENTATIONS OF THE GENERAL LINEAR GROUP.

One last thing worth mentioning is that the study of Lie algebra representations of $g\ell_n(\mathbb{C})$ is equivalent to the study of group representations of $G\ell(n,\mathbb{C})$. To obtain the equivalence, let $\phi$ be a Lie algebra representation of $g\ell_n(\mathbb{C})$ on V. Note that the map $a \to e^a$ is a one to one correspondence between the matrices in $g\ell_n(\mathbb{C})$ and those in $G\ell(n,\mathbb{C})$ which takes diagonals to diagonals. Define the group representation $\phi^*$ of $G\ell(n,\mathbb{C})$ on V by

$$\phi^*(e^a) = e^{\phi(a)} .$$

If $v_\gamma$ is a weight vector for $g\ell_n(\mathbb{C})$ in V with weight $\lambda^{(\gamma)}$ then $v_\gamma$ satisfies

$$\phi^*(\mathrm{diag}(x_1,\ldots,x_n))(v_\gamma) = (x_1^{\lambda_1^{(\gamma)}} x_2^{\lambda_2^{(\gamma)}} \ldots x_n^{\lambda_n^{(\gamma)}})(v_\gamma).$$

So when viewed in terms of the representation $\phi^*$ of $G\ell(n,\mathbb{C})$, the formal character $f_V(x_1,\ldots,x_n)$ is nothing more than the trace of $\phi^*(\delta)$ for $\delta$ a typical diagonal matrix in $G\ell(n,\mathbb{C})$. For a more complete discussion of the representations of $G\ell(n,\mathbb{C})$, see Macdonald [4] or Stanley [5].

REFERENCES

1. J. Humphreys, Introduction to Lie Algebras and Representation Theory, Springer (1972).

2. G. D. James, The Representation Theory of the Symmetric Groups, Springer Lecture Notes No. 682, Springer (1978).

3. W. Lederman, Introduction to Group Characters, Cambridge University Press (1977).

4. I. G. Macdonald, Lie groups and combinatorics, Contemporary Mathematics, Vol. 9 (1982), 73-83.

5. R. P. Stanley, $G\ell(n,\mathbb{C})$ for combinatorialists, Surveys in combinatorics, (E. K. Lloyd, editor), London Math. Soc. Lecture Notes No. 82, Cambridge University Press (1983).

Contemporary Mathematics
Volume 34, 1984

# THE CYCLOTOMIC IDENTITY

N. Metropolis
Gian Carlo Rota[1]

1. INTRODUCTION. One of the most useful identities in mathematics is the exponential identity

$$1 + \alpha z + \frac{\alpha^2 z^2}{2!} + \frac{\alpha^3 z^3}{3!} + \cdots = \lim_{n\to\infty} \left(1 + \frac{\alpha z}{n}\right)^n .$$

Over the years, the exponential identity has been variously interpreted and proved by combinatorial, probabilistic and algebraic methods.

By contrast, the cyclotomic identity

$$1 + \alpha z + \alpha^2 z^2 + \alpha^3 z^3 + \cdots = \prod_{n\geq 1} \left(\frac{1}{1 - z^n}\right)^{M(\alpha,n)}$$

where

$$M(\alpha,n) = \frac{1}{n} \sum_{d|n} \mu(\tfrac{n}{d})\, \alpha^d ,$$

and $\mu$ is the classical Möbius function, is less well known, though it plays a role which is sometimes analogous to the exponential identity.

Our objective is to give a bijective proof of the cyclotomic identity. In contrast to other proofs previously given, the proof given below is natural. It is, in fact, the natural proof, that is, the only proof in which set-theoretic operations naturally correspond to algebraic operations on formal power series. The present proof is a simplified version of the proof we have given elsewhere. Its extreme simplicity results from a pedestrian specification of some elementary algorithms relating permutations to linear orders that are usually passed over in silence.

---

1980 Mathematics Subject Classification. 05A19, 05A15, 05A10.

(*)Partially supported by NSF Grant No. MCS 8104855

2. LINEAR ORDERS, PARTITIONS, PERMUTATIONS. Some of the algorithms described in this Section have been known for a long time; others have been stated for present purposes. It will be prudent to begin with fundamentals.

Let S be a finite non-empty set of $|S|$ elements.

A linear order on S is a binary relation $\lambda$ on S with the following properties:

(1) $\lambda(a,a)$ holds for every $a \in S$;

(2) if $a \neq b$, then either $\lambda(a,b)$ or $\lambda(b,a)$ holds, but not both;

(3) if $a \neq b$, and $\lambda(a,b)$ holds, then either $\lambda(x,b)$ or $\lambda(a,x)$ holds for every $x \in S$.

The listing of the linear order $\lambda$ on S is a display $e_1 < e_2 < \cdots$ of the consecutive elements of S. The notation $<$ for the linear order $\lambda$ will be used when no confusion is possible.

If E is a subset of S, the linear order $\lambda$ on S defines a linear order on E, called the restriction of the linear order $\lambda$.

A partition q of S is a set of non-empty subsets of S, called blocks, with the following properties:

(1) If $I, J \in q$ and $I \neq J$, then I and J are disjoint.

(2) Every element of S belongs to some block.

Let q be a partition of S. A function $J \rightarrow e_J$ from q to S is called a set of representatives of the partition q when $e_I \neq e_J$ if $I \neq J$, and $e_J \in J$ for all $I, J \in q$.

Given a partition q, a set of representatives $\{e_J; J \in q\}$ of q, and a linear order $\lambda$ on the set of representatives $\{e_J; J \in q\}$, one defines a linear order $\lambda'$ on q by setting $\lambda'(I,J)$ whenever $\lambda(e_I, e_J)$ holds. The order $\lambda'$ is said to be induced on the partition q by the linear order $\lambda$ and by the set of representatives $\{e_J; J \in q\}$.

A linear partition $q_\lambda$ of S is a set of ordered pairs $(J, \lambda_J)$, where J ranges over the blocks of a partition q of S, and $\lambda_J$ is a linear order on each block J of q. We shall usually omit the subscript $\lambda$ and denote partitions and linear partitions by the same letter q, leaving it to the context to indicate which is meant.

The natural set of representatives of a *linear* partition q is the set $\{e_J; J \in q\}$, where $e_J$ is the first element in the linear order $\lambda_J$.

Given a linear partition q of S and a linear order $\lambda$ on S, the induced linear order $\lambda(q)$ on the set q is the linear order induced on q by $\lambda$ and by the natural set of representatives of q.

A partition q of S is said to be j-uniform when all blocks of q are of size j. Thus, the integer j divides $|S|$.

ALGORITHM 1. Let an integer $j > 0$ be given which divides $|S|$. A bijection between the set of all ordered pairs $(q,\lambda')$, where $q$ is a $j$-uniform linear partition of $S$, and $\lambda'$ is a linear order on the set $q$, and the set of all linear orders on $S$.

DESCRIPTION.

Step 1. Given: $(q,\lambda')$.

Define a linear order $\lambda(a,b)$ on $S$ as follows.

Case 1. $a$ and $b$ belong to the same block $J$ of $q$.

Set $\lambda(a,b)$ if $\lambda_J(a,b)$.

Case 2. $a$ and $b$ belong to different blocks $I$ and $J$ of $q$, respectively.

Set $\lambda(a,b)$ if $\lambda'(I,J)$.

Step 2. Given: a linear order $\lambda$ on $S$.

Let $e_1 < e_2 < \cdots < e_{|S|}$ be the listing of the linear order $\lambda$. Let

$J_1 = \{e_1, \cdots, e_j\}$, $J_2 = \{e_{j+1}, \cdots, e_{2j}\}$, $\cdots$ $J_k = \{e_{kj-j+1}, \cdots, e_{|S|}\}$.

On $J_i$, let $\lambda_i$ be the restriction of the linear order $\lambda$. The set $q = \{J_1, J_2, \cdots, J_k\}$ is a linear partition of $S$. The linear order $\lambda(q)$ defined on the set $q$ is the linear order induced by $\lambda$. The constructions in Steps 1 and 2 are clearly inverses and define the desired bijection, q.e.d.

A permutation $\pi$ of $S$ is a bijection of $S$. A permutation is cyclic if for an arbitrary choice of an element $e$ of $S$, the sequence $e$, $\pi(e)$, $\pi^2(e)$, $\pi^3(e)$, $\cdots$ contains all elements of $S$.

Given a permutation $\pi$, there exists a unique partition $q$ of $S$ such that the restriction of $\pi$ to each block of $q$ is a cyclic permutation. Such a restriction of $\pi$ is called a cycle of $\pi$. If $\mu$ is a cycle of $\pi$, we denote by $\bar{\mu}$ the block of $q$ on which the cycle $\mu$ is defined.

ALGORITHM 2. A bijection between the set of all ordered pairs $(\lambda_1,\lambda_2)$, where $\lambda_1,\lambda_2$ are linear orders on $S$, and the set of all pairs $(\lambda_1, \pi)$, where $\pi$ is a permutation of $S$.

DESCRIPTION. Let the permutation $\pi$ map $\lambda_1(a)$ to $\lambda_2(a)$. Conversely, given $(\lambda_1,\pi)$, define a linear order $\lambda_2$ by setting $\lambda_2(\pi(e), \pi(f))$ whenever $\lambda_1(e,f)$ for $e, f \in S$.

ALGORITHM 3. A bijection between the set of all ordered pairs $(e,\pi)$, when $e$ is an element of $S$ and $\pi$ is a cyclic permutation of $S$, and the set of all linear orders on $S$.

DESCRIPTION. Given $(e,\pi)$, set $e < \pi(e) < \pi^2(e) < \cdots < \pi^{|S|-1}(e)$ to obtain the listing of a linear order $\lambda$. Conversely, given $\lambda$, let $e_1 < e_2 < \cdots < e_{|S|}$ be the listing of $\lambda$, and set $e = e_1$, and $\pi(e_1) = e_2$, $\pi(e_2) = e_3$, $\cdots$, $\pi(e_{|S|-1}) = e_{|S|}$, $\pi(e_{|S|}) = e_1$.

3. NECKLACES AND PLACEMENTS. Given a set A of $\alpha$ element, which we call an alphabet, and whose elements we call letters or colors. A word is a finite juxtaposition of letters of the alphabet A, that is, an element of the free monoid generated by A. The length of a word is the number of letters; the product of two or more words is juxtaposition.

Two words w and w′ are said to be conjugate when w = uv and w′ = vu, where u and v are words. The empty word is the identity in the monoid of words; thus, every word is conjugate to itself. The relationship of conjugacy between words is an equivalence relation, and an equivalence class of words under the equivalence relation of conjugacy will be called a necklace. Conjugate words have the same length; thus, the notion of length is well-defined for necklaces.

If $w = u^i$, then i divides the length n of the word w. We say then that the word w has period n/i. Thus, the period in a representation of a word w in the form $u^i$, is the length of the word u. The smallest j such that $w = v^{n/j}$ for some word v (of length j) is called the primitive period of the word w. For example, w has primitive period 1 if and only if all letters of w are identical.

Clearly, every period of the word w is a multiple of the primitive period of w.

Conjugate words have the same primitive period and hence the same periods. A word of primitive period n is said to be aperiodic. An equivalence class of aperiodic words will be called a primitive necklace.

If $\gamma$ is a necklace of length n and if w is a word belonging to the equivalence class $\gamma$, we say that the word w is obtained by cutting open the necklace $\gamma$.

Let w be a word obtained by cutting open the necklace $\gamma$, and let $w = v^{n/j}$, where j is the primitive period of the word w. Let $\delta$ be the necklace to which the word v belongs. We shall say that the necklace $\gamma$ is of period $\delta$.

One verifies that the period $\delta$ of a necklace $\gamma$ is well-defined, that is, that it is independent of the choice of the word w in the equivalence class $\gamma$.

A placement is a generalization of a permutation. Given a set S of $|S|$ objects ("places") and a set A of $\alpha$ objects ("colors") labeled a, b, $\cdots$, c, or $a_1$, $a_2$, $\cdots$, we define a placement with domain S and range A ("of colors to places") to be a pair $(f,\pi)$, where f is a function from S to A, and $\pi$ is a permutation of the set S.

Suppose that $S = N = \{1, 2, \cdots, n\}$, the set of the first n integers in the natural order $\lambda_{nat}$. By Algorithm 2, we associate to the pair $(\lambda_{nat}, \pi)$ a pair $(\lambda_{nat}, \lambda(\pi))$ of linear orders. In this way, a placement can be visualized as a sequential assignment of the colors to the places.

The set A of colors will remain fixed throughout.

There are evidently $\alpha^n n!$ distinct placements of a set of $\alpha$ colors to a set of n places, that is, of placements with domain N and range A.

Let $(f,\pi)$ be a placement with domain N, and with a cyclic permutation $\pi$, or <u>cyclic placement</u> for short. Choose an element e of N arbitrarily. By Algorithm 3, the pair $(e,\pi)$ defines a linear order $\lambda_e$ on the set S, listed as $e = e_1 < e_2 < \cdots < e_n$. We thus obtain a word $w(e) = f(e_1)\, f(e_2) \cdots f(e_n)$. If $e' \neq e$, the words $w(e)$ and $w(e')$ are conjugate. Therefore the placement $(f,\pi)$ defines a necklace $\gamma$, the conjugacy class of the word $w(e)$. If the period of the necklace $\gamma$ is the primitive necklace $\delta$, we say that the placement $(f,\pi)$ has <u>period</u> $\delta$. We stress the fact that the period of a cyclic placement is not a number, but a primitive necklace. The length of the necklace $\delta$, written $|\delta|$, divides n.

Next, let $(f,\pi)$ be a placement with non-empty domain N and range A, where no restriction is placed on the permutation $\pi$. As $\mu$ ranges over all cycles of the permutation $\pi$, the sets $\bar{\mu}$ are the blocks of a partition q of the set N. The restriction of the placement $(f,\pi)$ to a block $\bar{\mu}$ of the partition q is a cyclic placement with domain $\bar{\mu}$ and range A. It therefore has a period which we shall write $\delta(\mu)$.

In this way, given a placement $(f,\pi)$, we associate a primitive necklace $\delta(\mu)$ to every cycle $\mu$ of the permutation $\pi$ of N. This assignment of primitive necklaces to every block of the partition q can and will be viewed as a generalization of the cycle-decomposition of a permutation.

We say that the placement $(f,\pi)$ is of <u>period $\delta$</u> whenever, for every cycle $\mu$ of the permutation $\pi$, one has $\delta(\mu) = \delta$, that is, when all the periods $\delta(\mu)$ coincide as $\mu$ ranges over the cycles of the permutation $\pi$.

Note that even when the placement $(f,\pi)$ is of period $\delta$, the permutation $\pi$ need not be cyclic. Nonetheless, we shall see that placements of period $\delta$ play a role similar to that of cyclic permutations.

4. PERIODIC PLACEMENTS. We consider throughout this Section placements of period $\delta$ with non-empty domain $N = \{1, 2, \cdots, n\}$ and range A. Choose a primitive necklace $\delta$ which will remain fixed throughout this Section. Its length $|\delta| = j$ divides n, and $n = kj$. Our objective is to give a bijective algorithm associating a placement of period $\delta$ to each permutation of the set N.

Let $w = a_1 a_2 \cdots a_j$ be a word obtained by cutting open the necklace $\delta$, where the letters $a_i$ belong to the alphabet A. The letters $a_i$ need not be distinct. The word w is primitive, since the necklace $\delta$ is primitive, and it will remain fixed throughout this Section.

ALGORITHM 4. A bijection between the set of all placements of N of period $\delta$ and the set of all pairs $(q,\sigma)$ where q is a j-uniform linear partition of N, and $\sigma$ is a permutation of q.

DESCRIPTION.

Step 1. Given: a placement $(f,\pi)$ of period $\delta$ with domain N and range A.

Let $\mu$ be a cycle of the permutation $\pi$, defined on the set $\bar{\mu}$. The family of sets $\bar{\mu}$, as $\mu$ ranges over all cycles of $\pi$, is a partition of N.

Let $f_\mu$ be the restriction of the function f to the set $\bar{\mu}$. Then $(f_\mu, \mu)$ is a placement of period $\delta$. Say the set $\bar{\mu}$ has jr elements. The integer r depends on $\mu$, but this dependency will not be explicitly displayed. The subset $E(\mu)$ of $\bar{\mu}$ consisting of all elements e of $\bar{\mu}$ such that $f(e) = a_1$ is unequivocally defined, since the word w is primitive. Choose arbitrarily an element of $E(\mu)$, call it e. Since $\mu$ is a cyclic permutation, Algorithm 3 yields from the pair $(e,\mu)$ a linear order $\lambda_\mu$ on the set $\bar{\mu}$. Let $e_1 < e_2 < \cdots < e_{rj}$ be the listing of the linear order $\lambda_\mu$.

Let $J_{\mu 1} = \{e_1, e_2, \cdots, e_j\}$, $J_{\mu 2} = \{e_{j+1}, \cdots, e_{2j}\}$, $\cdots$, $J_{\mu r} = \{e_{rj-j+1}, \cdots, e_{rj}\}$. Mapping $J_{\mu 1}$ to $J_{\mu 2}$, $J_{\mu 2}$ to $J_{\mu 3}$, $\cdots$, $J_{\mu r}$ to $J_{\mu 1}$ we define a cyclic permutation $\sigma(\mu)$ of the set $\{J_{\mu 1}, J_{\mu 2}, \cdots, J_{\mu r}\}$. As $\mu$ ranges over all cycles of $\pi$, the sets $J_{\mu 1}, J_{\mu 2}, \cdots, J_{\mu r}$ define a j-uniform linear partition q of N. The cyclic permutations $\sigma(\mu)$, as $\mu$ ranges over all cycles of $\pi$, define a permutation $\sigma$ of q, as desired.

Step 2. Given: a j-uniform linear partition q of N and a permutation $\sigma$ of q.

Let $\tau$ be a cycle of the permutation $\sigma$, defined on the subset $\bar{\tau}$ of q. Choose an element $J \in \bar{\tau}$ arbitrarily. Algorithm 3 associates with the pair $(J, \tau)$ a linear order on the elements of $\bar{\tau}$. Let $J_1 < J_2 < \cdots < J_r$ be a listing of the elements of $\bar{\tau}$ in this linear order. The integer r depends on $\tau$, but this dependency will not be explicitly displayed. Let $\bar{\bar{\tau}} = J_1 \cup J_2 \cup \cdots \cup J_r$. Thus, $\bar{\bar{\tau}}$ is the subset of N containing those elements of N which belong to one of $J_1, J_2, \cdots, J_r$. In other words, $\bar{\tau}$ is a j-uniform partition of $\bar{\bar{\tau}}$. Let $\lambda_\tau$ be the linear order on the set $\bar{\bar{\tau}}$ defined by Algorithm 1. Let $e_1$ be the first element of the linear order $\lambda_\tau$. Algorithm 3 yields from the pair $(e_1, \lambda_\tau)$ a cyclic permutation $\pi(\tau)$ of the set $\bar{\bar{\tau}}$. As $\tau$ ranges over all cycles of the permutation $\sigma$, the cyclic permutations $\pi(\tau)$ define a permutation $\pi$ of N. The cyclic permutation $\pi(\tau)$ is independent of the choice of J.

Let $e_1 < e_2 < \cdots < e_{jr}$ be the listing of the linear order $\lambda_\tau$ defined on $\bar{\bar{\tau}}$.

Define a function $f_\tau$ with domain $\bar{\bar{\tau}}$ and range A by setting

$f_\tau(e_1) = a_1,\ f_\tau(e_2) = a_2,\ \cdots,\ f_\tau(e_j) = a_j,$

$f_\tau(e_{j+1}) = a_1,\ \cdots$

$\cdots \qquad\qquad\qquad\qquad f_\tau(e_{jr}) = a_j.$

The function $f_\tau$ is independent of the choice of J. The placement $(f_\tau, \pi(\tau))$ is of period $\delta$. Define $f(e) = f_\tau(e)$ if $e \in \bar{\bar{\tau}}$. The placement $(f,\pi)$ is the desired placement.

The algorithms described in Steps 1 and 2 are inverses to each other, and define the required bijection, q.e.d.

ALGORITHM 5. A bijection between the set of all permutations of N and the set of all $(q,\sigma)$, where q is a j-uniform linear partition of N and $\sigma$ is a permutation of q.

DESCRIPTION.

Recall that $\lambda_{nat}$ is the natural linear order of the integers in N.

Step 1. Given: $(q,\sigma)$.

Let $\lambda_{nat}(q)$ be the linear order induced on q by $\lambda_{nat}$. Algorithm 2 yields from the pair $(\lambda_{nat}(q), \sigma)$ a pair $(\lambda_{nat}(q), \lambda(\sigma))$ of linear orders, where $\lambda(\sigma)$ is a linear order on q. Algorithm 1 yields a linear order $\lambda$ on N from the ordered pair $(q, \lambda(\sigma))$. Algorithm 2 yields a pair $(\lambda_{nat}, \nu)$ from the pair $(\lambda_{nat}, \lambda)$ of linear orders on N, where $\nu$ is the desired permutation of the set N.

Step 2. Given: a permutation $\nu$ of N.

The ordered pair $(\lambda_{nat}, \nu)$ yields, by Algorithm 2, an ordered pair of linear orders $(\lambda_{nat}, \lambda(\nu))$ on N. Let $e_1 < e_2 < \cdots < e_n$ be the listing of the linear order $\lambda(\nu)$. Let $J_1 = \{e_1, \cdots, e_j\}$, $J_2 = \{e_{j+1} \cdots, e_{2j}\}$, $\cdots$, $J_k = \{e_{kj-j+1}, \cdots, e_n\}$, and let q be the j-uniform linear partition of N whose blocks are $J_1, J_2, \cdots, J_k$. Let $\bar{\lambda}(\nu)$ be the linear order induced on q by the linear order $\lambda(\nu)$, and let $J_1 < J_2 < \cdots < J_k$ be the listing of the linear order $\bar{\lambda}(\nu)$ on q. Let $\lambda_{nat}(q)$ be the linear order induced on q by the linear order $\lambda_{nat}$. The ordered pair $(\lambda_{nat}(q), \bar{\lambda}(\nu))$ yields, by Algorithm 2, an ordered pair $(\lambda_{nat}(q), \sigma(\pi))$, where $\sigma$ is a permutation of the set q. The pair $(q,\sigma)$ is as desired.

The constructions in Steps 1 and 2 are inverses to each other, and define the required bijection, q.e.d.

ALGORITHM 6. Let $|\delta|$ divide $|N|$. A bijection between the set of all permutations of the set N and the set of all placements of period $\delta$ with domain N and range A.

DESCRIPTION.

Algorithm 4 gives a bijection between the set of all placements $(f,\pi)$ of period $\delta$ and the set of all pairs $(q,\sigma)$. Algorithm 5 gives a bijection between the set of all pairs $(q,\sigma)$ and the set of all permutations $\nu$ of N, as desired.

As a by-product, we obtain an algorithm which associates to every permutation $\nu$ of N a permutation $\pi$ of N whose cycles have sizes equal to multiples of j. This correspondence is not bijective, but it becomes bijective when permutations $\pi$ are replaced by placements!

5. THE CYCLOTOMIC IDENTITY. Let $M(\alpha,n)$ be the number of primitive necklaces of length n on an alphabet A of $\alpha$ letters. The particular expression of $M(\alpha,n)$ will not concern us. The cyclotomic identity is the following identity between formal power series in z:

$$(*)\quad \frac{1}{1-\alpha z} = \prod_{n\geq 1}\left(\frac{1}{1-z^n}\right)^{M(\alpha,n)} .$$

We give a bijective proof of this identity. The left side expands as

$$\frac{1}{1-\alpha z} = \sum_{n\geq 0} \alpha^n \cdot n! \frac{z^n}{n!} ,$$

and is thus seen to be the exponential generating function of the set of placements with variable domain N and fixed range A.

Now let $\Delta$ be the set of all primitive necklaces of fixed range A. The right side of the cyclotomic identity can be rewritten as

$$\prod_{\delta\varepsilon\Delta}\left(\frac{1}{1-z^{|\delta|}}\right).$$

Now let $(f,\pi)$ be a placement with domain N and range A. Then there is a unique partition q of N such that the restriction of $(f,\pi)$ to a block B of q is a placement of period $\delta(B)$, and such that for distinct blocks B, B$'$ of q the periods $\delta(B)$ and $\delta(B')$ are distinct. Thus, if $g(\delta,z)$ is the exponential generating function of the set of all placements of period δ of domain N and range A, as N varies we have

$$\frac{1}{1-\alpha z} = \prod_{\delta \varepsilon \Delta} g(\delta, z) \quad .$$

But by Algorithm 6 we have

$$g(\delta,z) = \sum_{n \geq 1} (|\delta|n)! \frac{z^{|\delta|n}}{(|\delta|n)!} = \frac{1}{1 - z^{|\delta|}} \quad .$$

This concludes the proof of the cyclotomic identity.

BIBLIOGRAPHY

[1] N. Metropolis and Gian-Carlo Rota, Witt vectors and the algebra of necklaces, Adv. in Math. 50 (1983), 95-125.

[2] N. Bourbaki, Algèbre commutative, Chapitres 8 et 9, Masson (Paris), 1983.

N. METROPOLIS
LOS ALAMOS NATIONAL LABORATORY
LOS ALAMOS, NEW MEXICO 87545

Theoretical Division, T-DOT, MS B210
Los Alamos National Laboratory
Los Alamos, New Mexico 87545

GIAN-CARLO ROTA
M.I.T. and LOS ALAMOS NATIONAL LABOATORY

Department of Mathematics
Massachusetts Institute of Technology
Cambridge, MA 02139

Contemporary Mathematics
Volume 34, 1984

# ARRANGEMENTS OF HYPERPLANES AND DIFFERENTIAL FORMS

Peter Orlik, Louis Solomon and Hiroaki Terao

1. INTRODUCTION. Let $\mathbb{K}$ be a field and let $V$ be a vector space of dimension $\ell$ over $\mathbb{K}$. A hyperplane in $V$ is a vector subspace of codimension 1. An arrangement $\mathcal{A} = (\mathcal{A}, V)$ in $V$ is a finite set of hyperplanes. Let $L(\mathcal{A})$ be the collection of all intersections of elements of $\mathcal{A}$. We partially order $L(\mathcal{A})$ by the reverse of inclusion, so that $X \leq Y$ means $X \supseteq Y$. Then $L(\mathcal{A})$ is a geometric lattice called the intersection lattice of $\mathcal{A}$. In this paper we study a certain algebra $R(\mathcal{A})$ of exterior differential forms, and show how combinatorial properties of $L(\mathcal{A})$, for example its characteristic polynomial and the matroid operations of deletion and contraction, are reflected in the algebraic properties of $R(\mathcal{A})$.

Let $F$ be the field of rational functions on $V$, and let $\Omega(V) = \bigoplus_{p \geq 0} \Omega^p(V)$ be the $F$-algebra of rational exterior differential forms on $V$. If $x_1, \ldots, x_\ell$ is a $\mathbb{K}$-basis for $V^*$ then $F = \mathbb{K}(x_1, \ldots, x_\ell)$ and the elements of $\Omega^p(V)$ have the shape

$$\omega = \sum_{i_1 < \cdots < i_p} f_{i_1 \cdots i_p} dx_{i_1} \cdots dx_{i_p}$$

where $f_{i_1 \cdots i_p} \in F$. Choose for each $H \in \mathcal{A}$ a linear form $\alpha_H \in V^*$ with kernel $H$. This linear form is determined uniquely by $H$ up to a constant multiple. The differential form $\omega_H = d\alpha_H/\alpha_H \in \Omega^1(V)$ is determined uniquely

---

This work was supported in part by the National Science Foundation. The third author held a Japanese Mathematical Scientist Research Fellowship.

by H. Our object of study is the $\mathbb{K}$-subalgebra $R(\mathcal{A})$ of $\Omega(V)$ generated by the $\omega_H$, $H \in \mathcal{A}$, and the identity. It has finite dimension over $\mathbb{K}$ and is graded by $R(\mathcal{A}) = \bigoplus_{p \geq 0} R_p(\mathcal{A})$ where $R_p(\mathcal{A}) = R(\mathcal{A}) \cap \Omega^p(V)$. If $U = \bigoplus_{p\geq 0} U_p$ is any graded vector space of finite dimension over $\mathbb{K}$ let

$$P(U,t) = \sum_{p\geq 0} \dim(U_p)\, t^p$$

be its Poincaré polynomial in the indeterminate $t$. If $X$ is a subspace of $V$ let $\operatorname{codim} X$ denote its codimension in $V$. Let $\mu$ be the Möbius function of $L(\mathcal{A})$ [1, p. 141] and write $\mu(X) = \mu(V,X)$.

(1.1) <u>Theorem</u>. Let $\mathcal{A} = (\mathcal{A}, V)$ be an arrangement of hyperplanes and let $L = L(\mathcal{A})$ be its intersection lattice. The Poincaré polynomial of the graded $\mathbb{K}$-algebra $R = R(\mathcal{A})$ of differential forms is

$$P(R,t) = \sum_{X \in L} \mu(X)(-t)^{\operatorname{codim} X}.$$

Since there is no obvious direct connection between differential forms and the Mobius function it may help the reader if we verify (1.1) in case $\dim V = 2$ by direct calculation. The arrangement $\mathcal{A}$ consists of $n$ lines in a plane. Say $\mathcal{A} = \{H_1, \ldots, H_n\}$. The Möbius function is given by

$$\mu(V) = 1, \quad \mu(H_i) = -1, \quad \mu(0) = n-1.$$

Thus we must check that $\dim R_1(\mathcal{A}) = n$ and $\dim R_2(\mathcal{A}) = n-1$. Let $x, y$ be a basis for $V^*$. Write $\alpha_i = \alpha_{H_i}$ and $\omega_i = \omega_{H_i}$. Say $\alpha_i = a_i x + b_i y$ with $a_i, b_i \in \mathbb{K}$. Then $\omega_i = (a_i/\alpha_i)dx + (b_i/\alpha_i)dy$. The 1-forms $\omega_1, \ldots, \omega_n$ span $R_1(\mathcal{A})$ over $\mathbb{K}$ by definition of $R(\mathcal{A})$ and they are linearly independent over $\mathbb{K}$ because the rational functions $1/\alpha_1, \ldots, 1/\alpha_n$ are linearly independent over $\mathbb{K}$. Thus $\dim R_1(\mathcal{A}) = n$. Since $\omega_i^2 = 0$ and $\omega_i\omega_j = -\omega_j\omega_i$, the space $R_2(\mathcal{A})$ is spanned over $\mathbb{K}$ by the $\omega_i\omega_j$ with $i < j$. We have

$$d\alpha_i\, d\alpha_j = (a_i b_j - b_i a_j)dxdy$$

and thus for any $i, j, k$ we have

$$\alpha_k d\alpha_i d\alpha_j + \alpha_i d\alpha_j d\alpha_k + \alpha_j d\alpha_k d\alpha_i = \det \begin{bmatrix} a_i & a_j & a_k \\ b_i & b_j & b_k \\ \alpha_i & \alpha_j & \alpha_k \end{bmatrix} dx\,dy = 0$$

because the third row is a linear combination of the first two. Thus multiplying by $1/\alpha_i\alpha_j\alpha_k$ we get

(1.2) $$\omega_i\omega_j + \omega_j\omega_k + \omega_k\omega_i = 0.$$

In particular we have $\omega_i\omega_j = -\omega_j\omega_n + \omega_i\omega_n$ if $1 \le i < j < n$ so that $R_2(\mathcal{A})$ is spanned over $\mathbf{K}$ by the $n-1$ elements $\omega_1\omega_n, \ldots, \omega_{n-1}\omega_n$. We must prove that these elements are linearly independent over $\mathbf{K}$. Define an F-linear map $\partial : \Omega^2(V) \to \Omega^1(V)$ by $\partial(f dx dy) = fx dy - fy dx$. Then

(1.3) $$\partial(\omega_i\omega_j) = \omega_j - \omega_i.$$

If $\sum_{i=1}^{n-1} c_i\omega_i\omega_n = 0$ with $c_i \in \mathbf{K}$ then applying $\partial$ gives $\sum_{i=1}^{n-1} c_i(\omega_n - \omega_i) = 0$. Since $\omega_1, \ldots, \omega_n$ are linearly independent over $\mathbf{K}$ we get $c_1 = \ldots = c_{n-1} = 0$ and the verification is complete.

This calculation contains two features of the general argument: the formula (1.2) which expresses the linear dependence of $\alpha_i, \alpha_j, \alpha_k$ in terms of $R(\mathcal{A})$, and the map $\partial$ used in (1.3). These features appear early in Section 2 in altered form: the analogue of (1.2) is (2.7) and the analogue of (1.3) is (2.9.i). The map $\partial$ reappears in (4.19) in its original context. It is not used in the proof of (1.1) for reasons described in the last paragraph of Section 4: the linear independence of $\omega_1, \ldots, \omega_n$ amounts to the assertion $R_1(\mathcal{A}) = \bigoplus_{i=1}^{n} \mathbf{K}\omega_i$ which is a special case of the direct sum decomposition (4.20).

One feature of the general argument does not appear in the preceding calculation. This is an exact sequence for $R$ corresponding to the matroid operations of deletion and contraction in $\mathcal{A}$. Suppose $\mathcal{A}$ is non void. (We allow the void arrangement, an empty set $\mathcal{A}$ of hyperplanes for which $L(\mathcal{A})$ consists of $V$ alone and $R(\mathcal{A}) = \mathbf{K}$). Choose $H_0 \in \mathcal{A}$ and hold it fixed. The subarrangement $\mathcal{A}' = \mathcal{A} - \{H_0\}$ is called the deletion of $H_0$

from $\mathcal{A}$. The arrangement $\mathcal{A}'' = \{H_0 \cap H \mid H \in \mathcal{A}'\}$ in $H_0$ is called the restriction of $\mathcal{A}$ to $H_0$, or in matroid terminology the contraction of $\mathcal{A}$ by $H_0$. Call $(\mathcal{A}, \mathcal{A}', \mathcal{A}'')$ a triple of arrangements.

(1.4) Theorem. Let $(\mathcal{A}, \mathcal{A}', \mathcal{A}'')$ be a triple of arrangements and let $R, R', R''$ be the corresponding graded $\mathbb{K}$-algebras of differential forms. There is an exact sequence

$$0 \to R' \xrightarrow{i} R \xrightarrow{j} R'' \to 0$$

in which $i$ is inclusion and $j$ is a $\mathbb{K}$-linear map which is homogeneous of degree $-1$.

The simplest case of (1.4) is $\ell = 1$ where $\mathcal{A}'$ and $\mathcal{A}''$ are void and $\mathcal{A} = \{H\}$ for a single hyperplane $H$, the origin. Let $x \in V^*$ be non zero. Then $R' = \mathbb{K} = R''$ and $R = \mathbb{K} + \mathbb{K}\frac{dx}{x}$. The exact sequence (1.4) is

$$0 \to \mathbb{K} \to \mathbb{K} + \mathbb{K}\frac{dx}{x} \to \mathbb{K} \to 0.$$

Since $i$ is the inclusion the only natural choice for $j$ is $j(a + b\frac{dx}{x}) = b$ where $a, b \in \mathbb{K}$. In case $\mathbb{K} = \mathbb{C}$ note that $b$ is the residue of the 1-form $b\frac{dx}{x}$. This is a key observation. The source of the map $j$ corresponding to the matroid operation of contraction is Leray's notion of residue for a differential form with a simple pole [7, Chap. III].

Here is an outline of the paper. In Section 2 we introduce an algebra $A$ defined by generators and relations and prove an analogue of (1.1) for $A$. This algebra was defined in [6, Sect. 2] in case $\mathbb{K} = \mathbb{C}$ is the complex field and the analogue of (1.1) was proved there; the argument here is much the same. In Section 3 we construct an exact sequence for $A$ analogous to (1.4). In Section 4 we prove that there is an isomorphism $A \simeq R$ of graded algebras and deduce (1.1) and (1.4) from the analogous theorems for $A$. The precise flow diagram for the argument is more complicated. At some places we have indicated possible alternate routes to the theorems; the present exposition is the best we can do at the moment. In Section 5 we show that if $L(\mathcal{A})$ contains a maximal chain of modular elements then

Stanley's factorization theorem [8] for the characteristic polynomial of $L(\mathcal{G})$ is reflected in a factorization of the algebra $R(\mathcal{G})$ or $A(\mathcal{G})$ as a tensor product of subspaces.

It is possible to increase the degree of generality in Sections 2, 3 and 5 by replacing the lattice of an arrangement by any geometric lattice. The arguments are precisely the same with appropriate changes in notation. In particular, there is an exact sequence

$$0 \to A' \to A \to A'' \to 0$$

for deletion and contraction in any geometric lattice $L$ where $A = A(L)$ is the algebra defined in [6]. Since Section 4 cannot be done in this generality we chose to write the paper in terms of the lattice of an arrangement.

To read this paper one needs only some familiarity with the Möbius function of a geometric lattice and elementary facts about exterior algebra and chain complexes. It would be misleading though to avoid mention of the deeper topological undercurrent in case $\mathbb{K} = \mathbb{C}$ which led to the algebraic arguments given here. Arnold [2] studied the cohomology of the manifold

$$M = \{(z_1, \ldots, z_\ell) \in \mathbb{C}^\ell \mid z_j \neq z_k \text{ for } 1 \leq j < k \leq \ell\}$$

and found that the cohomology ring of $M$ has generators

$$\omega_{jk} = \frac{1}{2\pi i} \frac{dz_j - dz_k}{z_j - z_k}$$

and relations

$$\omega_{jk}\omega_{km} + \omega_{km}\omega_{mj} + \omega_{mj}\omega_{jk} = 0.$$

The corresponding lattice is the partition lattice. This led to the definition of $R$ in [6] and a proof of (1.1) for $\mathbb{K} = \mathbb{C}$ by topological argument. These ideas are outside the scope of this paper. Cartier's Séminaire Bourbaki talk [4] contains a summary of this work and further references.

## 2. THE ALGEBRA $A(\mathcal{G})$.

In this section we associate to each arrangement $\mathcal{G} = (\mathcal{G}, V)$ a graded anticommutative $\mathbb{K}$-algebra $A = A(\mathcal{G})$ with identity, whose generators are in one-to-one correspondence with the hyperplanes of $\mathcal{A}$. This algebra is constructed in the form $A(\mathcal{A}) = E(\mathcal{A})/I(\mathcal{A})$ where $E(\mathcal{A})$ is an exterior algebra and the ideal $I(\mathcal{A})$ is determined by the dependence relations between the hyperplanes of $\mathcal{A}$. The main results are Theorem 2.26 which gives a direct sum decomposition of $A$, Theorem 2.37 which computes the dimensions of the summands and Theorem 2.38 which computes the Poincaré polynomial of $A$ in terms of the lattice $L = L(\mathcal{A})$. In (2.41) we compute an example which illustrates some of the main points in the argument. Let $\bigoplus_{H\in\mathcal{A}} \mathbb{K}e_H$ be a vector space over $\mathbb{K}$ which has a $\mathbb{K}$-basis consisting of elements $e_H$ in one-to-one correspondence with the hyperplanes $H \in \mathcal{A}$, and let

$$E = E(\mathcal{A}) = \Lambda(\bigoplus_{H\in\mathcal{A}} \mathbb{K}e_H)$$

be its exterior algebra. We write $uv$ rather than $u \wedge v$ for the multiplication in $E$. Thus $e_H^2 = 0$ and $e_H e_K = -e_K e_H$ for all $H, K \in \mathcal{A}$. Let $n = |\mathcal{A}|$. The algebra $E$ is graded by $E = \bigoplus_{p=0}^{n} E_p$ where $E_0 = E_0(\mathcal{A}) = \mathbb{K}$ and $E_p = E_p(\mathcal{A})$ is spanned over $\mathbb{K}$ by all $e_{H_1} \cdots e_{H_p}$ with $H_k \in \mathcal{A}$.

(2.1) <u>Definition</u>. Define a $\mathbb{K}$-linear map $\partial : E \to E$ by $\partial(1) = 0$ and

$$\partial(e_{H_1} \cdots e_{H_p}) = \sum_{k=1}^{p} (-1)^{k-1} e_{H_1} \cdots \hat{e}_{H_k} \cdots e_{H_p}$$

for all $H_1, \ldots, H_p \in \mathcal{A}$.

In case $p = 1$ we agree that this means $\partial e_H = 1$ for $H \in \mathcal{A}$. Sometimes we write $\partial = \partial_E$. The mapping $\partial$ has the properties

(2.2) $$\partial^2 = 0$$

(2.3) $$\partial(uv) = (\partial u)v + (-1)^p u(\partial v) \qquad u \in E_p, \quad v \in E.$$

These are known facts about the exterior algebra which have nothing to do with arrangements. Since the map $\partial$ is homogeneous of degree $-1$ we see that $(E, \partial)$ is a chain complex.

Let $\underline{\underline{S}}_p = \underline{\underline{S}}_p(\mathcal{A})$ denote the set of all p-tuples $(H_1, \ldots, H_p)$ of elements of $\mathcal{G}$ and let $\underline{\underline{S}} = \underline{\underline{S}}(\mathcal{G}) = \bigcup_{p\geq 0} S_p$. In case $p = 0$ we agree that $S = (\ )$ is the empty tuple. We abuse notation slightly and write $H_k \in S$. If $K = (T_1,\dots,T_q) \in \underline{\underline{S}}$ we write $(S,T) = (H_1,\dots,H_p, K_1,\dots,K_q) \in \underline{\underline{S}}$. Define $e_S = e_{H_1} \cdots e_{H_p} \in E_p$. Thus $e_S e_T = e_{(S,T)}$. If $p = 0$ we agree that $e_S = 1$.

(2.4) Lemma. If $S \in \underline{\underline{S}}$ and $H \in S$ then $e_S = e_H(\partial e_S)$.

Proof. If $H \in S$ then $e_H e_S = 0$ so that $0 = \partial(e_H e_S) = e_S - e_H(\partial e_S)$. □

If $S = (H_1,\dots,H_p) \in \underline{\underline{S}}$ define $\cap S = H_1 \cap \dots \cap H_p \in L$. If $p = 0$ we agree that $\cap S = V$. The rank function $r$ on the geometric lattice $L$ is given by $r(X) = \operatorname{codim} X$, the codimension of $X$ in $V$. If $r(\cap S) < p$ we say that $S$ is dependent. If $r(\cap S) = p$ we say that $S$ is independent.

(2.5) Definition. Let $I = I(\mathcal{G})$ be the ideal of $E$ generated by all elements $\partial e_S$ such that $S \in \underline{\underline{S}}$ is dependent.

(2.6) Definition. Let $A = A(\mathcal{G}) = E/I$. Let $\varphi : E \to A$ be the natural homomorphism and let $A_p = \varphi(E_p)$. If $H \in \mathcal{G}$ let $a_H = \varphi(e_H)$ and if $S \in \underline{\underline{S}}$ let $a_S = \varphi(e_S)$.

Since $I$ is generated by elements which are homogeneous with respect to the grading in $E$ we have $I = \bigoplus_{p=0}^{n} I_p$ where $I_p = I \cap E_p$ and thus $A = \bigoplus_{p=0}^{n} A_p$. Since the elements of $\underline{\underline{S}}_1$ are independent we have $I_0 = 0$ and thus $A_0 = \mathbb{K}$. The dependent elements of $\underline{\underline{S}}_2$ have the form $S = (H, H)$ where $H \in \mathcal{G}$. Since $e_S = e_H^2 = 0$ it follows that $I_1 = 0$. Thus the elements $a_H$, $H \in \mathcal{G}$, are linearly independent over $\mathbb{K}$ and we have $A_1 = \bigoplus_{H \in \mathcal{G}} \mathbb{K} a_H$. If $p > \ell$ then any tuple $S = (H_1,\dots,H_p)$ is dependent so (2.4) implies $e_S \in I$ and hence $a_S = 0$. Thus $A_p = 0$ if $p > \ell$.

(2.7) Lemma. If $S = (H_1,\dots,H_p) \in \underline{\underline{S}}$ is dependent then

$$\sum_{k=1}^{p} (-1)^{k-1} a_{H_1} \cdots \hat{a}_{H_k} \cdots a_{H_p} = 0 .$$

Proof. Since $\partial e_S \in I$ this follows from (2.1). □

(2.8) Lemma. $\partial I \subseteq I$.

Proof. It follows from the definition of $I$ that it is spanned over $\mathbb{K}$ by elements of the form $e_T(\partial e_S)$ where $S \in \underline{\underline{S}}$ is dependent and $T \in \underline{\underline{S}}$. From (2.2) and (2.3) we have

$$\partial(e_T \partial e_S) = (\partial e_T)(\partial e_S) \pm e_T(\partial^2 e_S) = (\partial e_T)(\partial e_S) \in I. \qquad \square$$

(2.9) Lemma. There exists a $\mathbb{K}$-linear map $\partial_A : A \to A$ such that

(i) $\partial_A(a_S) = \sum_{k=1}^{p} (-1)^{k-1} a_{H_1} \cdots \hat{a}_{H_k} \cdots a_{H_p}$ if $S = (H_1, \ldots, H_p) \in \underline{\underline{S}}$ .

(ii) $\partial_A^2 = 0$ .

(iii) $\partial_A(ab) = (\partial_A a)b + (-1)^p a(\partial_A b)$ $\quad a \in A_p$, $b \in A$ .

(iv) If $\mathcal{A}$ is non void then the chain complex $(A, \partial_A)$ is acyclic.

Proof. It follows from (2.8) that there exists a $\mathbb{K}$-linear map $\partial_A : A \to A$ such that $\partial_A \varphi = \varphi \partial_E$. The properties (i) - (iii) follow from (2.1) - (2.3). If $S$ is the empty tuple then (i) is to be interpreted as $\partial_A(1) = 0$. To prove (iv) note that since $\mathcal{A}$ is non void we may choose $v \in E$ such that $v$ is homogeneous of degree 1 and $\partial_E v = 1$. Let $b = \varphi(v)$. Now let $a \in A$ be arbitrary. Choose $u \in E$ with $\varphi(u) = a$. Then $\partial_E(vu) = (\partial_E v)u - v(\partial_E u) = u - v\partial_E u$. Since $\varphi\partial_E = \partial_A \varphi$ and $\varphi$ is a $\mathbb{K}$-algebra homomorphism we have $a = \partial_A(ba) + b\,\partial_A a$ for all $a \in A$. Thus $\mathrm{im}(\partial_A) \supseteq \ker(\partial_A)$. □

(2.10) Proposition. Let $\mathcal{A}$ be a nonvoid arrangement. Then

$$\sum_{p \geq 0} (-1)^p \dim A_p = 0 .$$

Proof. Since the chain complex $(A, \partial_A)$ is acyclic, its Euler characteristic is zero. □

The formula of Proposition (2.10) is the punch line in the proof of Theorem (2.37). Henceforth in this section $\partial$ means $\partial_E$.

(2.11) Definition. Let $J = J(\mathcal{A})$ be the subspace of $E$ spanned over $\mathbb{K}$ by all $e_S$, $S \in \underline{\underline{S}}$, such that $S$ is dependent.

(2.12) Lemma. (i) $J$ is an ideal of $E$. (ii) $I = J + \partial J$.

Proof. If $T \in \underline{\underline{S}}$ is dependent and $S \in \underline{\underline{S}}$ then $(S,T)$ is dependent. Thus $e_S e_T = e_{(S,T)} \in J$. This proves that $J$ is an ideal. Formula (2.4) shows that $J \subseteq I$. We have $\partial J \subseteq I$ by definition of $J$ and $I$, and thus $J + \partial J \subseteq I$. Since $J + \partial J$ contains the generators $\partial e_S$ ($S$ dependent) of $I$ the proof will be complete if we show that $J + \partial J$ is an ideal. Since $J$ is an ideal it suffices to show that $e_H \partial e_S \in J + \partial J$ when $H \in \mathcal{A}$ and $S \in \underline{\underline{S}}$ is dependent. This follows from the formula $e_H \partial e_S = e_S - \partial(e_H e_S) = e_S - \partial e_{(H,S)}$ because $(H, S)$ is dependent. □

If $X \in L$ define a subset $\underline{\underline{S}}_X$ of $\underline{\underline{S}}$ and a subspace $E_X$ of $E$ by

(2.13) $$\underline{\underline{S}}_X = \underline{\underline{S}}_X(\mathcal{A}) = \{S \in \underline{\underline{S}} \mid \cap S = X\}$$

(2.14) $$E_X = E_X(\mathcal{A}) = \sum_{S \in \underline{\underline{S}}_X} \mathbb{K} e_S .$$

Note that $e_S \in E_{\cap S}$ for all $S \in \underline{\underline{S}}$. Since $\underline{\underline{S}} = \bigcup_{X \in L} \underline{\underline{S}}_X$ is a disjoint union we have

(2.15) $$E = \bigoplus_{X \in L} E_X .$$

We will prove that the algebra $A$ has an analogous direct sum decomposition. If $X \in L$ let $\pi_X$ be the projection of $E$ on $E_X$ which annihilates all $E_Y$ for $Y \in L$ and $Y \neq X$. Thus

(2.16) $$\pi_X e_S = \begin{cases} e_S & \text{if } \cap S = X \\ 0 & \text{otherwise} . \end{cases}$$

If $F$ is a subspace of $E$ write $F_X = F \cap E_X$.

(2.17) Lemma. If $F$ is a subspace of $E$ and $\pi_X(F) \subseteq F$ for all $X \in L$ then $\pi_X(F) = F_X$ and $F = \bigoplus_{X \in L} F_X$.

Proof. This follows easily from (2.15). □

(2.18) Lemma. $J = \bigoplus_{X \in L} J_X$.

Proof. By definition $J$ is spanned by the elements $e_S$, $S \in \underline{\underline{S}}$, such that $S$ is dependent. It follows from (2.16) that $\pi_X(J) \subseteq J$ for all $X \in L$. The lemma follows from (2.17). □

(2.19) Definition. Let $J' = J'(\mathcal{G})$ be the subspace of $E$ spanned by all $e_S$, $S \in \underline{\underline{S}}$, such that $S$ is independent. Thus $E = J \oplus J'$. Let $\pi = \pi_{\mathcal{G}}$ be the projection of $E$ on $J'$ which annihilates $J$.

(2.20) Definition. Let $K = K(\mathcal{G}) = \pi(\partial J)$.

(2.21) Lemma. $I = J \oplus K$.

Proof. Since $1 - \pi$ is the projection of $E$ on $J$ which annihilates $J'$ and $I \supseteq J$ by (2.12), we have $(1-\pi)I = J$. Lemma 2.12 also shows that $\pi(I) = \pi(J + \partial J) = \pi(\partial J) = K$. Since $J \subseteq I$ we have $(1-\pi)I \subseteq I$ and hence $\pi(I) \subseteq I$. Thus $I = (1-\pi)I \oplus \pi I = J \oplus K$. □

(2.22) Lemma. $K = \bigoplus_{X \in L} K_X$.

Proof. In view of (2.17) it suffices to show that $\pi_X(K) \subseteq K$ for all $X \in L$. By (2.18) we have $K = \pi(\partial J) = \sum_{Y \in L} \pi(\partial J_Y)$. The space $\partial J_Y$ is spanned by elements $\partial e_S$ with $S$ dependent and $\cap S = Y$. Write $S = (H_1, \ldots, H_p)$ and let $S_k = (H_1, \ldots, \widehat{H}_k, \ldots, H_p)$. Then $\partial e_S = \sum_{k=1}^{p} (-1)^{k-1} e_{S_k}$. If $S_k$ is independent then $\cap S_k = \cap S = Y$ because $S$ is dependent. Thus $\pi(\partial e_S) \in E_Y$ so $\pi(\partial J_Y) \subseteq E_Y$. Since $\pi_X(E_Y) = 0$ for $X \neq Y$ we have

$\pi_X(K) = \pi(\partial J_X) \subseteq \pi(\partial J) = K.$ □

(2.23) Lemma. If $X \in L$ then $K_X = \pi(\partial J_X)$.

Proof. We showed in the last line of the preceding argument that $\pi_X(K) = \pi(\partial J_X)$. □

(2.24) Proposition. $I = \bigoplus_{X \in L} I_X$.

Proof. From (2.21) we have $I = J+K$. We showed $\pi_X(J) \subseteq J$ in (2.18) and $\pi_X(K) \subseteq K$ in (2.22). Thus $\pi_X(I) \subseteq I$. The proposition follows from (2.17). □

(2.25) Definition. If $X \in L$ let $A_X = \varphi(E_X)$.

(2.26) Theorem. Let $\mathcal{A}$ be an arrangement and let $A = A(\mathcal{A})$. Then

$$A = \bigoplus_{X \in L} A_X.$$

Proof. Since $E = \sum E_X$ we have $A = \sum A_X$. Proposition (2.24) shows that the sum is direct. □

(2.27) Corollary. $A_p = \bigoplus_{\substack{X \in L \\ r(X) = p}} A_X$.

Proof. It follows from (2.26) that we may replace $\oplus$ by $\sum$ in the statement. Suppose $a \in A_X$ where $X \in L$ and $r(X) = p$. Write $a = \varphi(u)$ where $u \in E_X$. Write $u = \sum_{S \in \underline{\underline{S}}_X} c_S e_S$ where $c_S \in \mathbb{K}$. If $S \in \underline{\underline{S}}_X$ is dependent then $e_S \in I$ and $\varphi(e_S) = 0$. If $S$ is independent then $r(\cap S) = r(X) = p$ implies $e_S \in E_p$ and $\varphi(e_S) \in A_p$. Thus $a = \varphi(u) \in A_p$. In the opposite direction suppose $a \in A_p$ and write $a = \varphi(u)$ with $u \in E_p$. Write $u = \sum_{S \in \underline{\underline{S}}_p} c_S e_S$ where $c_S \in \mathbb{K}$. If $S \in \underline{\underline{S}}_p$ is dependent then $\varphi(e_S) = 0$. If $S \in \underline{\underline{S}}_p$ is independent then $r(\cap S) = p$ and $e_S \in E_{\cap S}$ implies $\varphi(e_S) \in A_{\cap S}$. Thus $a = \varphi(u)$ lies in the sum of the $A_X$ with $r(X) = p$. □

(2.28) Definition. Let $T = T(\mathcal{G}) = \bigcap_{H \in \mathcal{G}} H$ denote the maximal element of L. Define the rank of $\mathcal{G}$ by $r(\mathcal{G}) = r(T(\mathcal{G}))$.

(2.29) Definition. If $X \in L$ let $\mathcal{G}_X = \{H \in \mathcal{G} \mid H \supseteq X\} = \{H \in \mathcal{G} \mid H \leq X\}$.

Our next aim is to prove that $\dim(A_X) = (-1)^{r(X)}\mu(X)$. We will do this by induction on $r(\mathcal{G})$ after several lemmas. Here it is important to make explicit the dependence on $\mathcal{G}$ of all the spaces we have constructed. If $\mathcal{B}$ is a subarrangement of $\mathcal{G}$ we view $E(\mathcal{B})$ as a subalgebra of $E(\mathcal{G})$ and $L(\mathcal{B})$ as a sublattice of $L(\mathcal{G})$. Also $\underline{\underline{S}}(\mathcal{B}) \subseteq S(\mathcal{G})$ and an element $S \in \underline{\underline{S}}(\mathcal{B})$ is dependent viewed in $\underline{\underline{S}}(\mathcal{B})$ if and only if it is dependent in $\underline{\underline{S}}(\mathcal{G})$. We write $E = E(\mathcal{G})$, $I = I(\mathcal{G})$, $J = J(\mathcal{G})$, $K = K(\mathcal{G})$ and $A = A(\mathcal{G})$. If $X \in L(\mathcal{G})$ we write $\underline{\underline{S}}_X = \underline{\underline{S}}_X(\mathcal{G})$, $K_X = K_X(\mathcal{G})$ and $A_X = A_X(\mathcal{G})$. Note that $\partial_{E(\mathcal{B})}$ is the restriction of $\partial_{E(\mathcal{G})}$ to $E(\mathcal{B})$. Since $J(\mathcal{B}) \subseteq J(\mathcal{G})$ and $J'(\mathcal{B}) \subseteq J'(\mathcal{G})$ the projection $\pi_{\mathcal{B}}$ of $E(\mathcal{B})$ onto $J'(\mathcal{B})$ is the restriction to $E(\mathcal{B})$ of the projection $\pi_{\mathcal{G}}$ of $E(\mathcal{G})$ onto $J'(\mathcal{G})$. For simplicity we let $\partial$ denote both $\partial_{E(\mathcal{G})}$ and $\partial_{E(\mathcal{B})}$, and we let $\pi$ denote both $\pi_{\mathcal{G}}$ and $\pi_{\mathcal{B}}$. Thus $K(\mathcal{B}) = \pi(\partial J(\mathcal{B})) \subseteq \pi(\partial J(\mathcal{G})) = K(\mathcal{G})$. It follows that $K_X(\mathcal{B}) \subseteq K_X(\mathcal{G})$ for any $X \in L(\mathcal{G})$.

Clearly $I(\mathcal{G}) \cap E(\mathcal{B}) \supseteq I(\mathcal{B})$ for any subarrangement $\mathcal{B}$ of $\mathcal{G}$. Lemma 2.14 of [6] asserts that

$$I(\mathcal{G}) \cap E(\mathcal{B}) = I(\mathcal{G}) \tag{2.30}$$

for any subarrangement $\mathcal{B}$ of $\mathcal{G}$. The proof given there is correct in case $\mathcal{B} = \mathcal{G}_X$ for some $X \in L(\mathcal{G})$ and, fortunately, this is the only case used in the rest of [6]. The mistake in [6, Lemma 2.14] lies in the assertion that $E_X(\mathcal{G}) \subseteq E(\mathcal{B})$ for all $X \in L(\mathcal{B})$[1]. In this section we prove (2.30) in case $\mathcal{B} = \mathcal{G}_X$ using the argument in [6] and use it to deduce the formula for $\dim A_X$. In Section 3 we construct an exact sequence which allows us to prove in (3.15) that (2.30) is in fact correct for all $\mathcal{B}$. Construction of the

[1] The first two authors would like to thank George Glauberman, W. A. M. Janssen and Nguyen Viet Dung for this remark.

exact sequence uses the truth of (2.30) in case $\mathcal{B} = \mathcal{A}_X$. We have no direct argument for (2.30). To isolate the difficulty we prove a lemma about any subarrangement $\mathcal{B}$ of $\mathcal{A}$ which points in the direction of (2.30).

(2.31) Lemma. Let $\mathcal{B}$ be a subarrangement of $\mathcal{A}$. Then

$$I(\mathcal{A}) \cap E(\mathcal{B}) = J(\mathcal{B}) \oplus \Big( \bigoplus_{Y \in L(\mathcal{B})} (K_Y(\mathcal{A}) \cap E(\mathcal{B})\Big).$$

Proof. Linearly order the elements of $\mathcal{A}$. Say that $S = (H_1, \ldots, H_p) \in \underline{\underline{S}}(\mathcal{A})$ is standard if $H_1 < \cdots < H_p$ in the linear order. The elements $e_S$ with $S \in \underline{\underline{S}}$ standard are a $\mathbb{K}$-basis for $E(\mathcal{A})$. Thus every element $u \in E(\mathcal{A})$ may be written uniquely as $u = \sum c_S e_S$, sum over standard $S \in \underline{\underline{S}}$. Define the support supp(u) of u to be the set of all standard S for which $c_S \neq 0$. Define the support supp(F) of a subspace F of $E(\mathcal{A})$ to be the set of supports of its elements. Let $F_1, F_2, \ldots$ be subspaces of $E(\mathcal{A})$ with pairwise disjoint supports. If $u_k \in F_k$ for $k = 1, 2, \ldots$ and $\sum u_k \in E(\mathcal{B})$ then $u_k \in E(\mathcal{B})$ for $k = 1, 2, \ldots$. Thus

$$(\sum F_k) \cap E(\mathcal{B}) = \sum (F_k \cap E(\mathcal{B})). \tag{2.32}$$

Note that supp $J(\mathcal{A})$ consists of dependent tuples. If $X \in L(\mathcal{A})$ then supp $K_X(\mathcal{A})$ consists of independent tuples and the supports of the various $K_X(\mathcal{A})$, $X \in L(\mathcal{A})$, are pairwise disjoint because $K_X(\mathcal{A}) = K(\mathcal{A}) \cap E_X(\mathcal{A}) \subseteq E_X(\mathcal{A})$. Thus we may apply (2.32) to the direct sum decomposition of $I(\mathcal{A})$ given by (2.21) and (2.22) and conclude that

$$I(\mathcal{A}) \cap E(\mathcal{B}) = (J(\mathcal{A}) \cap E(\mathcal{B})) \oplus \bigoplus_{X \in L(\mathcal{A})} (K_X(\mathcal{A}) \cap E(\mathcal{B})). \tag{2.33}$$

It follows from the definition of $J(\mathcal{A})$ and $J(\mathcal{B})$ that $J(\mathcal{A}) \cap E(\mathcal{B}) = J(\mathcal{B})$. Suppose $X \in L(\mathcal{A})$ and $K_X(\mathcal{A}) \cap E(\mathcal{B}) \neq 0$. Then $E_X(\mathcal{A}) \cap E(\mathcal{B}) \neq 0$. If u is a nonzero element in $E_X(\mathcal{A}) \cap E(\mathcal{B})$ then there exists $S \in$ supp(u) with $S \in \underline{\underline{S}}_X(\mathcal{A}) \cap \underline{\underline{S}}(\mathcal{B})$ and thus $X \in L(\mathcal{B})$. Thus the lemma follows from (2.33). □

(2.34) Lemma. If $X \in L(\mathcal{A})$ then $I(\mathcal{A}) \cap E(\mathcal{A}_X) = I(\mathcal{A}_X)$.

Proof. Let $\mathcal{B} = \mathcal{A}_X$. Suppose $Y \in L(\mathcal{B})$. If $S \in \underline{\underline{S}}_Y(\mathcal{A})$ then $S = (H_1, \ldots, H_p)$ with $\cap H_k = Y$. Thus $H_k \supseteq Y$ for $k = 1, \ldots, p$. Since $Y \in L(\mathcal{A}_X)$ we have $Y \supseteq X$ so $H_k \supseteq X$. Thus $S \in \underline{\underline{S}}_Y(\mathcal{B})$. This shows that $\underline{\underline{S}}_Y(\mathcal{A}) \subseteq \underline{\underline{S}}_Y(\mathcal{B})$ and thus, since $\mathcal{B} \subseteq \mathcal{A}$ we have $\underline{\underline{S}}_Y(\mathcal{A}) = \underline{\underline{S}}_Y(\mathcal{B})$. Thus $E_Y(\mathcal{A}) = E_Y(\mathcal{B})$ and $J_Y(\mathcal{A}) = J_Y(\mathcal{B})$. It follows from (2.23) that $K_Y(\mathcal{A}) = \pi(\partial J_Y(\mathcal{A})) = \pi(\partial J_Y(\mathcal{B})) = K_Y(\mathcal{B})$. Since $K_Y(\mathcal{B}) \subseteq E(\mathcal{B})$ it follows from (2.31) that

$$I(\mathcal{A}) \cap E(\mathcal{B}) = J(\mathcal{B}) \oplus (\bigoplus_{Y \in L(\mathcal{B})} K_Y(\mathcal{B})).$$

Now (2.21) and (2.22) applied to the arrangement $\mathcal{B}$ show that the right hand side of the last equation is $I(\mathcal{B})$. □

Let $\mathcal{B}$ be a subarrangement of $\mathcal{A}$. Since $I(\mathcal{A}) \cap E(\mathcal{B}) \supseteq I(\mathcal{B})$ the inclusion $E(\mathcal{B}) \subseteq E(\mathcal{A})$ induces a $\mathbb{K}$-algebra homomorphism

$$i : A(\mathcal{B}) \to A(\mathcal{A}) \tag{2.35}$$

such that

$$i(e_H + I(\mathcal{B})) = e_H + I(\mathcal{A}) \qquad H \in \mathcal{A}. \tag{2.36}$$

Note that $i$ is a monomorphism precisely when (2.30) holds. Thus (2.34) shows that $i$ is a monomorphism when $\mathcal{B} = \mathcal{A}_X$.

(2.37) Theorem. Let $\mathcal{A}$ be an arrangement and let $A = A(\mathcal{A})$. If $X \in L(\mathcal{A})$ then $\dim(A_X) = (-1)^{r(X)}\mu(X)$.

Proof. We argue by induction on the rank $r(\mathcal{A})$. If $r(\mathcal{A}) = 0$ then $\mathcal{A}$ is void so $A = \mathbb{K}$ and the assertion is clear. If $r(\mathcal{A}) = 1$ then $\mathcal{A} = \{H\}$ so $A = \mathbb{K} + \mathbb{K}a_H$ and the assertion is clear. Suppose $r(\mathcal{A}) \geq 2$. Fix $X \in L$ and let $i : A(\mathcal{A}_X) \to A(\mathcal{A})$ be the homomorphism corresponding to the subarrangement $\mathcal{A}_X$. Lemma 2.34 shows that $i$ is a monomorphism. The space $A_X(\mathcal{A}) = \varphi(E_X(\mathcal{A}))$ is spanned over $\mathbb{K}$ by all elements $e_S + I(\mathcal{A})$ with $S \in \underline{\underline{S}}_X(\mathcal{A})$. Similarly $A_X(\mathcal{A}_X)$ is spanned over $\mathbb{K}$ by all elements $e_S + I(\mathcal{A}_X)$ with $S \in \underline{\underline{S}}_X(\mathcal{A}_X)$. Since $S_X(\mathcal{A}) = S_X(\mathcal{A}_X)$ we have

$i(A_X(\mathcal{A}_X)) = A_X(\mathcal{A})$. Since $i$ is a monomorphism it follows that $\dim A_X(\mathcal{A}_X) = \dim A_X(\mathcal{A})$. Suppose $X \neq T(\mathcal{A})$. Then $r(\mathcal{A}_X) < r(\mathcal{A})$. Let $\mu_X$ be the Möbius function of $L(\mathcal{A}_X)$. By induction $\dim A_X(\mathcal{A}_X) = (-1)^{r(X)}\mu_X(X) = (-1)^{r(X)}\mu(X)$ so $\dim A_X(\mathcal{A}) = (-1)^{r(X)}\mu(X)$. Let $T = T(\mathcal{A})$. It follows from (2.10) and (2.27) that

$$0 = \sum_{p\geq 0} (-1)^p \dim A_p(\mathcal{A}) = \sum_{X\in L} (-1)^{r(X)} \dim A_X(\mathcal{A})$$

$$= (-1)^{r(T)} \dim A_T(\mathcal{A}) + \sum_{\substack{X\in L\\ X\neq T}} \mu(X) .$$

Since $\sum_{X\in L} \mu(X) = 0$ this shows that $\mu(T) = (-1)^{r(T)} \dim A_T(\mathcal{A})$. □

(2.38) Theorem. Let $\mathcal{A}$ be an arrangement, let $L = L(\mathcal{A})$ and let $A = A(\mathcal{A})$. The Poincaré polynomial of the graded $\mathbb{K}$-algebra $A$ is

$$P(A, t) = \sum_{X\in L} \mu(X)(-t)^{\operatorname{codim} X} .$$

Proof. This follows at once from (2.27) and (2.37). □

The characteristic polynomial of the lattice $L = L(\mathcal{A})$ is by definition [1, p. 155]

(2.39)
$$\chi(L, t) = \sum_{X\in L} \mu(X)\, t^{r(L)-r(X)}$$

where $r(L) = r(\mathcal{A})$ is the rank of $L$. Since $r(X) = \operatorname{codim} X$ it follows from (2.38) and (2.39) that the Poincaré polynomial of $A$ and the characteristic polynomial of $L$ are related by the formula

(2.40)
$$P(A, t) = (-t)^{r(L)} \chi(L, -t^{-1}).$$

(2.41) Example. Let $\dim V = 3$ and let $x_1, x_2, x_3$ be a basis for $V^*$. Let $H_0 = \ker(x_1 + x_2)$ and for $m = 1, 2, 3$ let $H_m = \ker(x_m)$. Let $\mathcal{A} = \{H_0, H_1, H_2, H_3\}$. Write $e_m = e_{H_m}$ and $a_m = e_m + I$ for $m = 0, 1, 2, 3$. The ideal $I$ is generated by $\partial(e_0 e_1 e_2)$ and $\partial(e_0 e_1 e_2 e_3)$.

The generators $a_m$ are thus subject to the relations

$$a_1a_2 - a_0a_2 + a_0a_1 = 0$$

$$a_1a_2a_3 - a_0a_2a_3 + a_0a_1a_3 - a_0a_1a_2 = 0.$$

We have $J = \mathbb{K}e_0e_1e_2 + \mathbb{K}e_0e_1e_2e_3$. Let $S = (H_0, H_1, H_2)$ and let $T = (H_0, H_1, H_2, H_3)$. The nonzero subspaces $K_X$ are $K_{H_1 \cap H_2}$ spanned by $\pi(\partial e_S) = \partial e_S = e_1e_2 - e_0e_2 + e_0e_1$ and $K_{H_1 \cap H_2 \cap H_3}$ spanned by $\pi(\partial e_T) = e_1e_2e_3 - e_0e_2e_3 + e_0e_1e_3$. Thus $\dim I = \dim J + \dim K = 4$ and $\dim A = \dim E - \dim I = 12$. The Poincaré polynomials of $J$ and $K$ are $P(J,t) = t^3 + t^4$ and $P(K,t) = t^2 + t^3$. Since $I = J \oplus K$ we have $P(I,t) = t^2 + 2t^3 + t^4 = t^2(t+1)^2$. Thus

$$P(A,t) = P(E,t) - P(I,t) = (1+t)^4 - t^2(1+t)^2 = (1+t)^2(1+2t).$$

On the other hand direct computation of the Mobius function shows that $\chi(L,t) = (t-1)^2(t-2)$ as in (2.40). The algebra $A$ has a $\mathbb{K}$-basis

$$1, a_0, a_1, a_2, a_3, a_0a_1, a_0a_2, a_0a_3, a_1a_3, a_2a_3, a_0a_2a_3, a_1a_2a_3.$$

## 3. THE EXACT SEQUENCE OF A TRIPLE $(\mathcal{A}, \mathcal{A}', \mathcal{A}'')$.

Let $\mathcal{A} = (\mathcal{A}, V)$ be a non void arrangement. Choose $H_0 \in \mathcal{A}$ and hold it fixed throughout this section.

(3.1) Definition. The subarrangement $\mathcal{A}' = \mathcal{A} - \{H_0\}$ is called the deletion of $H_0$ from $\mathcal{A}$. The arrangement $\mathcal{A}'' = \{H_0 \cap H \mid H \in \mathcal{A}'\}$ is called the restriction of $\mathcal{A}$ to $H_0$. We say that $(\mathcal{A}, \mathcal{A}', \mathcal{A}'')$ is a triple of arrangements.

Note that $\mathcal{A}'$ is an arrangement in $V$ and $\mathcal{A}''$ is an arrangement in $H_0$. Our aim in this section is to construct $\mathbb{K}$-linear maps $i = i_A : A(\mathcal{A}') \to A(\mathcal{A})$ and $j = j_A : A(\mathcal{A}) \to A(\mathcal{A}'')$ such that the sequence

(3.2) $$0 \to A(\mathcal{A}') \xrightarrow{i} A(\mathcal{A}) \xrightarrow{j} A(\mathcal{A}'') \to 0$$

is exact. It seems appropriate to comment on our notation and terminology which is not standard. The restriction $\mathcal{A}''$ has been called the induced arrangement on $H_0$ [10, p. 16]. We use the term restriction because our ultimate interest is in the restriction of linear forms or differential forms on $V$ to $H_0$. In matroid terminology $\mathcal{A}''$ is the contraction of $\mathcal{A}$ by $H_0$ usually written $\mathcal{A}/H_0$. The $\mathcal{A}, \mathcal{A}', \mathcal{A}''$ notation is appropriate here for its homological flavor and because it is simpler in our context. Thus we write $L = L(\mathcal{A})$, $L' = L(\mathcal{A}')$ and $L'' = L(\mathcal{A}'')$ for the corresponding lattices, we write $A = A(\mathcal{A})$, $A' = A(\mathcal{A}')$ and $A'' = A(\mathcal{A}'')$ for the corresponding algebras, and we use notation like $E, E', E''$ or $\underline{\underline{S}}, \underline{\underline{S}}', \underline{\underline{S}}''$ or $J, J', J''$ inherited from Section 2, without comment.

(3.3) Proposition. Let $(\mathcal{A}, \mathcal{A}', \mathcal{A}'')$ be a triple of arrangements and let $A, A', A''$ be the corresponding algebras. Then

$$P(A,t) = P(A',t) + tP(A'',t) .$$

Proof. Suppose first that $r(L') < r(L)$. Then $H_0$ is a separator of $L$ in the sense of geometric lattices [1, Thm. 2.45.iii]. By [1, Thm. 2.45.iv] we have $L \simeq L' \times L_0$, direct product of lattices, where $L_0$ is the lattice consisting of the two subspaces $V$ and $H_0$. Furthermore the mapping $X \to X \cap H_0$, $X \in L'$, is a lattice isomorphism from $L'$ to $L''$. Thus $\chi(L,t) = \chi(L_0,t)\chi(L',t) = (t-1)\chi(L',t)$ and $P(A',t) = P(A'',t)$. It follows from these formulas and (2.40) that

$$P(A,t) = (-t)^{r(L)}\chi(L,-t^{-1}) = (1+t)(-t)^{r(L')}\chi(L',-t^{-1})$$

$$= (1+t)P(A',t) = P(A',t) + tP(A'',t) .$$

This proves the assertion in case $r(L') < r(L)$. Suppose that $r(L') = r(L)$. Then $H_0$ is not a separator of $L$. A theorem of Brylawski [3, Thm. 4.2] on matroids states, when written in the language of geometric lattices, that if $H_0$ is not a separator then

$$(-1)^{r(L)}\chi(L,t) = (-1)^{r(L')}\chi(L',t) + (-1)^{r(L'')}\chi(L'',t) . \qquad (3.4)$$

In Brylawski' terminology (3.4) says that $(-1)^{r(L)}\chi(L,t)$ is a Tutte-

Grothendieck invariant. Since $r(L'') = r(L) - 1$ the Proposition follows directly from (3.4) and (2.40). □

The formula (3.4) was first applied in the context of arrangements by Zaslavsky [10, Lemma 4A4].

(3.5) Corollary. Let $(\mathcal{A}, \mathcal{A}', \mathcal{A}'')$ be a triple of arrangements and let $A, A', A''$ be the corresponding algebras. Then

$$\dim A = \dim A' + \dim A''.$$

Proof. Let $t = 1$ in (3.3). □

Let $i : A' \to A$ be the $\mathbb{K}$-algebra homomorphism defined in (2.35) which is induced by the inclusion $E' \subset E$. If $H \in \mathcal{A}$ write $a_H = e_H + I$ as in Section 2. If $H \in \mathcal{A}'$ it is important to distinguish between $a_H$ and $e_H + I'$; we cannot identify the two elements because we do not know that $i$ is a monomorphism. If $S = (H_1, \ldots, H_p) \in \underline{\underline{S}}$ write $a_S = a_{H_1} \cdots a_{H_p}$ as in Section 2. If $S \in \underline{\underline{S}}'$ then $a_S \in iA'$. The hyperplanes of $\mathcal{A}''$ have the form $H_0 \cap H$ where $H \in \mathcal{A}'$. We write the corresponding generators of $E''$ and $A''$ as $e_{H_0 \cap H}$ and $a_{H_0 \cap H}$. If $S = (H_1, \ldots, H_p) \in \underline{\underline{S}}$ and $\sigma$ is a permutation of $1, \ldots, p$ let $\sigma S = (H_{\sigma 1}, \ldots, H_{\sigma p})$. To define a $\mathbb{K}$-linear map $\theta$ from $E$ to some vector space over $\mathbb{K}$ it suffices to prescribe the values $\theta(e_S)$ for $S \in \underline{\underline{S}}$ and check that $\theta(e_{\sigma S}) = \operatorname{sgn}(\sigma)\,\theta(e_S)$. We apply this remark to $E$ and $E''$ in Lemmas (3.6) and (3.8). For convenience we agree that if $H_0 \in S$ then $H_0$ is the first element of the tuple $S$ and write $S = (H_0, H_1, \ldots, H_p)$ where $H_1, \ldots, H_p \in \mathcal{A}'$.

(3.6) Lemma. There exists a surjective $\mathbb{K}$-linear map $\theta : E \to E''$ such that

$$\theta(e_{H_1} \cdots e_{H_p}) = 0$$

$$\theta(e_{H_0} e_{H_1} \cdots e_{H_p}) = e_{H_0 \cap H_1} \cdots e_{H_0 \cap H_p}$$

for all $(H_1, \cdots, H_p) \in \underline{\underline{S}}'$. This map satisfies $\theta I \subseteq I''$.

Proof. Since $E = E' \oplus e_{H_0} E'$ the remarks of the preceding paragraph allow us to define $\theta$ by the formulas in the lemma. It is understood that $\theta(1) = 0$ and $\theta(e_{H_0}) = 1$. For convenience we define $\lambda : \mathcal{A}' \to \mathcal{A}''$ by $\lambda H = H_0 \cap H$ for $H \in \mathcal{A}'$. Define $\lambda : \underline{\underline{S}}' \to \underline{\underline{S}}''$ by $\lambda(H_1, \cdots, H_p) = (\lambda H_1, \cdots, \lambda H_p)$. In case $S = (\ )$ we agree that $\lambda S = (\ )$. Thus $\theta(e_S) = 0$ and $\theta(e_{H_0} e_S) = e_{\lambda S}$ for $S \in \underline{\underline{S}}'$. Since $I = J + \partial J$ it suffices to show that $\theta(e) \in I''$ and $\theta(\partial e) \in I''$ for any $e \in J$. Since $\theta(e_S) = 0$ for $S \in \underline{\underline{S}}'$ we may assume that $e = e_{H_0} e_S$ with $S \in \underline{\underline{S}}'$ dependent. Then $(H_0, S)$ is dependent and hence so is $\lambda S$. Thus $\theta(e_{H_0} e_S) = e_{\lambda S} \in J'' \subseteq I''$. We also have $\theta(\partial(e_{H_0} e_S)) = \theta(e_S - e_{H_0} \partial e_S) = -\partial(e_{\lambda S}) \in \partial J'' \subseteq I''$. □

(3.7) Corollary. There exists a surjective $\mathbb{K}$-linear map $j : A \to A''$ such that the diagram

$$\begin{array}{ccc} E & \xrightarrow{\theta} & E'' \\ \varphi \downarrow & & \downarrow \varphi'' \\ A & \xrightarrow{j} & A'' \end{array}$$

commutes. In particular

$$j(a_{H_1} \cdots a_{H_p}) = 0$$

$$j(a_{H_0} a_{H_1} \cdots a_{H_p}) = a_{H_0 \cap H_1} \cdots a_{H_0 \cap H_p}$$

for all $(H_1, \cdots, H_p) \in \underline{\underline{S}}'$.

As in the proof of (3.6) define $\lambda : \mathcal{A}' \to \mathcal{A}''$ by $\lambda H = H_0 \cap H$ for $H \in \mathcal{A}'$. Since $\lambda$ is surjective we may choose a map $\nu : \mathcal{A}'' \to \mathcal{A}'$ such that $\lambda \nu K = \nu K \cap H_0 = K$ for all $K \in \mathcal{A}''$. The map $\nu$ is not canonically defined. Define a map $\nu : \underline{\underline{S}}'' \to \underline{\underline{S}}'$ by $\nu(K_1, \cdots, K_p) = (\nu K_1, \cdots, \nu K_p)$. In case $S = (\ )$ is the empty tuple in $\underline{\underline{S}}''$ we agree that $\nu S = (\ )$ is the empty tuple in $\underline{\underline{S}}'$.

(3.8) Definition. Let $\rho : E'' \to E$ be the $\mathbb{K}$-linear map defined by

$$\rho(e_S) = e_{H_0} e_{\nu S} \qquad S \in \underline{\underline{S}}''.$$

(3.9) Lemma. Let $e \in E$. Then $e - \rho\theta(e) \in E' + I$.

Proof. It suffices to prove that $e_S - \rho\theta(e_S) \in E' + I$ for any $S \in \underline{\underline{S}}$. If $S \in \underline{\underline{S}}'$ then $\theta(e_S) = 0$ by definition so $e_S \in E' \subseteq E' + I$. Thus it suffices to prove the assertion for tuples of the form $(H_0, S)$ where $S \in \underline{\underline{S}}'$. Since

$$e_{(H_0, S)} - \rho\theta(e_{(H_0, S)}) = e_{H_0}(e_S - e_{\nu\lambda S})$$

it suffices to prove that $e_{H_0}(e_S - e_{\nu\lambda S}) \in E' + I$ for all $S \in \underline{\underline{S}}'$. We argue by induction on the length $p$ of the tuple $S$. If $p = 0$ we have $e_{H_0} - \rho\theta(e_{H_0}) = e_{H_0} - e_{H_0} = 0$. If $p = 1$ we have $S = (H)$ where $H \in \mathcal{A}'$. Since $H_0 \cap H = \lambda H = H_0 \cap \nu\lambda H$ it follows that $(H_0, H, \nu\lambda H)$ is dependent. Thus

$$e_H e_{\nu\lambda H} - e_{H_0} e_{\nu\lambda H} + e_{H_0} e_H = \partial e_{(H_0, H, \nu\lambda H)} \in I.$$

Since $H \in \mathcal{A}'$ and $\nu\lambda H \in \mathcal{A}'$ we have $e_H e_{\nu\lambda H} \in E'$. Thus

$$e_{H_0} e_S - e_{H_0} e_{\nu\lambda S} = e_{H_0} e_H - e_{H_0} e_{\nu\lambda H} \in E' + I.$$

For $p \geq 2$ write $S = (H, T)$ where $H \in \mathcal{A}'$ and $T \in \underline{\underline{S}}'$ has length $p - 1$. Write

$$e_{H_0}(e_S - e_{\nu\lambda S}) = e_{H_0}(e_H - e_{\nu\lambda H}) e_T - e_{\nu\lambda H} e_{H_0}(e_T - e_{\nu\lambda T}).$$

Since $I$ is an ideal of $E$ and $E'$ is a subalgebra of $E$ we have $E'(E' + I) \subseteq E' + I$. Since $e_T$ and $e_{\nu\lambda H}$ are in $E'$ the assertion follows from the case $p = 1$ and the induction hypothesis. □

(3.10) Lemma. $\rho(I'') \subseteq E' + I$.

Proof. By Lemma (2.12.ii) we have $I'' = J'' + \partial J''$. Thus it suffices to show that $\rho(e_T)$ and $\rho(\partial e_T)$ are in $E' + I$ for dependent $T \in \underline{\underline{S}}''$. By definition $\rho(e_T) = e_{H_0} e_{\nu T} = e_{(H_0, \nu T)}$. Since $T$ is dependent so is $(H_0, \nu T)$ and thus $\rho(e_T) \in J \subseteq I \subseteq E' + I$.

Since $e_{\nu T} \in E'$ and $(H_0, \nu T)$ is dependent we have

$$\rho(\partial e_T) = e_{H_0}\,\partial(e_{\nu T}) = e_{\nu T} - \partial(e_{(H_0, \nu T)}) \in E' + I. \qquad \square$$

(3.11) Proposition. The sequence

$$A' \xrightarrow{i} A \xrightarrow{j} A'' \longrightarrow 0$$

is exact.

Proof. We showed in (3.7) that $j$ is surjective and that $j(a_{H_1} \cdots a_{H_p}) = 0$ for all $(H_1, \cdots, H_p) \in \underline{\underline{S}}'$ where $a_H = e_H + I \in A$. Let $a'_H = e_H + I' \in A'$. Then $i(a'_H) = a_H$ so $ji(a'_{H_1} \cdots a'_{H_p}) = 0$. Thus $\operatorname{im} i \subseteq \ker j$. Next we show that $\ker j \subseteq \operatorname{im} i$. Let $a \in \ker j$. Choose $e \in E$ such that $\varphi(e) = a$. It follows from (3.7) that $\varphi''\theta(e) = j\varphi(e) = ja = 0$. Thus $\theta(e) \in \ker \varphi'' = I''$ so by (3.10) we have $\rho\theta(e) \in E' + I$. By (3.9) we have $e - \rho\theta(e) \in E' + I$ and hence

$$e = (e - \rho\theta(e)) + \rho\theta(e) \in E' + I.$$

Thus

$$a = \varphi(e) \in \varphi(E' + I).$$

Since $\varphi(I) = 0$ it follows from the commutative diagram

$$\begin{array}{ccc} E' & \xrightarrow{i} & E \\ {\scriptstyle \varphi'}\downarrow & & \downarrow{\scriptstyle \varphi} \\ A' & \xrightarrow{i} & A \end{array}$$

that $\varphi(E' + I) \subseteq \operatorname{im} I$. $\square$

(3.12) Lemma. The map $i : A' \to A$ is injective.

Proof. Since $\ker j = iA'$ we have

$$\dim A = \dim iA' + \dim A''.$$

By (3.5) it follows that $\dim iA' = \dim A'$ so $i$ is injective. $\square$

(3.13) Theorem. Let $\mathcal{A}$ be an arrangement. Let $H_0 \in \mathcal{A}$ and let $(\mathcal{A}, \mathcal{A}', \mathcal{A}'')$ be the corresponding triple. Let $i = i_A : A(\mathcal{A}') \to A(\mathcal{A})$ be the natural homomorphism and let $j = j_A : A(\mathcal{A}) \to A(\mathcal{A}'')$ be the $\mathbb{K}$-linear map defined by

$$j(a_{H_1} \cdots a_{H_p}) = 0$$

$$j(a_{H_0} a_{H_1} \cdots a_{H_p}) = a_{H_0 \cap H_1} \cdots a_{H_0 \cap H_p}$$

for $(H_1, \ldots, H_p) \in \underline{\underline{S}}(\mathcal{G}')$. Then the sequence

$$0 \to A(\mathcal{G}') \xrightarrow{i} A(\mathcal{G}) \xrightarrow{j} A(\mathcal{G}'') \to 0$$

is exact.

Proof. This follows at once from (3.11) and (3.12). □

(3.14) Remark. This argument is the best we can do at the moment. Note that (3.11) is proved entirely within the framework of this paper while (3.12) depends on a dimension count which uses the Tutte-Grothendieck invariance (3.4) of the characteristic polynomial. We have remarked in the Introduction that this Section could be written with any geometric lattice $L$ in place of $L(\mathcal{G})$, using precisely the same arguments with appropriate changes in notation. If this is done there is no naturally given coefficient field $\mathbb{K}$ for the algebra $A$. The field comes in our case from the underlying vector space $V$. Thus it might be useful to define $A$ and study its properties over any commutative ring $\mathbb{K}$, in particular over $\mathbb{K} = \mathbb{Z}$. Then one can no longer argue (3.12) with a dimension count as we have done; one is forced to prove it directly. If this is done then Brylawski's theorem (3.4) follows from existence of the exact sequence (3.13).

(3.15) Proposition. Let $\mathcal{G}$ be an arrangement and let $\mathcal{B}$ be a subarrangement of $\mathcal{G}$. Then the natural homomorphism $i : A(\mathcal{B}) \to A(\mathcal{G})$ is a monomorphism.

Proof. We have remarked following (2.36) that $i$ is a monomorphism if and only if $I(\mathcal{G}) \cap E(\mathcal{B}) = I(\mathcal{B})$. It follows from (3.12) that $I(\mathcal{G}) \cap E(\mathcal{B}) = I(\mathcal{B})$ in case $|\mathcal{G}| - |\mathcal{B}| = 1$. Using this formula we get $I(\mathcal{G}) \cap E(\mathcal{B}) = I(\mathcal{B})$ for all $\mathcal{B}$ by induction on $|\mathcal{G}| - |\mathcal{B}|$. □

## 4. THE ALGEBRA OF DIFFERENTIAL FORMS.

Let $\mathbb{K}$ be a field and let $V$ be a vector space of dimension $\ell$ over $\mathbb{K}$. Let $S$ be the symmetric algebra of $V^*$ and let $F$ be the quotient field of $S$. It will sometimes be convenient to indicate the dependence of $S$ and $F$ on $V$ in which case we write $S = \mathbb{K}[V]$ and $F = \mathbb{K}(V)$. We view $F \otimes V^* = F \otimes_{\mathbb{K}} V^*$ as vector space over $F$ by defining

$$f(g \otimes \alpha) = fg \otimes \alpha \qquad f, g \in F \qquad \alpha \in V^* .$$

There exists a unique $\mathbb{K}$-linear map $d : F \to F \otimes V^*$ such that

$$d(fg) = f(dg) + g(df) \qquad f, g \in F$$

and $d\alpha = 1$ for $\alpha \in V^*$. If $x_1, \ldots, x_\ell$ is a basis for $V^*$ then $S = \mathbb{K}[x_1, \ldots, x_\ell]$ is a polynomial algebra on $x_1, \ldots, x_\ell$ and $F = \mathbb{K}(x_1, \ldots, x_\ell)$ is the field of rational functions. In terms of this basis the differential $df$ is given by the formula

$$df = \sum_{i=1}^{\ell} \frac{\partial f}{\partial x_i} \otimes x_i = \sum_{i=1}^{\ell} \frac{\partial f}{\partial x_i} dx_i$$

Note that $F \otimes V^* = Fdx_1 \oplus \cdots \oplus Fdx_\ell$.

(4.1) Definition. Let $\Omega(V)$ be the exterior algebra of the $F$-vector space $F \otimes V^*$ graded by $\Omega(V) = \bigoplus_{p=0}^{\ell} \Omega^p(V)$ where

$$\Omega^p(V) = \bigoplus_{1 \le i_1 < \cdots < i_p \le \ell} Fdx_{i_1} \wedge \cdots \wedge dx_{i_p} .$$

If $\omega, \eta \in \Omega(V)$ we often write $\omega\eta = \omega \wedge \eta$. Thus $dx_{i_1} \wedge \cdots \wedge dx_{i_p} = dx_{i_1} \cdots dx_{i_p}$. We agree to identify $\Omega^0(V)$ with $F$. The elements of $\Omega(V)$ are called (rational) differential p-forms on $V$. We recall the definition and properties of the exterior derivative $d : \Omega(V) \to \Omega(V)$.

(4.2) Lemma. The map $d : F \to F \otimes V^*$ may be extended in a unique way to a $\mathbb{K}$-linear map $d : \Omega(V) \to \Omega(V)$ with the properties (i) $d^2 = 0$ and

(ii) $d(\omega\eta) = (d\omega)\eta + (-1)^p \omega(d\eta) \qquad \omega \in \Omega^p(V), \quad \eta \in \Omega(V)$.

If $x_1,\ldots,x_\ell$ is a basis for $V^*$ and $\omega = \sum f_{i_1\ldots i_p} dx_{i_1}\ldots dx_{i_p}$ where $1 \le i_1 < \ldots < i_p \le \ell$ and $f_{i_1\ldots i_p} \in F$ then

(iii) $$d\omega = \sum_{j=1}^{\ell} \sum (\partial f_{i_1\ldots i_p}/\partial x_j)\, dx_j\, dx_{i_1}\ldots dx_{i_p}.$$

<u>Proof</u>. To prove uniqueness suppose $\omega = \sum f_{i_1\ldots i_p} dx_{i_1}\ldots dx_{i_p}$. Then (ii) implies

$$d\omega = \sum (df_{i_1\ldots i_p})\, dx_{i_1}\ldots dx_{i_p} + \sum f_{i_1\ldots i_p}\, d(dx_{i_1}\ldots dx_{i_p}).$$

The second sum vanishes by (i) and (ii). This shows that $d\omega$ is given by (iii) and proves uniqueness. To prove existence define $d\omega$ by (iii) and check (i) and (ii) by direct calculation. □

The main object of study in this section is a $\mathbb{K}$-subalgebra $R = R(\mathcal{A})$ of $\Omega(V)$ defined by an arrangement $\mathcal{A}$ in $V$. If $H \in \mathcal{A}$ choose a nonzero linear form $\alpha_H \in V^*$ such that $\ker(\alpha_H) = H$. Note that $\alpha_H$ is determined uniquely up to a constant multiple.

(4.3) <u>Definition</u>. If $H \in \mathcal{A}$ let $\omega_H = d\alpha_H/\alpha_H \in \Omega^1(V)$. Note that although the linear form $\alpha_H$ is only determined up to a constant multiple, the differential form $\omega_H$ is uniquely determined by $H$.

(4.4) <u>Definition</u>. If $(\mathcal{A}, V)$ is an arrangement, let $R = R(\mathcal{A})$ be the $\mathbb{K}$-subalgebra of $\Omega(V)$ generated by all $\omega_H$, $H \in \mathcal{A}$, and the identity. In case $\mathcal{A}$ is the void arrangement we agree that $R(\mathcal{A}) = \mathbb{K}$.

Let $R_p = R \cap \Omega^p(V)$. Since the generators $\omega_H$ are homogeneous, the algebra $R$ is naturally graded by $R = \bigoplus_{p=0}^{\ell} R_p$.

(4.5) <u>Lemma</u>. There exists a surjective homomorphism $\gamma : A(\mathcal{A}) \to R(\mathcal{A})$ of graded $\mathbb{K}$-algebras such that $\gamma(a_H) = \omega_H$ for all $H \in \mathcal{A}$.

Proof. Define a $\mathbb{K}$-algebra homomorphism $\nu : E(\mathcal{C}) \to R(\mathcal{C})$ by $\nu(e_H) = \omega_H$. To show that $\nu(I(\mathcal{C})) = 0$ we need to show that if $S = (H_1, \ldots, H_p)$ is dependent then $\nu(\partial e_S) = 0$. First note that $\nu(J(\mathcal{C})) = 0$ because $(H_1, \ldots, H_p)$ dependent implies $d\alpha_{H_1} \ldots d\alpha_{H_p} = 0$ and thus $\omega_{H_1} \ldots \omega_{H_p} = 0$. We prove that $\nu(\partial e_S) = 0$ by induction on $p$. Let $\alpha_i = \alpha_{H_i}$ and let $\omega_i = \omega_{H_i}$. For $p = 2$ we have $S = (H, H)$ for some $H \in \mathcal{C}$ so $\partial e_S = 0$ and the assertion is clear. In general let $S_k = (H_1, \ldots, \hat{H}_k, \ldots, H_p)$ for $k = 1, \ldots, p$. If $S_k$ is dependent for some $k$ then $\nu(e_{S_k}) \in \nu(J(\mathcal{C})) = 0$. We may assume by induction that $\nu(\partial e_{S_k}) = 0$. Since $e_S = (-1)^{k-1} e_{H_k} e_{S_k}$ it follows from (2.3) that $(-1)^{k-1}\partial(e_S) = \partial(e_{H_k} e_{S_k}) = e_{S_k} - e_{H_k} \partial(e_{S_k})$ so $\nu(\partial e_S) = 0$ and we are done. Thus we may assume that no proper subset of $\alpha_1, \ldots, \alpha_p$ is linearly dependent so there exist $c_i \in \mathbb{K}$, all nonzero, with $\sum_{i=1}^{p} c_i \alpha_i = 0$. If we replace $\alpha_i$ by $c_i \alpha_i$ then $\omega_i$ is unchanged. Thus we may assume that $\sum \alpha_i = 0$. Suppose $1 \le j \le p-1$. Then $0 = \sum d\alpha_i$ implies $0 = (\sum d\alpha_i)(d\alpha_1 \ldots \widehat{d\alpha_j}\widehat{d\alpha_{j+1}} \ldots d\alpha_p)$ so $d\alpha_1 \ldots \widehat{d\alpha_{j+1}} \ldots d\alpha_p = -d\alpha_1 \ldots \widehat{d\alpha_j} \ldots d\alpha_p$. Define $\eta_j$ by $\alpha_j \eta_j = (-1)^{j-1} \omega_1 \ldots \hat{\omega}_j \ldots \omega_p$. Then $\alpha_1 \ldots \alpha_p \eta_{j+1} = (-1)^j d\alpha_1 \ldots \widehat{d\alpha_{j+1}} \ldots d\alpha_p = (-1)^{j-1} d\alpha_1 \ldots \widehat{d\alpha_j} \ldots d\alpha_p = \alpha_1 \ldots \alpha_p \eta_j$. Let $\eta$ be the common value of the $\eta_j$. Then

$$\nu(\partial e_S) = \sum_{i=1}^{p} (-1)^{i-1} \omega_1 \ldots \hat{\omega}_i \ldots \omega_p = \sum_{i=1}^{p} \alpha_i \eta_i = (\sum_{i=1}^{p} \alpha_i)\eta = 0.$$

Thus $\nu(I(\mathcal{C})) = 0$ and $\nu$ induces $\gamma : A(\mathcal{C}) \to R(\mathcal{C})$ such that $\gamma(a_H) = \omega_H$. □

We will prove that $\gamma : A(\mathcal{C}) \to R(\mathcal{C})$ is an isomorphism of $\mathbb{K}$-algebras. Then the theorems of Section 2 may be translated into theorems about the structure of $R(\mathcal{C})$ and give us the main results of this paper. Assume that $\mathcal{C}$ is non void and choose $H_0 \in \mathcal{C}$. Let $(\mathcal{C}, \mathcal{C}', \mathcal{C}'')$ be the corresponding triple. We will construct $\mathbb{K}$-linear maps $i = i_R : R(\mathcal{C}') \to R(\mathcal{C})$ and $j = j_R : R(\mathcal{C}) \to R(\mathcal{C}'')$ and then prove that there is an exact sequence for $R(\mathcal{C})$

analogous to (3.2) and a commutative diagram

$$(4.6)\qquad \begin{array}{ccccccccc} 0 & \to & A(\mathcal{C}') & \xrightarrow{i} & A(\mathcal{C}) & \xrightarrow{j} & A(\mathcal{C}'') & \to & 0 \\ & & \downarrow \gamma' & & \downarrow \gamma & & \downarrow \gamma'' & & \\ 0 & \to & R(\mathcal{C}') & \xrightarrow{i} & R(\mathcal{C}) & \xrightarrow{j} & R(\mathcal{C}'') & \to & 0 \end{array}$$

where the maps $\gamma', \gamma''$ are defined on $A(\mathcal{C}')$ and $A(\mathcal{C}'')$ just as $\gamma$ is on $A(\mathcal{C})$. It is this diagram which allows us to argue by induction and prove that $A(\mathcal{C}) \simeq R(\mathcal{C})$. Note that $R(\mathcal{C}')$ and $R(\mathcal{C})$ are both subalgebras of $\Omega(V)$ and that $R(\mathcal{C}') \subseteq R(\mathcal{C})$. We choose $i : R(\mathcal{C}') \to R(\mathcal{C})$ to be the inclusion map. In Section 2, injectivity of $i$ was a subtle point because we had no analogue of $\Omega(V)$. The injectivity of $i = i_R$ is clear.

To define $j$ we use the Leray residue map on differential forms. The definition of the residue and the preliminary Lemmas (4.8) and (4.10) are adapted from Pham's discussion [7, Chap. III] of the residue in case $\mathbb{K} = \mathbb{C}$ and the forms are holomorphic. Let $\alpha_0 = \alpha_{H_0}$ and let $S_0$ be the localization of $S$ at $\alpha_0$. By definition $S_0$ is the subring of $F$ consisting of all $f/g$ such that $f, g \in S$ and $g$ is prime to $\alpha_0$. Let $\rho : V^* \to H_0^*$ be the restriction map and let $y_i = \rho(x_i)$. We may extend $\rho$ uniquely to a $\mathbb{K}$-algebra homomorphism $\rho : S_0 \to \mathbb{K}(H_0)$. Both existence and uniqueness are shown with the formula $\rho(f/g) = f(y_1,\dots,y_\ell)/g(y_1,\dots,y_\ell)$. Note that $g(y_1,\dots,y_\ell) \neq 0$ because $g$ is prime to $\alpha_0$. Define a $\mathbb{K}$-subalgebra $\Omega_0$ of $\Omega(V)$ by

$$\Omega_0 = \bigoplus_{p=0}^{\ell} \bigoplus_{i_1 < \dots < i_p} S_0\, dx_{i_1} \dots dx_{i_p}.$$

This subalgebra does not depend on the basis $x_1,\dots,x_\ell$.

(4.7) <u>Lemma</u>. The map $\rho : S_0 \to \mathbb{K}(H_0)$ may be extended in a unique way to a $\mathbb{K}$-linear map $\rho : \Omega_0 \to \Omega(H_0)$ with the properties

(i) $\rho(\omega\eta) = \rho(\omega)\rho(\eta) \qquad \omega, \eta \in \Omega_0$

(ii) $\rho(f\omega) = \rho(f)\rho(\omega) \qquad f \in S_0,\ \omega \in \Omega_0$

(iii) $\rho(d\alpha) = d\rho(\alpha) \qquad \alpha \in V^*$.

If $x_1, \ldots, x_\ell$ is a basis for $V^*$ and $\omega = \sum f_{i_1 \ldots i_p} dx_{i_1} \ldots dx_{i_p}$ then

(iv) $\rho(\omega) = \sum f_{i_1 \ldots i_p}(y_1, \ldots, y_\ell) dy_{i_1} \ldots dy_{i_p}$.

<u>Proof</u>. If $\omega = \sum f_{i_1 \ldots i_p} dx_{i_1} \ldots dx_{i_p}$ and $\rho$ has the properties (i)-(iii) then

$$\rho(\omega) = \sum \rho(f_{i_1 \ldots i_p}) \rho(dx_{i_1}) \ldots \rho(dx_{i_p}) = \sum \rho(f_{i_1 \ldots i_p}) dy_{i_1} \ldots dy_{i_p}.$$

This shows $\rho(\omega)$ is given by (iv) and proves uniqueness. To prove existence define $\rho(\omega)$ by (iv) and then (i)-(iii) are clear. □

(4.8) <u>Lemma</u>. Suppose $\alpha \in V^*$ and $\alpha \neq 0$. If $\omega \in \Omega_0$ and $d\alpha \wedge \omega = 0$ then there exists $\psi \in \Omega_0$ with $\omega = d\alpha \wedge \psi$.

<u>Proof</u>. Choose a basis $x_1, \ldots, x_\ell$ for $V^*$ such that $\alpha = x_1$. We may assume that $\omega \in \Omega^p(V)$ for some $p$. Write $\omega = \sum f_{i_1 \ldots i_p} dx_{i_1} \ldots dx_{i_p}$ where $f_{i_1 \ldots i_p} \in S_0$ and the sum is over all $1 \leq i_1 < \ldots < i_p \leq \ell$. Then

$$0 = dx_1 \wedge \omega = \sum{}' f_{i_1 \ldots i_p} dx_1 dx_{i_1} \ldots dx_{i_p}$$

where the sum is over all $2 \leq i_1 < \ldots < i_p \leq \ell$. Thus $f_{i_1 \ldots i_p} = 0$ if $i_1 \geq 2$. □

(4.9) <u>Definition</u>. Say that $\varphi \in \Omega(V)$ has at most a simple pole along $H_0$ if $\alpha_0 \varphi \in \Omega_0$.

(4.10) <u>Lemma</u>. Suppose $\varphi \in \Omega(V)$ has at most a simple pole along $H_0$ and that $d\varphi = 0$. Then there exist $\psi, \theta \in \Omega_0$ such that $\varphi = (d\alpha_0/\alpha_0) \wedge \psi + \theta$. The form $\rho(\psi) \in \Omega(H_0)$ is uniquely determined by $\varphi$.

<u>Proof</u>. For simplicity write $\alpha = \alpha_0$. Since $d\varphi = 0$ it follows from (4.2.ii) that $d(\alpha\varphi) = d\alpha \wedge \varphi + \alpha \wedge d\varphi = d\alpha \wedge \varphi$. Since $\alpha\varphi \in \Omega_0$ by hypothesis and $d\Omega_0 \subseteq \Omega_0$, it follows from (4.8) that there exists $\theta \in \Omega_0$ such that

$d(\alpha\varphi) = d\alpha \wedge \theta$. Thus $d\alpha \wedge \varphi = d\alpha \wedge \theta$, which implies $d\alpha \wedge \alpha(\varphi - \theta) = 0$. since $\alpha(\varphi - \theta) \in \Omega_0$ it follows from (4.8) that there exists $\psi \in \Omega_0$ such that $\alpha(\varphi - \theta) = d\alpha \wedge \psi$. This proves the existence of $\theta$ and $\psi$.

To prove the uniqueness of $\rho(\psi)$ it suffices to show that if $\psi, \theta \in \Omega_0$ and $(d\alpha/\alpha)\wedge\psi + \theta = 0$ then $\rho(\psi) = 0$. First note that $d\alpha \wedge \theta = 0$. It follows from (4.8) that there exists $\theta' \in \Omega_0$ such that $\theta = d\alpha \wedge \theta'$. Now $d\alpha \wedge (\psi + \alpha\theta') = d\alpha \wedge \psi + \alpha\theta = 0$. Since $\psi + \alpha\theta' \in \Omega_0$ we may apply (4.8) again to conclude that there exists $\theta'' \in \Omega_0$ with $\psi + \alpha\theta' = d\alpha \wedge \theta''$. Since $\rho(\alpha) = 0$ it follows from (4.7) that $\rho(\alpha\theta') = 0$ and $\rho(d\alpha \wedge \theta'') = 0$. Thus $\rho(\psi) = 0$. □

(4.11) Definition. The uniquely determined form $\rho(\psi)$ is called the residue of $\varphi$ along $H_0$ and we denote it $\mathrm{res}(\varphi)$.

If $H \in \mathcal{C}$ then $d\omega_H = d(d\alpha_H/\alpha_H) = d\alpha_H \wedge d\alpha_H + (1/\alpha_H)d(d\alpha_H) = 0$ so $d(\omega_{H_1} \cdots \omega_{H_p}) = 0$ for all $H_1, \ldots, H_p \in \mathcal{C}$ and thus $d\varphi = 0$ for all $\varphi \in R(\mathcal{C})$. It is clear from the definition that each $\varphi \in R(\mathcal{C})$ has at most a simple pole along $H_0$. Thus $\mathrm{res}(\varphi)$ is defined for all $\varphi \in R(\mathcal{C})$.

(4.12) Lemma. Suppose $H_1, \ldots, H_p \in \mathcal{C}'$. Then

(i) $\mathrm{res}(\omega_{H_1} \cdots \omega_{H_p}) = 0$

(ii) $\mathrm{res}(\omega_{H_0}\omega_{H_1} \cdots \omega_{H_p}) = \omega_{H_0 \cap H_1} \cdots \omega_{H_0 \cap H_p}$.

(iii) $\mathrm{res}\, R(\mathcal{C}) \subseteq R(\mathcal{C}'')$.

Proof. In case $p = 0$ formulas (i) and (ii) are to be interpreted as $\mathrm{res}(1) = 0$ and $\mathrm{res}(\omega_{H_0}) = 1$. Let $\varphi = \omega_{H_1} \cdots \omega_{H_p}$. We may choose $\psi = 0$ and $\theta = \omega_{H_1} \cdots \omega_{H_p}$ in (4.10). This shows that $\mathrm{res}(\varphi) = 0$ and proves (i). Now let $\varphi = \omega_{H_0}\omega_{H_1} \cdots \omega_{H_p}$. We may choose $\psi = \omega_{H_1} \cdots \omega_{H_p}$ and $\theta = 0$ in (4.10). This shows that $\mathrm{res}(\varphi) = \rho(\omega_{H_1} \cdots \omega_{H_p})$. By (4.7.i) we have $\rho(\omega_{H_1} \cdots \omega_{H_p}) = \rho(\omega_{H_1}) \cdots \rho(\omega_{H_p})$.

Thus we must show $\rho(\omega_{H_i}) = \omega_{H_0 \cap H_i}$. If $H \in \mathcal{C}'$ then it follows from (4.7.i) and (4.7.iii) that $\rho(\omega_H) = \rho(d\alpha_H/\alpha_H) = d\rho(\alpha_H)/\rho(\alpha_H)$. Since $\rho(\alpha_H)$ is a linear form on $H_0$ which defines the hyperplane $H_0 \cap H$ we have $\rho(\omega_H) = \omega_{H_0 \cap H}$. This proves (ii). To prove (iii) note that since $\omega_{H_0}^2 = 0$ it follows from the definition of $R(\mathcal{C})$ and $R(\mathcal{C}')$ that $R(\mathcal{C}) = R(\mathcal{C}') + \omega_{H_0} R(\mathcal{C}')$. Thus (iii) follows from (i) and (ii). □

(4.13) Theorem. Let $(\mathcal{C}, V)$ be an arrangement of hyperplanes and let $R(\mathcal{C})$ be the algebra of differential forms generated by the forms $\omega_H = d\alpha_H/\alpha_H$ and the identity. There exists an isomorphism $\gamma : A(\mathcal{C}) \to R(\mathcal{C})$ of graded $\mathbb{K}$-algebras such that $\gamma(a_H) = \omega_H$.

(4.14) Theorem. Let $(\mathcal{C}, V)$ be a non void arrangement and let $(\mathcal{C}, \mathcal{C}', \mathcal{C}'')$ be the triple defined by a hyperplane $H_0 \in \mathcal{C}$. Let $i = i_R : R(\mathcal{C}') \to R(\mathcal{C})$ be the inclusion map and let $j = j_R : R(\mathcal{C}) \to R(\mathcal{C}'')$ be defined by $j(\varphi) = \mathrm{res}(\varphi)$ for $\varphi \in R(\mathcal{C})$, where $\mathrm{res}(\varphi)$ is the residue of $\varphi$ along $H_0$. Then the sequence

$$0 \to R(\mathcal{C}) \xrightarrow{i} R(\mathcal{C}') \xrightarrow{j} R(\mathcal{C}'') \to 0$$

is exact.

Proof. We prove (4.13) and (4.14) simultaneously by induction on $|\mathcal{C}|$. If $\mathcal{C}$ is void then $A(\mathcal{C}) = \mathbb{K} = R(\mathcal{C})$ and (4.13) is clear. If $|\mathcal{C}| = 1$ then $\mathcal{C}'$ and $\mathcal{C}''$ are void. Say $\mathcal{C} = \{H\}$. Then $R(\mathcal{C}) = \mathbb{K} + \mathbb{K}\omega_H$ and $R(\mathcal{C}') = \mathbb{K} = R(\mathcal{C}'')$ so both (4.13) and (4.14) are clear. If $|\mathcal{C}| > 1$ (4.12.iii) shows that $j_R$ maps $R(\mathcal{C})$ into $R(\mathcal{C}'')$, from (4.12.ii) that $j_R$ is surjective and from (4.12.i) that $j_R i_R = 0$. To prove that $\ker j_R \subseteq \operatorname{im} i_R$ consider the diagram (4.6). The diagram is commutative. This is clear for the left hand square by definition of $i_A$ and $i_R$. For the right hand square it follows from (3.7) and (4.12). The top row is exact by (3.13). We may assume by the induction hypothesis in (4.13) that $\gamma'$

and $\gamma''$ are isomorphisms. Now a diagram chase shows that $\ker j_R \subseteq \operatorname{im} i_R$. This proves that the second row of (4.6) is exact. It follows now from the Five Lemma that $\gamma$ is an isomorphism. □

(4.15) Theorem. Let $(\mathcal{A}, V)$ be an arrangement of hyperplanes, let $L = L(\mathcal{A})$ be its intersection lattice and let $R = R(\mathcal{A})$ be the algebra of differential forms generated by the forms $\omega_H = d\alpha_H/\alpha_H$ and the identity. The Poincaré polynomial of $R$ is

$$P(R, t) = \sum_{X \in L} \mu(X)(-t)^{\operatorname{codim} X}.$$

Proof. This follows at once from (2.38) and (4.14). □

(4.16) Remark. The argument in (4.13) - (4.14) which proves that $\ker j_R \subseteq \operatorname{im} i_R$ does not use the fact, known to us by the induction hypothesis, that $\gamma'$ is an isomorphism. This point is worth inspection. Suppose given a commutative diagram of abelian groups and homomorphisms

$$\begin{array}{ccccc} A' & \xrightarrow{i_A} & A & \xrightarrow{j_A} & A'' \\ \gamma' \downarrow & & \gamma \downarrow & & \gamma'' \downarrow \\ R' & \xrightarrow{i_R} & R & \xrightarrow{j_R} & R'' \end{array}$$

where $\gamma$ is surjective, $\gamma''$ is injective and $\ker j_A \subseteq \operatorname{im} i_A$. Then $\ker j_R \subseteq \operatorname{im} i_R$, and nothing has been assumed about $\gamma'$. If in addition both $i_R$ and $\gamma'$ are injective then $i_A$ is injective. This closer analysis shows first, that only (3.11) and not (3.12) is used in the proof of (4.13) - (4.14) and second, that (3.12) is a consequence of (4.14). Thus the injectivity of $i_R$, which is obvious, implies the injectivity of $i_A$, which is not. Note that (2.38) is also a consequence of (4.14) because (2.38) is not used in the proof of (3.11). We have chosen our manner of exposition because Sections 2 and 3 may be done without additional effort for an arbitrary geometric lattice $L$ where the differential forms and the map $i_R$ are not available.

We have seen in (2.8.ii) that $\partial_A$ is a $\mathbb{K}$-derivation of $A$ which is homogeneous of degree $-1$. Since $\gamma : A \to R$ is an isomorphism of graded $\mathbb{K}$-algebras it follows that $\gamma\partial_A\gamma^{-1}$ is a $\mathbb{K}$-derivation of $R$ and that $(R, \gamma\partial_A\gamma^{-1})$ is a chain complex which, by (2.9.iv), is acyclic when $\mathcal{C}$ is non void. We show now how to construct this complex and prove its acyclicity without using $A$ as an intermediary. To do this we use a close relative of the Koszul complex; the precise connection with the Koszul complex is described below. Recall from (4.1) that $\Omega(V)$ is by definition the exterior algebra of the $F$ vector space $F \otimes V^*$ where $\otimes = \otimes_{\mathbb{K}}$. Thus $\Omega(V) = F \otimes \wedge V^*$ where $\wedge V^*$ is the exterior algebra of the $\mathbb{K}$-vector space $V^*$ and $\Omega^p(V) = F \otimes \wedge^p V^*$ for $0 \le p \le \ell$. Choose a basis $x_1, \ldots, x_\ell$ for $V^*$. There is a unique $F$-linear map $\partial = \partial_\Omega : F \otimes \wedge^p V^* \to F \otimes \wedge^{p-1} V^*$ for $1 \le p \le \ell$ such that

$$\partial(f \otimes x_{i_1} \wedge \ldots \wedge x_{i_p}) = \sum_{k=1}^{p} (-1)^{k-1} x_{i_k} f \otimes (x_{i_1} \wedge \ldots \hat{x}_{i_k} \wedge \ldots x_{i_p}) \tag{4.17}$$

if $1 \le p \le \ell$, and $\partial(f \otimes 1) = 0$. Direct computation shows that $\partial^2 = 0$. Thus $(\Omega(V), \partial)$ is a chain complex. Note that $S \otimes V^*$ is stable under $\partial$ and thus $(S \otimes V^*, \partial)$ is a subcomplex. This subcomplex differs from the Koszul complex only in that when $p = 0$ the map $f \otimes 1 \to 0$ is replaced, in the Koszul complex, by $f \otimes 1 \to \varepsilon(f)$ where $\varepsilon : S \to \mathbb{K}$ is the natural augmentation. The map $\partial$ is a $\mathbb{K}$-derivation of $\Omega$:

$$\partial(\omega\eta) = (\partial\omega)\eta + (-1)^p \omega(\partial\eta) \qquad \omega \in \Omega^p(V), \quad \eta \in \Omega(V). \tag{4.18}$$

If $p = 0$ this is true because $\partial(f \otimes 1) = 0$. To check it for $p \ge 1$ it suffices, because $\partial$ is an $F$-linear map, to consider the cases $\omega = dx_{i_1} \ldots dx_{i_p}$, and $\eta = 1$ or $dx_{j_1} \ldots dx_{j_q}$.

(4.19) <u>Proposition</u>. Let $(\mathcal{C}, V)$ be an arrangement and let $\partial = \partial_\Omega$. Then $\partial R(\mathcal{C}) \subseteq R(\mathcal{C})$. If $\partial_R$ denotes the restriction of $\partial$ to $R(\mathcal{C})$ then (i) $(R(\mathcal{C}), \partial_R)$ is a chain complex which is acyclic if $\mathcal{C}$ is non void and (ii) $\partial_R = \gamma\partial_A\gamma^{-1}$.

Proof. It follows from (4.18) that for $(H_1,\ldots,H_p) \in \underline{\underline{S}}(\mathcal{Q})$ we have

$$\partial(\omega_{H_1}\cdots\omega_{H_p}) = \sum_{k=1}^{p} (-1)^{k-1}\omega_{H_1}\cdots\hat{\omega}_{H_k}\cdots\omega_{H_p}.$$

Thus $\partial R(\mathcal{Q}) \subseteq R(\mathcal{Q})$, so $(R(\mathcal{Q}), \partial_R)$ is a chain complex. If $\mathcal{Q}$ is non void choose $H \in \mathcal{Q}$. Let $\alpha = \alpha_H$ and let $\omega = \omega_H$. Write $\alpha = \sum c_k x_k$ for suitable $c_k \in \mathbb{K}$. It follows from (4.17) that $\partial(dx_k) = \partial(1 \otimes x_k) = x_k \otimes 1$. Since $\partial$ is F-linear we have

$$\partial\omega = \frac{1}{\alpha}\partial(d\alpha) = \frac{1}{\alpha}\partial(\sum c_k dx_k) = \frac{1}{\alpha}\sum c_k x_k \otimes 1 = 1 \otimes 1 = 1$$

where, in the last equality we identify $F \otimes \mathbb{K}$ with $F$. Now let $\eta \in R(\mathcal{Q})$ be arbitrary. From (4.18) we get $\partial(\omega\eta) = (\partial\omega)\eta - \omega(\partial\eta) = \eta - \omega(\partial\eta)$. If $\eta \in \ker \partial$ then $\eta = \partial(\omega\eta)$. This proves that $(R(\mathcal{Q}), \partial_R)$ is acyclic because $R(\mathcal{Q})$ is an algebra and thus $\omega\eta \in R(\mathcal{Q})$. To prove (ii) note that $\gamma\partial_A\gamma^{-1}\omega_H = \gamma\partial_A a_H = \gamma 1 = 1 = \partial_R\omega_H$ for any $H \in \mathcal{Q}$. Since both $\gamma\partial_A\gamma^{-1}$ and $\partial_R$ are $\mathbb{K}$-derivations of $R(\mathcal{Q})$ which agree on the generators $\omega_H$ of $R(\mathcal{Q})$ they are equal. □

For $X \in L = L(\mathcal{Q})$ let $R_X(\mathcal{Q}) = \sum \mathbb{K}\omega_{H_1}\cdots\omega_{H_p}$ where the sum is over all $(H_1,\ldots,H_p)$ in $\underline{\underline{S}}_X = \underline{\underline{S}}_X(\mathcal{Q})$. Then

$$R_p = \sum_{\substack{X \in L \\ r(X)=p}} R_X. \tag{4.20}$$

Corollary (2.27) shows that (4.20) is a direct sum because $\gamma(A_X) = R_X$ and $\gamma$ is an isomorphism. If one can prove that (4.20) is a direct sum without using the isomorphism $\gamma : A \to R$ then acyclicity of $(R, \partial_R)$ and an argument like that in (2.37) - (2.38) proves the formula for $P(R,t)$ entirely within the context of differential forms.

## 5. MODULAR ELEMENTS AND FACTORIZATION OF $A(\mathcal{A})$.

Recall that an element $Y$ in a geometric lattice $L$ is modular if $r(X)+r(Y) = r(X \vee Y) + r(X \wedge Y)$ for all $X \in L$. [1, p. 58]. If $\mathcal{A}$ is an arrangement and $Y \in L(\mathcal{A})$ recall our notation $\mathcal{A}_Y = \{H \in \mathcal{A} \mid H \leq Y\}$ defined in (2.29). Define $\mathcal{A}^Y = \{H \cap Y \mid H \cap Y \neq Y\}$. Then $\mathcal{A}^Y$ is an arrangement in $Y$. If $Y = H_0$ is a hyperplane of $\mathcal{A}$ then $\mathcal{A}^Y = \mathcal{A}''$ in our earlier notation.

(5.1) <u>Lemma</u>. Let $\mathcal{A}$ be an arrangement. Suppose there exists a modular element $Y \in L(\mathcal{A})$ such that $r(Y) = r(\mathcal{A}) - 1$. If $H \in \mathcal{A}$ and $H \cap Y \neq Y$ then there is a $\mathbb{K}$-algebra isomorphism $\tau : A(\mathcal{A}_Y) \to A(\mathcal{A}^H)$ such that $\tau(a_K) = a_{H\cap K}$ for all $K \in \mathcal{A}_Y$.

<u>Proof</u>. It follows from [1, Prop. 2.42] that the mapping $\lambda : X \to X \vee H = X \cap H$ is a lattice isomorphism from $L(\mathcal{A}_Y)$ to $L(\mathcal{A}^H)$. If $S = (H_1, \cdots, H_p) \in \underline{\underline{S}}(\mathcal{A}_Y)$ define $\lambda S = (\lambda H_1, \cdots, \lambda H_p) \in \underline{\underline{S}}(\mathcal{A}^H)$. The $\mathbb{K}$-algebra isomorphism from $E(\mathcal{A}_Y)$ to $E(\mathcal{A}^H)$ which sends $e_S$ to $e_{\lambda S}$ maps $I(\mathcal{A}_Y)$ to $I(\mathcal{A}^H)$ and hence induces a $\mathbb{K}$-algebra homomorphism $\tau : A(\mathcal{A}_Y) \to A(\mathcal{A}^H)$ such that $\tau a_S = a_{\lambda S}$. The inverse of $\lambda$ gives the inverse of $\tau$. □

Proposition (3.15) allows us to identify $A(\mathcal{A}_Y)$ with the $\mathbb{K}$-subalgebra of $A(\mathcal{A})$ generated by the elements $a_H = e_H + I(\mathcal{A})$ for $H \in \mathcal{A}_Y$.

(5.2) <u>Proposition</u>. Let $\mathcal{A}$ be an arrangement. Suppose there exists a modular element $Y \in L(\mathcal{A})$ such that $r(Y) = r(\mathcal{A}) - 1$. Let $\mathcal{B} = \mathcal{A} - \mathcal{A}_Y$. Then

$$A(\mathcal{A}) = A(\mathcal{A}_Y) \oplus \Big( \bigoplus_{H \in \mathcal{B}} A(\mathcal{A}_Y)\, a_H \Big).$$

<u>Proof</u>. Let $Q = \sum_{H\in\mathcal{B}} A(\mathcal{A}_Y) a_H$. We show first that if $H, K \in \mathcal{B}$ then $a_H a_K \in Q$. If $H = K$ this is clear since $a_H^2 = 0$. Suppose $H \neq K$. We have remarked in the proof of Lemma (5.2) that the mapping $X \to X \cap H$ is a lattice isomorphism from $L(\mathcal{A}_Y)$ to $L(\mathcal{A}^H)$. In particular there exists

$M \in L(\mathcal{C}_Y)$ such that $M \cap H = K \cap H$. Since $K \neq H$ we have $r(K \cap H) = 2$ so $r(M \cap H) = 2$ and thus $M \in \mathcal{C}_Y$. Since $r(M \cap H \cap K) = 2$ it follows that $(M, H, K)$ is dependent and thus $a_H a_K - a_M a_K + a_M a_H = 0$. This shows $a_H a_K \in A(\mathcal{C}_Y) a_H + A(\mathcal{C}_Y) a_K \subseteq Q$ as desired.

Since $A(\mathcal{C}_Y)$ is a $\mathbb{K}$-subalgebra of $A(\mathcal{C})$ containing the identity element, it follows that $Q$ is closed under multiplication and $A(\mathcal{C}_Y)Q \subseteq Q$. Thus $A(\mathcal{C}_Y) + Q$ is a $\mathbb{K}$-subalgebra of $A(\mathcal{C})$. But $\mathcal{C} = \mathcal{C}_Y \cup \mathcal{B}$ so $A(\mathcal{C}_Y) + Q$ contains all the generators $a_H$, $H \in \mathcal{C}$ of $A(\mathcal{C})$. Thus

$$A(\mathcal{C}) = A(\mathcal{C}_Y) + Q = A(\mathcal{C}_Y) + \sum_{H \in \mathcal{B}} A(\mathcal{C}_Y) a_H .$$

We must prove that this is a direct sum. First we show that $A(\mathcal{C}_Y) \cap Q = 0$. By Theorem (2.26) we have $A = \bigoplus_{X \in L} A_X$ where $A_X = \varphi(E_X) = \sum_{\cap S = X} \mathbb{K} a_S$. Let $\pi_X$ be the natural projection of $A$ onto $A_X$. Then $\pi_X(a_S) = a_S$ if $\cap S = X$ and $\pi_X(a_S) = 0$ otherwise. We show that $\pi_X(A(\mathcal{C}_Y) \cap Q) = 0$ for all $X \in L$ by showing that (i) $\pi_X(A(\mathcal{C}_Y)) = 0$ if $X \not\leq Y$ and (ii) $\pi_X(Q) = 0$ if $X \leq Y$. Since $A(\mathcal{C}_Y) = \sum_{\cap S \leq Y} \mathbb{K} a_S$ (i) is immediate. To prove (ii) we observe that if $H \in \mathcal{B}$ and $X \leq Y$ then $\pi_X(a_S a_H) = \pi_X(a_{(S, H)}) = 0$ because $(\cap S) \cap H \not\leq Y$. Thus $A(\mathcal{C}_Y) \cap Q = 0$.

It remains to prove that the sum $\sum_{H \in \mathcal{B}} A(\mathcal{C}_Y) a_H$ is direct. Fix $H_0 \in \mathcal{B}$ and use the notation of Section 3. Let $\lambda : \underline{\underline{S}}' \to \underline{\underline{S}}''$ and let $j : A \to A''$ be the maps defined in (3.6-7). Since $H_0 \in \mathcal{B}$ we have $\underline{\underline{S}}(\mathcal{C}_Y) \subseteq \underline{\underline{S}}'$. It follows from (3.7) that for $S \in \underline{\underline{S}}(\mathcal{C}_Y)$ and $H \in \mathcal{B}$ we have $j(a_H a_S) = a_{\lambda S}$ if $H = H_0$ and $j(a_H a_S) = 0$ otherwise. By Lemma (5.1) with $H$ replaced by $H_0$, there exists a $\mathbb{K}$-algebra isomorphism $\tau : A(\mathcal{C}_Y) \to A''$ with $\tau(a_S) = a_{\lambda S}$. Thus $j(a_H a_S) = \tau(a_S)$ if $H = H_0$ and $j(a_H a_S) = 0$ otherwise. It follows that for $u \in A(\mathcal{C}_Y)$ we have

$$j(a_H u) = \begin{cases} \tau(u) & \text{if } H = H_0 \\ 0 & \text{if } H \in \mathcal{B} - \{H_0\} . \end{cases}$$

Suppose $\sum_{H \in \mathcal{B}} a_H u_H = 0$ where $u_H \in A(\mathcal{C}_Y)$. Then

$$0 = j\Big(\sum_{H \in \mathcal{B}} a_H u_H\Big) = \tau(u_{H_0})$$

and thus $u_{H_0} = 0$. Since $H_0$ was an arbitrary hyperplane of $\mathcal{B}$ it follows that $u_H = 0$ for all $H \in \mathcal{B}$. □

(5.3) Theorem. Let $\mathcal{A}$ be an arrangement and let $A = A(\mathcal{A})$. Suppose there exists a maximal chain $Y_0 < Y_1 < \cdots < Y_r$, $r = r(\mathcal{A})$, of elements of $L(\mathcal{A})$ such that $Y_{i-1}$ is a modular element of $L(\mathcal{A}_{Y_i})$ for $i = 1, \cdots, r$. Let $\mathcal{B}_i = \mathcal{A}_{Y_i} - \mathcal{A}_{Y_{i-1}}$ for $i = 1, \cdots, r$ and let $B_i = \sum_{H \in \mathcal{B}_i} \mathbb{K} a_H$. Then the $\mathbb{K}$-linear map

$$(\mathbb{K} + B_1) \otimes \cdots \otimes (\mathbb{K} + B_r) \to A$$

defined by multiplication in $A$ is an isomorphism of graded vector spaces. In particular

$$P(A, t) = (1 + b_1 t) \cdots (1 + b_r t)$$

where $b_i = |\mathcal{B}_i|$.

Proof. Let $Y = Y_{r-1}$. Then $Y$ is a modular element of $L(\mathcal{A})$ such that $r(Y) = r(\mathcal{A}) - 1$. Let $\mathcal{B} = \mathcal{A} - \mathcal{A}_Y$ and let $B = \sum_{H \in \mathcal{B}} \mathbb{K} a_H$. Proposition (3.2) shows that the $\mathbb{K}$-linear map $A(\mathcal{A}_Y) \otimes (\mathbb{K} + B) \to A(\mathcal{A})$ defined by the multiplication in $A(\mathcal{A})$ is an isomorphism of graded vector spaces. The theorem follows by induction on $r$. □

The reader should note that the isomorphism in Theorem (5.3) is not an isomorphism of $\mathbb{K}$-algebras. It follows from (2.40) that under the hypotheses of (5.3) the characteristic polynomial of $L = L(\mathcal{A})$ is

$$\chi(L, t) = (t - b_1) \cdots (t - b_r).$$

This factorization was proved by Stanley [8], [1, p. 177]. Thus Theorem (5.3) gives a factorization of the algebra $A(\mathcal{A})$ which corresponds to the factorization of $\chi(L, t)$ for a supersolvable lattice $L$. There is another algebraic source of the factorization of $\chi(L, t)$ for a supersolvable lattice $L$. It is shown in [5] that if $L(\mathcal{A})$ is supersolvable then $\mathcal{A}$ is a free arrangement in the sense of [9]. Then one may apply the factorization theorem [9] for the characteristic polynomial of a free arrangement.

Let $\mathcal{A}$ be the arrangement in $\mathbb{K}^\ell$ consisting of all hyperplanes $H_{jk} = \ker(x_j - x_k)$ where $x_1, \cdots, x_\ell$ are the coordinate functions. The lattice $L(\mathcal{A})$ is the partition lattice and satisfies the conditions of Theorem (5.3); see [1, p. 178]. Since $A(\mathcal{A}) \simeq R(\mathcal{A})$ we may apply Theorem (5.3) to $R(\mathcal{A})$. Let $\alpha_{jk} = x_j - x_k$ and let $\omega_{jk} = d\alpha_{jk}/\alpha_{jk}$. If $\mathbb{K} = \mathbb{C}$ this is Arnold's example cited in the Introduction; we have omitted the factor $1/2\pi i$ which has no combinatorial significance. We may choose $Y_0 = V$ and choose $Y_i$ to be the subspace defined by $x_1 = x_2 = \cdots = x_{i+1}$ for $i = 1, \ldots, \ell - 1$. Then $B_i = \mathbb{K}\omega_{1,i+1} + \mathbb{K}\omega_{2,i+1} + \cdots + \mathbb{K}\omega_{i,i+1}$. Theorem 5.2 says

$$R \simeq \bigotimes_{i=1}^{\ell-1} (\mathbb{K} + \mathbb{K}\omega_{1,i+1} + \cdots + \mathbb{K}\omega_{i,i+1})$$

and the Poincaré polynomial is

$$P(R, t) = (1+t)(1+2t) \ldots (1+(\ell-1)t) .$$

One sees these formulas in Corollaries 3 and 2 of Arnold's paper.

## BIBLIOGRAPHY


1. Aigner, M., Combinatorial Theory, Grundlehren der math. Wiss. Band 234, Springer Verlag, Berlin, 1979.

2. Arnold, V. I., The Cohomology Ring of the Colored Braid Group. Mat. Zametki 5 (1969), 227-231, Math. Notes. 5 (1969), 138-140.

3. Brylawski, T. H., A Decomposition for Combinatorial Geometries, Trans. Amer. Math. Soc., 171 (1972), 235-282.

4. Cartier, P., Les Arrangements d'Hyperplans: un Chapitre de Géométrie Combinatoire, Séminaire Bourbaki, 33e année 1980-1981, pp. 1-22, Springer Lecture Notes No. 901, Springer Verlag, Berlin, 1981.

5. Jambu, M. and Terao, H., Arrangements libres d'hyperplans et treillis hyper-résolubles, C. R. Acad. Sc. Paris, 296 (1983), 623-624.

6. Orlik, P. and Solomon, L., Combinatorics and Topology of Complements of Hyperplanes, Inventiones math., 56 (1980), 167-189.

7. Pham, F., Introduction a l'Étude Topologique des Singularités de Landau, Mémorial des Sci. Math. Fasc. CLXIV, Gauthier-Villars, Paris, 1967.

8. Stanley, R. P., Modular Elements of Geometric Lattices, Algebra Universalis, 1 (1971), 214-217.

9. Terao, H., Generalized Exponents of a Free Arrangement of Hyperplanes and Shephard-Todd-Brieskorn Formula, Invent. math. 63 (1981), 159-179.

10. Zaslavsky, T., Facing up to Arrangements: Face-Count Formulas for Partitions of Space by Hyperplanes, Amer. Math. Soc. Memoir, 154, (1975).

Peter Orlik and Louis Solomon
Department of Mathematics
University of Wisconsin, Madison, WI 53706

Hiroaki Terao
Department of Mathematics
International Christian University
Mitaka 181, Tokyo, Japan

Contemporary Mathematics
Volume 34, 1984

# DOUBLE CENTRALIZING THEOREMS FOR WREATH PRODUCT

Amitai Regev*

I would like to describe here some extensions of the famous theorem of I. Schur about the double centralization between the Symmetric group $S_n$ and the general Linear Lie Group $GL(k,\mathbb{C})$. In these generalizations, $S_n$ is replaced by a wreath-product $A \sim S_n$. I would also like to indicate how these extensions can be used to trivialize the representation theory of such wreath products, as well as to deduce some new results about them. I shall describe the main results here, but the proofs will be omitted, as they are given in [7] in full detail.

Our starting point is the celebrated theorem of Schur [9], which we now describe:

Let $W$ be a vector space, $\dim W = k$, and define $T^n(W) = W^{\otimes n}$. The symmetric group permutes $T^n(W)$: if $\sigma \in S_n$, then

$$\varphi(\sigma)(w_1 \otimes \cdots \otimes w_n) = w_{\sigma^{-1}(1)} \otimes \cdots \otimes w_{\sigma^{-1}(n)},$$

while $GL(k,\mathbb{C}) \equiv GL(W)$ acts on $T^n(W)$ "diagonally": if $s \in GL(W)$, then

$$\psi(s)(w_1 \otimes \cdots \otimes w_n) = s(w_1) \otimes \cdots \otimes s(w_n).$$

Let $\mathfrak{A} = \varphi(FS_n)$ and let $\mathfrak{B} = \mathrm{span}_F\, \psi(GL(W))$, let $E = \mathrm{End}(T^n(W))$ and let $C_E(-)$ denote centralizer in $E$, then Schur proved the following

1. THEOREM: (SCHUR). $C_E(\mathfrak{A}) = \mathfrak{B}$ and $C_E(\mathfrak{B}) = \mathfrak{A}$.

There is an important principal here that will repeatedly be used later: Whenever such a double centralizing property exists, the representations of one object can be easily deduced from those of the other object.

It is in this way that Schur deduced the representations of $GL(W)$ from

1980 Mathematics Subject Classification. 20C05
*Partially supported by an N.S.F. grant.

those of $S_n$ (which he already knew from Frobenious).

Related to the above theorem of Schur we also mention an important theorem of H. Weyl:

It is well known that $FS_n = \bigoplus_{\lambda \vdash n} I_\lambda$ where the $I_\lambda$'s are minimal two-sided ideals, explicitely given by the theory of Young tableaux. Denote by $\Lambda_k(n)$ the partitions with $\leq k$ parts:

$\Lambda_k(n) = \{\lambda \vdash n | \lambda_{k+j}=0 \ \ j=1,2,\ldots\}$. We can now state

2. (WEYL'S "STRIP") THEOREM. With the previous notations, $\varphi(FS_n) = \bigoplus_{\lambda \in \Lambda_k(n)} I_\lambda$.

Both Schur's and Weyl's theorems were extended in [2] in a direction which I now describe:

Schur's theorem has an equivalent formulation--as well as proof--when $GL(W)$ is properly replaced by the Lie algebra $g\ell(W)$. A "Lie superalgebra extension" of Schur's theorem was obtained, in [2], by introducing a "sign permutation" action of $S_n$ on $T^n(W)$, where $W = T \oplus U$. That action can be described as follows: first, $S_n$ does permute the coordinates of $T^n(W)$; in addition to that, it assigns a $\pm$ sign in the following way: when a $t \in T$ is commuted with either a $t \in T$ or with a $u \in U$, no sign is introduced; when two $u$'s ($\in U$) are permuted, a minus sign is introduced. This action yields a representation $\varphi^*: FS_n \to E = \text{End}(T^n(W))$ and a subalgebra $\mathfrak{A}^* = \varphi^*(FS_n)$. Let now $\dim T = k$, $\dim U = \ell$, $W = T \oplus U$, then the Lie superalgebra $p\ell(k,\ell)$ [2], [8] acts on $T^n(W)$ as superderivations, yielding the superalgebra homomorphism $D_n: p\ell(k,\ell) \to E$ and the associative subalgebra $\mathfrak{B}^* \subseteq E$ generated by $D_n(p\ell(k,\ell))$. (We remark here that the above two actions were also considered--independently--by few physicists working in the physics of elementary particles, [1], [3].) The two main results in [2] are

3. THEOREM (EXTENSION OF SCHUR'S THEOREM), [2,4.14]: $C_E(\mathfrak{A}^*) = \mathfrak{B}^*$ and $C_E(\mathfrak{B}^*) = \mathfrak{A}^*$.

4. THEOREM (EXTENSION OF WEYL'S "STRIP" THEOREM) [2,3.20]: Let $H(k,\ell;n) = \{\lambda \vdash n | \lambda_{k+j} \leq \ell, \ \ell=1,2,\ldots\}$, then $\varphi^*(FS_n) = \mathfrak{A}^* \cong \bigoplus_{\lambda \in H(k,\ell;n)} I_\lambda$.

My aim is to report here on some further extensions of the above theorems, extensions in which $S_n$ is replaced by wreath products $A \sim S_n$, $A$ a group or an associative algebra. This is done over a field $F$ of characteristic zero and which is algebraically closed.

Wreath products arise naturally in group theory. Moreover, all the classical Weyl groups (of types A, B, C, D) are wreath products. Thus the motivation for studying these groups and their representations is obvious. The first to consider such representations was A. Young [11] who obtained the representations of $\mathbb{Z}_2 \sim S_n$ (the Weyl groups of types B, C) from those of $S_n$. A general method for obtaining the irreducible representations of wreath products was later obtained by Specht [9]; the appendix in [6] contains a clear and a short account of Specht's method. See also [4] and [5].

Following Schur, we shall obtain the representations of $A \sim S_n$ from double centralizing theorems involving $A \sim S_n$. The first such theorem, which is given below as theorem 5, is an extension of Schur's theorem and it suffices for deducing the representations of $A \sim S_n$.

Given a semi-simple associative algebra A ($1 \in A$), a semi-simple left A-module $M$ and a vector space $W$, there is a natural action of $A \sim S_n$ on $T^n(M\otimes W)$ yielding the homomorphism $\varphi: A \sim S_n \to E = \mathrm{End}(T^n(M\otimes W))$, with $\mathfrak{A} = \varphi(A\sim S_n)$. Let $R_M = \mathrm{Hom}_A(M,M)$ be the endomorphism algebra of $M$, then

$$\mathrm{End}(M\otimes W) \supseteq \mathrm{Hom}_A(M\otimes W, M\otimes W) \equiv R_M \otimes \mathrm{End}(W),$$

and we define $u_{M,W}$, $u_{M,W} \subseteq R_M \otimes \mathrm{End}(W)$ to be the group of units in $R_M \otimes \mathrm{End}(W)$. Since $M$ and $A$ are semi-simple, $u_{M,W}$ is a direct product of general linear groups $GL(Z_i)$. Since $u_{M,W} \subseteq GL(M\otimes W)$, hence $u_{M,W}$ acts diagonally on $T^n(M\otimes W)$, yielding the group homomorphism $\psi: u_{M,W} \to W$ and we denote by $\mathfrak{B}$ the algebra generated by $\psi(u_{M,W})$ in $E$. We can now state

5. THEOREM. With the above notations, $C_E(\mathfrak{A}) = \mathfrak{B}$ and $C_E(\mathfrak{B}) = \mathfrak{A}$.

There is now a "multi-strip" theorem that extends theorem 2, but this will be a special case of the "multi-hook" theorem--to be described later. Before proceeding to a wreath-product generalization of theorem 3, let me briefly indicate how theorem 5 is applied to derive the irreducible representations of $A \sim S_n$:

Let $A \cong \bigoplus_{i=1}^{t} F_{r_i}$ ($F_r$ = the $r\times r$ matrices), choose $A = M$ and choose $W$ so that $k = \dim W \geq n$, then $u_{A,W} \cong \mathop{X}_{i=1}^{t} GL(r_i k)$. For a single $GL(r\cdot k)$, the irreducible representations (polynomial, homogeneous of degree $n$) are parametrized by $\lambda \in \Lambda_{kr}(n)$, denoted $\{M_\lambda | \lambda \in \Lambda_{kr}(n)\}$. Hence, those for $u_{A,W} \equiv \mathop{X}_{i=1}^{t} GL(r_i k)$ are

$$\{M_{<\lambda>} = \bigotimes_{i=1}^{t} M_{\lambda(i)} | \lambda(i) \in \Lambda_{kr_i}(m_i), \sum_{i=1}^{t} m_i = n\}.$$

By theorem 5, $N_{<\lambda>} = \mathrm{Hom}_{\mathfrak{B}}(M_{<\lambda>}, T^n(A \otimes W))$ are irreducible $A \sim S_n$ modules, pairwise non-equivalent. A dimension argument then proves

6. THEOREM. $\{N_{<\lambda>}\}$ is a complete set of the irreducible representations for $A \sim S_n$.

When $G$ is a finite group and $A_n$ is the alternating group, it is not too difficult to obtain the representation of $G \sim A_n$ by an application of theorem 5, combined with a classical method of Frobenius. The representation theory of wreath products thus "reaches the Mecca of triviality", if I am to use Professor Rota's words.

Theorem 5 can be viewed as a special case of a more general double centralizing theorem that involve $A \sim S_n$ and the Lie superalgebra $R_M \otimes p\ell(W)$, where $W = T \oplus U$. The set-up for that theorem is the following: As in [2], $W = T \oplus U$, $\dim T = k$, $\dim U = \ell$ and $S_n$ acts on $T^n(W)$ by the "sign permutation" action $*$. Let $A$ be an associative algebra and let $M$ be a semi-simple $A$-module, then this defines a corresponding module structure on $T^n(M \otimes W)$ over $A \sim S_n$, which yields the homomorphism

$$\varphi^*: A \sim S_n \to E = \mathrm{End}(T^n(M \otimes W))$$

and the algebra $\mathfrak{A}^* = \varphi^*(A \sim S_n)$. The centralizing action comes from the Lie superalgebra $R_M \otimes p\ell(W)$: as before,

$$\mathrm{End}(M \otimes W) \supseteq R_M \otimes p\ell(W) \qquad (p\ell(W) \equiv p\ell(k,\ell) \equiv \mathrm{End}(T \oplus U)).$$

The decomposition $M \otimes W \equiv (M \otimes T) \oplus (M \otimes U)$ identifies $\mathrm{End}(M \otimes W) \equiv p\ell(M \otimes W)$, with $R_M \otimes p\ell(W)$ as a Lie sub-superalgebra. The superderivation $D_n : p\ell(A \otimes W) \to E$ (see 3.), when restricted to $R_M \otimes p\ell(W)$, yields $\psi^*: R_M \otimes p\ell(W) \to E$ as well as the associative algebra $\mathfrak{B}^*$ which is generated by $\psi^*(R_M \otimes p\ell(W))$ in $E$. We now have

7. THEOREM. With the above notations, $C_E(\mathfrak{A}^*) = \mathfrak{B}^*$. If $A$ is semi-simple and finite dimensional then, also, $C_E(\mathfrak{B}^*) = \mathfrak{A}^*$.

In the special case $\ell = 0$ (i.e.: $W = T$, $U = 0$), $p\ell(k,0) = g\ell(k)$ is the general linear Lie algebra, $\mathfrak{A}^* = \mathfrak{A}$, and the above theorem becomes the Lie algebra version of theorem 5.

To apply theorem 7, return first to theorem 6. Since $\{N_{<\lambda>}\}$ form a complete set for the irreducible representations of $A \sim S_n$, hence $A \sim S_n = \bigoplus_{<\lambda>} I_{<\lambda>}$ where each $I_{<\lambda>}$ is a minimal two-sided ideal in $A \sim S_n$, defined by the property $\varphi(I_{<\lambda>})M_{<\lambda>} \neq 0$.

We now fix $W = T \oplus U$, $\dim T = k$, $\dim U = \ell$, choose $M = A$, so $R_M \cong A^{op}$, and let $n > 0$ be any integer. Consider now $\varphi^*$ of theorem 7: since each $I_{<\lambda>}$ is a matrix algebra, hence $\mathfrak{A}^* = \varphi^*(A \sim S_n) \cong \bigoplus_{<\lambda> \in \Gamma} I_{<\lambda>}$ where $\Gamma$ is a certain subset of triple partitions.

We now have

8. (THE "MULTI-HOOK") THEOREM. Let $A = \bigoplus_{i=1}^{t} A_i$, $A_i \cong F_{r_i}$, $W = T \oplus U$, $\dim T = k$, $\dim U = \ell$, $M = A$, $\varphi^*$ and $\Gamma$ as above, then

$$\Gamma = \{<\lambda>=<\lambda(1),\ldots,\lambda(t)> \mid \lambda(i) \in H(r_i k, r_i \ell; m_i),\ 1 \le i \le t \text{ and } \sum_{i=1}^{t} m_1 = n\}.$$

Note that the special case $\ell = 0$ is a "multi-strip" generalization of (Weyl's) theorem 2.

## REFERENCES

1. Bars I., Balantekin A., J. Math. Phys.: 22(6) June 1981, 1149-1162, 22(8) August 1981, 1810-1818, 23(4) April 1982, 486-489, 23(7) July 1982, 1239-1247.

2. Berele A., Regev A., Hook Young diagrams with applications to combinatorics and to representation theory of Lie superalgebras.

3. Dondi P.H., Jarvis P.D., Diagram and superfield techniques in the classical superalgebras, J. Phys. A. Math. Gen. 14(1981) 547-563.

4. James G., Kerber A., The representation theory of the symmetric group. Encyclopedia of Mathematics and its applications, Vol. 16, (1981), Addison-Wesley.

5. Kerber, A., Zur darstellungstheorie von kranzprodukten, Can. J. Math. 20(1968) 665-672.

6. MacDonald I.G., Polynomial functors and Wreath products, J. Pure Appl. Algebra, 18(1980) 173-204.

7. Regev A., The representations of Wreath products via double centralizing theorems, Preprint.

8. Scheunert M., The theory of Lie superalgebras, Springer Verlag Lecture Notes in Math. No. 716.

9. Schur I., Über die rationalen darstellungen der allgemeinen linearen gruppe (1927); in: I. Schur, Gesammelte Abhandlungen III, 68-85, Springer Berlin, 1973.

10. Specht W., Eine Verallgemeinerung der symmetrischen Gruppe, Schriften Math. Seminar Berlin 1(1932) 1-32.

11. Young A., Q.S.A. V. Proc. London Math. Soc. (2), 31, (1930), 273-288.

Department of Mathematics
The Pennsylvania State University
University Park, PA 16802

Contemporary Mathematics
Volume **34**, 1984

# ON THE CONSTRUCTION OF THE MAXIMAL WEIGHT VECTORS IN THE TENSOR ALGEBRA OF $gl_n(\mathbb{C})$.

Phil Hanlon

ABSTRACT. We study the decomposition of $T^p(gl_n(\mathbb{C}))$. The $p^{th}$ graded piece of the tensor algebra of $gl_n(\mathbb{C})$ considered as a $gl_n(\mathbb{C})$-module. In particular we construct all maximal weight vectors of a given weight in $T^p(gl_n(\mathbb{C}))$ and give a partial description of the linear dependencies which hold amongst them.

INTRODUCTION.

In this paper we describe a combinatorial answer to a question in the representation theory of the complex general linear group. We assume some background in representation theory. This background information can be obtained from James [3] (for the symmetric group) and from Hanlon [1], Macdonald [4], Serre [5] or Stanley [6] (for $Gl(n,\mathbb{C})$). The contents of this paper are described in more detail in Hanlon [2].

## Section 1: Preliminaries

Throughout this paper n is fixed and $\mathcal{G}$ denotes the Lie algebra $gl_n(\mathbb{C})$. We let $T(\mathcal{G})$ be the tensor algebra of $\mathcal{G}$ and $T^p(\mathcal{G})$ its $p^{th}$ graded piece. There is a natural representation $ad : \mathcal{G} \to Hom(\mathcal{G})$ of $\mathcal{G}$ on itself defined by $ad(a)(b) = [a,b]$. This gives an action (also called ad) of $\mathcal{G}$ on $T^p(\mathcal{G})$ by

$$ad(a)(b_1 \otimes \cdots \otimes b_p) = \sum_{i=1}^{p} (b_1 \otimes \cdots \otimes ad(a)(b_i) \otimes \cdots \otimes b_p).$$

The problem we consider in this paper is how to construct all occurrences of a given irreducible $\mathcal{G}$-module in $T^p(\mathcal{G})$. We obtain a partial solution for all p by constructing all maximal weight vectors with a given dominant weight $\lambda$ in $T^p(\mathcal{G})$. We give a complete solution to the problem in the case

1980 Mathematics Subject Classification 20G05

that n is large with respect to p and $\lambda$ by finding a basis for the vector space spanned by the maximal weight vectors of weight $\lambda$ in $T^p(\mathfrak{g})$.

Section 2: Dominant Weights and Maximal Weight Vectors

Let V be a finite dimensional complex vector space and let $\phi : \mathfrak{g} \to \mathrm{End}(V)$ be a representation of $\mathfrak{g}$ on V. Let $\{v_\lambda\}$ be a basis of weight vectors for V with $\lambda^{(\lambda)} \in \mathbb{C}^n$ the weight of $v_\lambda$.

By $y_{ij}$ we mean the matrix in $\mathfrak{g}$ with a one in the (i,j) entry and zeros elsewhere. Then $\mathfrak{g}$ has the decomposition $\mathfrak{g} = \mathfrak{g}_- \oplus H \oplus \mathfrak{g}_+$ where $\mathfrak{g}_+ = \langle y_{ij}: i < j\rangle$, $\mathfrak{g}_- = \langle y_{ij} : i > j\rangle$, and H is the set of diagonal matrices in $\mathfrak{g}$. Define the dominance order $<$ on $\mathbb{Z}^n$ by

$$(a_1, \ldots, a_n) < (b_1, \ldots, b_n)$$

if $\sum_{i=1}^{s} a_i \le \sum_{i=1}^{s} b_i$ for $s = 1, 2, \ldots, n$. Note that

$$\phi(\mathrm{diag}(u_1, \ldots, u_n))(\phi(y_{ij})v_\gamma)$$

$$= \phi(y_{ij})(\phi(\mathrm{diag}(u_1,\ldots,u_n))v_\gamma) + \phi([\mathrm{diag}(u_1,\ldots,u_n), y_{ij}])v_\gamma$$

$$= ((u_1,\ldots,u_n)\cdot\lambda^{(\gamma)})\phi(y_{ij})v_\gamma + \phi((u_i-u_j)y_{ij})v_\gamma$$

$$= ((u_1,\ldots,u_n)\cdot(\lambda^{(\gamma)} + e_i - e_j))\,(\phi(y_{ij})v_\gamma).$$

Thus $\phi(y_{ij})v_\gamma$ is a weight vector of V with weight $\lambda^{(\gamma)}+e_i-e_j$. When i is less than j, $\lambda^{(\gamma)} < \lambda^{(\gamma)} + e_i - e_j$ so we have the following observation:

(*) for $g \in \mathfrak{g}_+$ and v a weight vector of V, $\phi(g)(v)$ is a linear combination of weight vectors whose weights strictly dominate the weight of v.

Since V is finite dimensional, the set of weights with nonzero multiplicity in V is finite. So by (*), there exist weight vectors $v_\gamma$ in V with $\phi(\mathfrak{g}_+)v_\gamma = 0$.

DEFINITION 1: A nonzero weight vector $v_\gamma$ of weight $\lambda^{(\gamma)}$ is called maximal if $\phi(\mathfrak{g}_+)v_\gamma = 0$. Given a weight $\lambda$ we let

$M_\lambda(V)$ denote the vector space of maximal weight vectors of weight $\lambda$ (together with 0), and let $m_\lambda(V)$ denote the dimension of $M_\lambda(V)$. Recall that the irreducible $\mathcal{G}$-modules are indexed by weights $(\lambda_1, \ldots, \lambda_n)$ with $\lambda_1 \geq \lambda_2 \geq \cdots \geq \lambda_n$. Such a weight is called <u>dominant</u>.

THEOREM 1. Let V be a finite dimensional $\mathcal{G}$-module. Then

(A) $m_\lambda(V) = 0$ unless $\lambda$ is a dominant weight.

(B) If $\lambda$ is a dominant weight then $m_\lambda(V)$ is the multiplicity of the irreducible $\mathcal{G}$-module indexed by $\lambda$ in the module V.

Specialize now to the case where V is $T^p(\mathcal{G})$ considered as a $\mathcal{G}$-module under the adjoint action. A basis of weight vectors is the set of $y_{i_1 j_1} \otimes \cdots \otimes y_{i_p j_p}$ where each $i_s$ and $j_s$ is in the range 1 to n. The above weight vector has weight

$$\lambda = \sum_{s=1}^{p} (e_{i_s} - e_{j_s})$$

so $\lambda \cdot (1,1, \ldots, 1) = 0$ and $\lambda$ is in $\mathbb{Z}^n$. If $\lambda$ is also dominant, then there exist partiitions $\alpha$ and $\beta$ of the same integer r with $\ell(\alpha) + \ell(\beta) \leq n$ such that

$$\lambda = \left(\sum_{i=1}^{\ell(\alpha)} \alpha_i e_i\right) - \left(\sum_{i=1}^{\ell(\beta)} \beta_i e_{n+1-i}\right)$$

$$= (\alpha_1, \alpha_2, \cdots, \alpha_{\ell(\alpha)}, 0, \ldots, 0, -\beta_{\ell(\beta)}, \ldots, -\beta_2, -\beta_1).$$

DEFINITION 2: Let $\alpha$ and $\beta$ be partitions of r with $\ell(\alpha) + \ell(\beta) \leq n$. The <u>mixed tensor</u> representation $V_{\alpha,\beta,n}$ is the irreducible $\mathcal{G}$-module indexed by the dominant weight $[\alpha,\beta]_n$ $= (\alpha_1, \alpha_2, \ldots, \alpha_{\ell(\alpha)}, 0, \ldots, 0, -\beta_{\ell(\beta)}, \ldots, -\beta_1)$.

PROBLEM 1: Given $\alpha, \beta$ partitions of r, with $\ell(\alpha) + \ell(\beta) \leq n$, construct all maximal weight vectors of weight $[\alpha,\beta]_n$ in $T^p(\mathcal{G})$.

PROBLEM 2: If possible, determine the linear dependencies amongst these vectors and a basis for $M_{[\alpha,\beta]_n}(V)$. The dimension of $M_{[\alpha,\beta]_n}(V)$ will then be the multiplicity of $V_{\alpha,\beta,n}$ in $T^p(\mathcal{G})$.

Section 3. Construction of the maximal weight vectors

In this section $\alpha$ and $\beta$ are fixed partitions of a non-negative integer r. Our goal is to construct all maximal weight vectors in $T^p(\mathfrak{G})$ of weight $[\alpha,\beta]_n$. Note that the sum of the positive entries in $[\alpha,\beta]_n$ is r. So r is the smallest value of p for which there even exist weight vectors of weight $[\alpha,\beta]_n$ in $T^p(\mathfrak{G})$. We begin with this minimal value of p.

CASE 1: $p = r$.

Let $W_r$ and $H_r$ denote the vector spaces of weight vectors and maximal weight vectors of weight $[\alpha,\beta]_n$ in $T^r(\mathfrak{G})$. Then $W_r$ has as basis the set of $y_{i_1,j_1} \otimes \cdots \otimes y_{i_r,j_r}$ such that

$$(*)\begin{cases} \sum_{s=1}^{r} e_{i_s} = (\alpha_1,\alpha_2,\ldots,\alpha_{\ell(\alpha)},0,0,\ldots,0) \text{ and} \\ \sum_{s=1}^{r} - e_{j_s} = (0,0,\ldots,0,-\beta_{\ell(\beta)},\ldots,-\beta_2,-\beta_1). \end{cases}$$

It is easy to see that the equations (*) hold if and only if exactly $\alpha_u$ of the numbers $i_1, i_2, \ldots, i_r$ equal u and exactly $\beta_u$ of the numbers $j_1, \ldots, j_r$ equal $(n+1) - u$.

Let $M^\alpha$ and $M^\beta$ be the complex vector spaces spanned by the $\alpha$ and $\beta$ tabloids respectively. For T a tabloid and i an integer in T let $r_T(i)$ denote the row of T containing i.

DEFINITION 3: Define the linear map $\pi : M^\alpha \otimes M^\beta \to W_r$ by $\pi(T \otimes S) = \bigotimes_{i=1}^{r} y_{r_T(i),(n+1)-r_S(i)}$. By the comments above, $\pi$ is an isomorphism. The following theorem is proved using a result of Hodge (see [2]). This result gives spanning sets for the orthogonal complements to the Specht modules in $M^\alpha$ and $M^\beta$. Let $S^\alpha \subseteq M^\alpha$ and $S^\beta \subseteq M^\beta$ be the Specht modules.

THEOREM 2: The restriction of $\pi$ to $S^\alpha \otimes S^\beta$ is a (vector space) isomorphism from $S^\alpha \otimes S^\beta$ onto $H_r$.

$$\begin{array}{ccl} M^\alpha \otimes M^\beta & \xrightarrow{\ \pi\ } & W_r \text{ (weight vectors of weight } [\alpha,\beta]_n) \\ \Big| & & \Big| \\ S^\alpha \otimes S^\beta & \xrightarrow{\ \pi\ } & H_r \text{ (maximal weight vectors).} \end{array}$$

This completes the construction of the maximal weight vectors of weight $[\alpha,\beta]_n$ in $H_r$.

CASE 2:  $p > r$.

We would like to reduce this case to Case 1. There are two special constructions which will help in this reduction.

(1)  If z' is a maximal weight vector of weight $[\alpha,\beta]_n$ in $T^p(\mathcal{G})$ with p' > p, and $z_0$ is a maximal weight vector of weight 0 in $T^{p-p'}(\mathcal{G})$ then $z' \otimes z_0$ is in $H_p$.

(2)  If $\sigma$ is a permutation in $S_p$ and z is in $H_p$ then $\sigma(z)$ is in $H_p$ (where $\sigma(z)$ is the vector obtained from z by permuting tensor positions according to $\sigma$).

The construction of $H_p$ will be complete when we find all vectors in $H_r$ modulo the subspace of those constructed by rules (1) and (2).

DEFINITION 4:  Define the linear map $\partial : T^a(\mathcal{G}) \to T^{a+1}(\mathcal{G})$

by   (A)   $\partial(y_{ij}) = \sum_{u=1}^{n} (y_{iu} \otimes y_{uj})$

     (B)   $\partial$ is a derivation.

Define $Y(i,j,k) \in T^k(\mathcal{G})$ by $Y(i,j,k) = \partial^{k-1} y_{ij}$.

LEMMA 1:  For any a and b, we have
$$ad(y_{ab})\partial \;=\; \partial\, ad(y_{ab})$$
where the multiplication is composition of maps.

It follows that $Y(i,j,k)$ is a weight vector of weight $e_i - e_j$ in $T^k(\mathcal{G})$ which satisfies $ad(y_{ab})\,(Y(i,j,k)) = \delta_{bi}\, Y(a,j,k) - \delta_{aj}\, Y(i,b,k)$. So with respect to the adjoint action of $\mathcal{G}$, $Y(i,j,k)$ looks exactly like $y_{ij}$ except that it occupies k tensor positions instead of 1.

DEFINITION 5:  Let f be a function from $\underline{r}$ into the positive integers with $\sum_{i=1}^{r} f(i) = p$. Define the linear map

$\pi_f : M^\alpha \otimes M^\beta \to W_p$ by
$$\pi_f(T \otimes S) = \bigotimes_{i=1}^{r} Y(r_T(i),\, (n+1) - r_S(i),\, f(i)).$$

It is $\underline{not}$ the case that $\pi_f$ is onto $W_p$ for p > r. However, combining Theorem 1 with Lemma 1 gives

LEMMA 2: For $u \in M^{\alpha} \otimes M^{\beta}$ we have $\pi_f(u)$ is in $H_p$ if and only if $u$ is in $S^{\alpha} \otimes S^{\beta}$. The vectors $\pi_f(u)$ together with the constructions (1) and (2) give a complete spanning set for $H_p$.

THEOREM 3: Let $H_p$ be the set of maximal weight vectors of weight $[\alpha,\beta]_n$ in $T^p(\mathbb{C})$.

(A) Let $f$ be a function from $\underline{r}$ into $\mathbb{Z}^+$ with

$$\Sigma f(i) = p' < p.$$

Let $u$ be a vector in $S^{\alpha} \otimes S^{\beta}$.

Let $z_0$ be a maximal weight vector of weight 0 in $T^{p-p'}(\mathbb{C})$.

Let $\sigma$ be a permutation in $S_p$.

Then $\sigma(z_0 \otimes \pi(f)(a))$ is in $H_p$.

(B) The set of vectors defined in (A) span $H_p$.

Part (B) of this theorem was originally proved by the author for $p \leq n - \ell(\alpha) - \ell(\beta) + r$ only. However together with R. K. Gupta the proof has been extended to all $p$. This completes the solution to problem 1 and leaves the question of what linear dependencies hold amongst the constructed maximal weight vectors.

PROBLEM 2: Linear Dependencies

It is an interesting open question to describe the linear dependencies amongst the vectors $\sigma(z_o \otimes \pi_f(u))$ for general $\alpha,\beta$ and $p$. For $\alpha = \beta = \phi$, this is done by an elegant result of Schur (see [7], pg. 128). We consider here the case where $n$ is large with respect to $r$ and $p$.

The situation is simplest if we disregard those maximal weight vectors $\sigma(z_0 \otimes \pi_f(u))$ which contain a nontrivial $z_0$. Let $H_p^{(0)}$ denote the subspace of $H_p$ spanned by vectors $\sigma(z_o \otimes \pi_f(u))$ where $z_o$ is nontrivial. We begin by computing the dimension of $H_p/H_p^{(0)}$.

THEOREM 4: Suppose $p$ is less than $n + r - \ell(\alpha) - \ell(\beta)$.

Then $H_p/H_p^{(0)}$ has dimension $\binom{p-1}{r-1}\frac{p!}{r!} f_{\alpha} f_{\beta}$.

The idea behind this theorem is the following.

Suppose $u_1, \ldots, u_{f_\alpha f_\beta}$ is a basis for $S^\alpha \otimes S^\beta$. Note that $S^\alpha \otimes S^\beta$ is an $S_r$-module so for each $\tau$ in $S_r$ we can write $\tau u_i = \sum_j c_{i,j}^{(\tau)} u_j$.

Our initial guess for a basis of $H_p/H_p^{(0)}$ is the set of $\sigma(\pi(f)(u_i)) + H_p^{(0)}$ for $\sigma$ in $S_p$, $1 \leq i \leq f_\alpha f_\beta$, and f a function from $\underline{r}$ to $\mathbb{Z}^+$ satisfying $\Sigma f(i) = p$. In fact this is a spanning set for $H_p/H_p^{(0)}$ of size $\binom{p-1}{r-1}$ p! $f_\alpha f_\beta$ (note that $\binom{p-1}{r-1}$ is the number of functions $f : \underline{r} \to \mathbb{Z}^+$ which satisfy $\Sigma f(i) = p$). There are linear dependencies amongst these vectors. These come about as follows: for any $f : \underline{r} \to \mathbb{Z}^+$ with $\Sigma f(i) = p$ and any $\tau$ in $S_r$ we have that $f\tau^{-1}$ is again a function from $\underline{r}$ to $\mathbb{Z}^+$ satisfying $\Sigma f\tau^{-1}(i) = p$. Moreover for any $u_i$, $\pi(f\tau^{-1})(u_i) = \sigma_\tau(\pi(f)(u_i))$ where $\sigma_\tau$ is an appropriate permutation of tensor positions. So

$$\sigma(\sigma_\tau(\pi(f)(u_i))) = \sigma(\pi(f)(\tau^{-1}u_i)) = \sum_j c_{i,j}^{(\tau^{-1})} \sigma(\pi(f)(u_j))$$

which is a nontrivial linear dependence amongst the vectors $\sigma(\tau(f)(u_i))$. One can check (see [2]) that these dependencies generate all linear dependencies and that this forces a basis of $H_p/H_p^{(0)}$ to have dimension $\binom{p-1}{r-1}\frac{p!}{r!} f_\alpha f_\beta$.

The problem with extending this basis to all of $H_p$ is that for $z_0$ a maximal weight vector of weight 0 in $T^a(\mathfrak{G})$ and $\eta \in S_a$ we may have $\eta(z_0) = z_0$. This is in fact the case for the usual basis of maximal 0-weight vectors in $T^a(\mathfrak{G})$. So we can't allow arbitrary $\sigma$ applied to $z_0 \otimes \pi(f)(u)$ when forming a basis for $H_p$. This is easily remedied by taking $\sigma$ from a set of coset leaders in $S_p$ mod the subgroup of permutations of the tensor positions occupied by $z_0$.

DEFINITION 6: For $p' < p$, let $C(p', p)$ be a complete set of right coset representatives for the subgroup $S_{p-p'} \times S_{p'}$ in $S_p$ (we consider $S_{p-p'}$ as permuting the symbols 1, 2, ..., $p - p'$, and $S_{p'}$ as permuting the symbols $p-p' + 1, \ldots, p$).

For each a, let $B_a^{(0)}$ be a basis for the vector space of maximal weight vectors of weight 0 in $T^a(\mathfrak{G})$, and let $B_a^{(\alpha,\beta)}$ be a basis for the subspace $H_a/H_a^{(0)}$.

THEOREM 5: Suppose p is less than $n + r - \ell(\alpha) - \ell(\beta)$. Then a basis for $H_p$ is given by the set of $\sigma(z_0 \times z)$ where is in $C(p',p)$ for $p' < p$, where $z_0$ is in $B_{(p-p')}^{(0)}$ and where z is in $B_{(p')}^{(\alpha,\beta)}$.

This theorem has a very nice corollary which is the final result of this paper.

COROLLARY: Let $F^{(n)}_{(\alpha,\beta)}(q) = \sum_{p=0}^{\infty} \frac{(\dim H_p)}{p!} q^p$

where $\dim H_p$ refers to the dimension of the vector space $H_p$ in $T^p(g\ell_n(\mathbb{C}))$. Then

$$\lim_{n\to\infty} F^{(n)}_{(\alpha,\beta)}(q) = q^r \frac{f_\alpha f_\beta}{r!}(1-q)^{-(r+1)}.$$

PROOF: It is enough to show that for fixed n, the power series on the right agrees with $F^{(n)}_{(\alpha,\beta)}(q) \bmod q^{n+r-\ell(\alpha)-\ell(\beta)}$. For $p < n + r - \ell(\alpha) - \ell(\beta)$ we have

$$\frac{\dim H_p}{p!} = \frac{1}{p!} \sum_{p'=r}^{p} \binom{p}{p'} \left(\binom{p'-1}{r-1} \frac{f_\alpha f_\beta}{r!} (p')!\right) \left| B^0_{p-p'} \right| .$$

It follows from a result of Schur ([7]) that $\left|B_a^{(0)}\right| = a!$ for a less than n. Thus

$$\frac{\dim H_p}{p!} = \left(\frac{f_\alpha f_\beta}{r!}\right)\left(\sum_{p'=r}^{p}\binom{p'-1}{r-1}\right) = \frac{f_\alpha f_\beta}{r!} \left(\sum_{p'=0}^{p}\binom{p'-1}{r-1}\right) .$$

It is clear that this number is the coefficient of $q^p$ in $q_r \frac{f_\alpha f_\beta}{r!} (1-q)^{-1}(1-q)^{-r}$ which completes the result.

## BIBLIOGRAPHY

1. P. Hanlon, An introduction to the complex representations of the symmetric and general linear groups, Proc. of the 1983 Summer Research Conference in Combinatorics and Algebra, AMS Contemporary Mathematics series.

2. P. Hanlon, On the decomposition of the tensor algebra of the classical groups. (submitted to Advances in Math.).

3. G. D. James, The representation theory of the symmetric groups, Springer Lecture Notes No. 682, Springer (1978).

4. I. G. Macdonald, Lie groups and combinatorics, Contemporary Mathematics, Vol. 9 (1982), 73-83.

5. J. P. Serre, Lie algebras and Lie groups, Benjamin, Inc. (1965).

6. R. P. Stanley, Gℓ(n,C) for combinatorialists, Surveys in combinatorics, (E. K. Lloyd, editor), London Math. Soc. Lecture Notes No. 82, Cambridge. University Press (1983).

7. H. Weyl, The classical groups, Princeton University Press (1973).

Contemporary Mathematics
Volume 34, 1984

# THE Q-DYSON CONJECTURE, GENERALIZED EXPONENTS, AND THE INTERNAL PRODUCT OF SCHUR FUNCTIONS

Richard P. Stanley[1]

ABSTRACT. The q-Dyson conjecture is a combinatorial problem posed by G. Andrews in 1975. The conjecture can be formulated in terms of symmetric functions, and it is shown how the theory of symmetric functions can be used to prove a limiting form of the conjecture. The proof uses a new identity involving the internal product of Schur functions. The same techniques yield information about a limiting form of the "generalized exponents" of $SL(n,\mathbb{C})$, as defined by Kostant. Complete details will appear in a forthcoming paper in Linear and Multilinear Algebra.

1. THE Q-DYSON CONJECTURE. Let $a_1,\cdots, a_n$ be nonnegative integers. In 1962 Dyson [2] conjectured that when the product

$$\prod_{\substack{i,j=1\\ i\neq j}}^{n} \left(1-x_i x_j^{-1}\right)^{a_i}$$

is expanded as a Laurent polynomial in the variables $x_1,\cdots, x_n$, then the constant term is equal to the multinomial coefficient $(a_1+\cdots+a_n)!/a_1!\cdots a_n!$. This conjecture was proved in 1962 by Gunson [4] and Wilson [15], and in 1970 an exceptionally elegant proof was given by Good [3].

In 1975 G. Andrews [1, p. 216] formulated a "q-analogue" of the Dyson conjecture, which reduces to the original conjecture when $q = 1$. Write $(a)_n = (1-a)(1-aq)\cdots(1-aq^{n-1})$, so $(q)_n = (1-q)(1-q^2)\cdots(1-q^n)$.

Q-DYSON CONJECTURE. When the product

$$\prod_{1\le i<j\le n} (qx_i x_j^{-1})_{a_i} (x_j x_i^{-1})_{a_j}$$

is expanded as a Laurent polynomial in the variables $x_1,\cdots, x_n$, then the constant term is equal to the q-multinomial coefficient

---

1980 Mathematics Subject Classification. 22E46, 05A15, 20C30.

[1]Partially supported by a grant from the National Science Foundation.

$$(q)_{a_1+\cdots+a_n}/(q)_{a_1}\cdots(q)_{a_n}\ .$$

This conjecture was proved for $n \le 3$ by Andrews [1] and for $n = 4$ by Kadell [6]. It was also proved for $a_1 = \cdots = a_n = 1$, 2, or $\infty$ by Macdonald [12], who formulated a far-reaching generalization. Further work appears in [5]. Here we will establish, as a corollary to a more general result, the case $a_1 = a_2 = \cdots = \ell$ in the limit $n \longrightarrow \infty$. In the form stated above, the q-Dyson conjecture becomes meaningless when $n \longrightarrow \infty$. However, it can be restated to make sense in this limit. We will give a restatement in terms of representation theory due to Macdonald [12, Conj. 3.1']. Let $SL(n,\mathbb{C})$ denote the group of $n \times n$ complex matrices of determinant one, and $s\ell(n,\mathbb{C})$ its Lie algebra of all $n \times n$ complex matrices of trace zero. Let

$$\text{ad}\colon SL(n,\mathbb{C}) \longrightarrow GL(s\ell(n,\mathbb{C}))$$

denote the adjoint representation of $SL(n,\mathbb{C})$, defined by

$$(\text{ad } X)(A) = XAX^{-1},$$

for $X \in SL(n,\mathbb{C})$ and $A \in s\ell(n,\mathbb{C})$.

Q-DYSON CONJECTURE FOR $a_1 = \cdots = a_n = \ell$ (reformulated). The multiplicity of the trivial character of $SL(n,\mathbb{C})$ in the virtual character

$$\det(1-q\cdot\text{ad } X)(1-q^2\cdot\text{ad } X)\cdots(1-q^{\ell-1}\cdot\text{ad } X)$$

is equal to

$$\prod_{i=1}^{n-1}\prod_{j=1}^{\ell-1}\left(1-q^{\ell i+j}\right). \tag{1}$$

2. SYMMETRIC FUNCTIONS. The above conjecture can be formulated in terms of symmetric functions, which we will now briefly review. Let $\lambda = (\lambda_1,\lambda_2,\cdots)$ be a <u>partition</u>, i.e., a decreasing sequence $\lambda_1 \ge \lambda_2 \ge \cdots \ge 0$ of nonnegative integers with only finitely many $\lambda_i$ unequal to zero. If $\lambda_{n+1} = \lambda_{n+2} = \cdots = 0$ then we also write $\lambda = (\lambda_1,\cdots,\lambda_n)$. The number of nonzero $\lambda_i$ is the <u>length</u> of $\lambda$, denoted $\ell(\lambda)$. If $m = \lambda_1 + \lambda_2 + \cdots$ then we write $\lambda \vdash m$ or $|\lambda| = m$. The <u>conjugate</u> partition $\lambda' = (\lambda_1', \lambda_2', \cdots)$ to $\lambda$ has $\lambda_i - \lambda_{i+1}$ parts equal to $i$.

Let $\Lambda_n = \Lambda_n(x)$ denote the ring of all symmetric polynomials with rational coefficients in the variables $x = (x_1,\cdots,x_n)$, and let $\Omega_n$ denote $\Lambda_n$ modulo the ideal generated by $x_1x_2\cdots x_n - 1$. A vector space basis for $\Omega_n$ consists of all <u>Schur functions</u> $s_\lambda(x) = s_\lambda(x_1,\cdots,x_n)$, where $\lambda$ ranges over all partitions of length $\le n-1$. For the definition and basic properties of Schur functions, see [11].

If $\varphi: SL(n,\mathbb{C}) \longrightarrow GL(N,\mathbb{C})$ is a (polynomial) representation of $SL(n,\mathbb{C})$, then the <u>character</u> of $\varphi$ is the unique polynomial char $\varphi \in \Omega_n$ satisfying char $\varphi = \mathrm{tr}\, \varphi(X)$ for any $X \in SL(n,\mathbb{C})$ with eigenvalues $x_1, \cdots, x_n$. A basic theorem (e.g. [14]) on the representations of $SL(n,\mathbb{C})$ states that the irreducible (polynomial) characters of $SL(n,\mathbb{C})$ are precisely the Schur functions $s_\lambda(x) \in \Omega_n$, $\ell(\lambda) \le n-1$. Thus the problem of decomposing char $\varphi$ into irreducible characters is equivalent to expanding char $\varphi$ as a linear combination of Schur functions in the ring $\Omega_n$.

Sometimes it is convenient to work with symmetric functions (= formal power series) in infinitely many variables $x = (x_1, x_2, \cdots)$. We let $\Lambda = \Lambda(x)$ denote the ring of all symmetric formal power series of bounded degree with rational coefficients in the variables $x$. $\Lambda$ is the inverse limit of the rings $\Lambda_n$ in the category of <u>graded</u> rings. The Schur functions $s_\lambda(x)$, for all partitions $\lambda$, form a vector space basis for $\Lambda(x)$. The completion $\hat{\Lambda}$ of $\Lambda$ (with respect to the ideal of symmetric functions with zero constant term) consists of all symmetric formal power series with no restriction on the degree. $\hat{\Lambda}$ is the inverse limit of the rings $\Lambda_n$ in the category of rings. For further information, see [11, Ch. I.2]. Let us remark that in [11] the elements of $\Lambda_n$, $\Lambda$, and $\hat{\Lambda}$ have <u>integer</u> coefficients, but we will find it convenient to allow rational coefficients from the start.

Now suppose $X \in SL(n,\mathbb{C})$ has eigenvalues $x_1, \cdots, x_n$, i.e.,

$$\det(1-qX) = \prod_{i=1}^{n} (1-qx_i) .$$

Since ad $X$ has eigenvalues $x_i x_j^{-1}$ (once each for $i \neq j$) and 1 ($n-1$ times) (e.g., [14, eqn. (8)]), we have

$$\begin{aligned} &\det(1-q\cdot \mathrm{ad}\, X)\cdots(1-q^{\ell-1}\cdot \mathrm{ad}\, X) \\ &= (q)_{\ell-1}^{-1} \prod_{i,j=1}^{n} \prod_{k=1}^{\ell-1} (1-q^k x_i x_j^{-1}) . \end{aligned} \tag{2}$$

Q-DYSON CONJECTURE FOR $a_1 = \cdots = a_n = \ell - 1$ (again reformulated). When equation (2) is expanded in the ring $\Omega_n \otimes \mathbb{Q}[[q]]$ as a linear combination of Schur functions $s_\lambda(x)$ with $\ell(\lambda) \le n-1$, then the coefficient of the trivial Schur function $s_\phi(x)$ is given by (1).

The above conjecture makes sense as $n \longrightarrow \infty$. We might as well consider a much more general situation, and later specialize to the case at hand . Thus introduce two new sets $u = (u_1, u_2, \cdots)$ and $v = (v_1, v_2, \cdots)$ of variables, and write (in the ring $\Omega_n \otimes \mathbb{Q}[[u,v]]$)

$$\left[\prod_k \frac{1-u_k}{1-v_k}\right] \det \prod_k \frac{1-u_k\,\mathrm{ad}\,X}{1-v_k\,\mathrm{ad}\,X} = \sum_\lambda c_\lambda^n(u;v)s_\lambda(x), \tag{3}$$

where $\lambda$ ranges over all partitions with $\ell(\lambda) \leq n-1$. Clearly $c_\lambda^n(u;v) \in \mathbb{Z}[[u,v]]$, the ring of formal power series with integer coefficients in the variables $u = (u_1, u_2, \cdots)$ and $v = (v_1, v_2, \cdots)$.

We wish to consider $c_\lambda^n(u;v)$ as $n \to \infty$. To do so, one must vary $\lambda$ with $n$ or else the limit becomes zero or undefined. The correct way of passing to the limit was suggested by R. Gupta (in the somewhat less general context of Section 5). Given any two partitions $\alpha$ and $\beta$ of lengths $k$ and $\ell$ of the same integer $m$, and given $n \geq k+\ell$, define the partition

$$[\alpha, \beta]_n = (\beta_1+\alpha_1, \beta_1+\alpha_2, \cdots, \beta_1+\alpha_k, \underbrace{\beta_1, \cdots, \beta_1}_{n-k-\ell}, \beta_1-\beta_\ell, \beta_1-\beta_{\ell-1}, \cdots, \beta_1-\beta_2)$$

of length $\leq n-1$.

$[\alpha,\beta]_n \;=\; n$

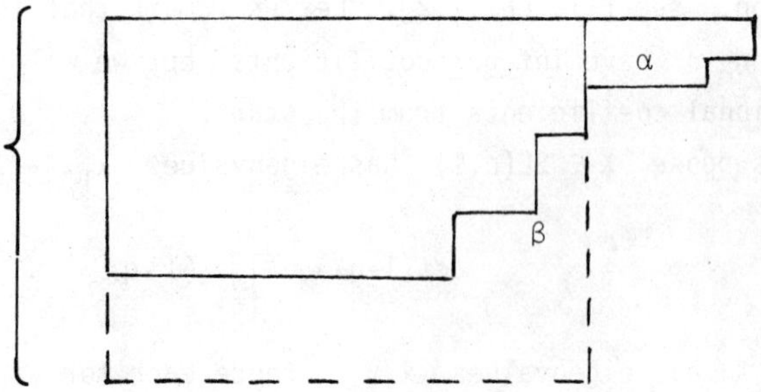


It follows from Gupta's work that

$$\lim_{n\to\infty} c^n_{[\alpha,\beta]_n}(u;v)$$

exists as a formal power series, which we denote by $c_{\alpha\beta}(u;v)$. Our main goal here is a formula for $c_{\alpha\beta}(u;v)$. The q-Dyson conjecture in the case $n \to \infty$ corresponds to taking $\alpha = \beta = \phi$ (the void or trivial partition) and $v_i = 0$, $u_1 = q$, $u_2 = q^2, \cdots, u_{\ell-1} = q^{\ell-1}$, $u_\ell = u_{\ell+1} = \cdots = 0$.

In order to state our result, we first review some more background from the theory of symmetric functions. The irreducible characters $\chi^\lambda$ of the symmetric group $S_m$ are indexed by partitions $\lambda$ of $m$. If $w \in S_m$, then define $\rho(w)$ to be the partition of $m$ whose parts are the cycle lengths of $w$. For any

$\lambda \vdash m$ with $\ell = \ell(\lambda)$, define the power-sum symmetric function $p_\lambda = p_{\lambda_1} p_{\lambda_2} \cdots p_{\lambda_\ell}$, where $p_n(x) = \sum_i x_i^n$. The Schur functions and power-sums are related by (e.g., [11, Ch. I.7])

$$s_\lambda = \frac{1}{m!} \sum_{w \in S_m} \chi^\lambda(w) p_{\rho(w)} . \tag{4}$$

Now let

$$\chi^\alpha \chi^\beta = \sum_\gamma g_{\alpha\beta\gamma} \chi^\gamma ,$$

where each $g_{\alpha\beta\gamma}$ is a nonnegative integer. It is an important open problem to obtain a nice combinatorial interpretation of $g_{\alpha\beta\gamma}$. D.E. Littlewood [10], in order to incorporate the Kronecker product $\chi^\alpha\chi^\beta$ into the theory of symmetric functions, defined an associative (and commutative) product $f*g$ on symmetric functions by

$$s_\alpha * s_\beta = \sum_\gamma g_{\alpha\beta\gamma} s_\gamma ,$$

called the <u>internal product</u>. (Littlewood uses the term "inner product". Since the product $f*g$ has nothing to do with the usual definition of inner product in linear algebra, we have followed a suggestion of I.G. Macdonald in calling it the internal product. Littlewood uses the notation $f\circ g$ for internal product. Since we are adhering to the notation of [11], where $f\circ g$ denotes plethysm , we have introduced the new notation $f*g$.) Note that $s_\alpha * s_m = s_\alpha$ and $s_\alpha * s_{1^m} = s_{\alpha'}$, where $\alpha'$ denotes the conjugate partition to $\alpha$.

In terms of the power-sums we have the expansion

$$s_\alpha * s_\beta = \frac{1}{m!} \sum_{w \in S_m} \chi^\alpha(w) \chi^\beta(w) p_{\rho(w)} . \tag{5}$$

The following basic property of the internal product is due to Schur [13] (p.69 of Dissertation; p.65 of GA):

2.1. PROPOSITION. We have

$$\prod_{i,j,k} (1 - x_i y_j v_k)^{-1} = \sum_{\alpha,\beta} s_\alpha * s_\beta(v) s_\alpha(x) s_\beta(y). \tag{6}$$

Now define a scalar product $\langle f,g \rangle$ in the ring $\Lambda$ by letting the Schur functions form an orthonormal basis, i.e.,

$$\langle s_\lambda , s_\mu \rangle = \delta_{\lambda\mu} .$$

Given partitions $\alpha,\beta$, define a symmetric function $s_{\alpha/\beta} \in \Lambda$, called a <u>skew Schur function</u>, by the rule

$$\langle s_{\alpha/\beta}, s_\gamma\rangle = \langle s_\alpha, s_\beta s_\gamma\rangle .$$

In other words, multiplication by $s_\beta$ is adjoint to the linear transformation sending $s_\alpha$ to $s_{\alpha/\beta}$. It is not difficult to show that $s_{\alpha/\beta} = 0$ unless $\beta \leq \alpha$, i.e., $\beta_i \leq \alpha_i$ for all $i$. For further information, see [11, Ch.I.5].

Let us remark that the Schur function $s_{[\alpha,\beta]_n}(x)$ was considered by D.E. Littlewood [8], who essentially showed that in the ring $\Omega_n$ we have

$$s_{[\alpha,\beta]_n}(x) = \sum_\lambda (-1)^{|\lambda|} s_{\alpha/\lambda}(x) s_{\beta/\lambda'}(1/x) ,$$

where $x = (x_1,\dots, x_n)$ and $1/x = (1/x_1,\dots, 1/x_n)$. For instance, the adjoint representation of $SL(n,\mathbb{C})$ corresponds to the partition $[1,1]_n$, with character

$$\begin{aligned} s_{[1,1]_n}(x) &= s_1(x)s_1(1/x) - 1 \\ &= (x_1 +\dots+ x_n)(x_1^{-1} +\dots+ x_n^{-1}) - 1 \\ &= n - 1 + \sum_{i\neq j} x_i x_j^{-1} . \end{aligned}$$

3. A FORMULA FOR $c_{\alpha\beta}(u;v)$. In order to evaluate $c_{\alpha\beta}(u;v)$ we first obtain a formula for the generating function

$$C(x,y) = \sum_{\alpha,\beta} c_{\alpha\beta}(u;v) s_\alpha(x) s_\beta(y) \in \mathbb{Q}[[u,v]] \otimes \hat{\Lambda}(x) \otimes \hat{\Lambda}(y) .$$

It will be more convenient to work with

$$C_0(x,y) = \sum_{\alpha,\beta} c_{\alpha\beta}(0;v) s_\alpha(x) s_\beta(y) ,$$

and later to apply a standard trick to obtain $C(x,y)$ from $C_0(x,y)$. (Here $c_{\alpha\beta}(0;v)$ denotes the substitution $u_k = 0$ in $c_{\alpha\beta}(u;v)$.)

3.1. LEMMA. We have

$$C_0(x,y) = \sum_{\lambda,\mu,\alpha} s_\lambda * s_\mu(v) s_{\lambda/\alpha}(x) s_{\mu/\alpha}(y)$$

<u>Sketch of proof</u>. One begins by setting $y_j = x_j^{-1}$ in (6) so that the left-hand side of (6) coincides with the left-hand side of (3) when $u_k = 0$. The proof then proceeds by standard manipulations of symmetric functions, which we omit. □

Next we find a more tractable expression for $C_0(x,y)$ by establishing the following symmetric function identity, which apparently is new.

3.2. LEMMA. We have

$$\prod_{i,j}\prod_{r>0}\prod_{a_1,\dots,a_r}(1-x_iy_jv_{a_1}\cdots v_{a_r})^{-1}$$

$$=\Big[\prod_{k\geq 1}(1-p_k(v))\Big]\sum_{\lambda,\mu,\alpha}s_\lambda * s_\mu(v)s_{\lambda/\alpha}(x)s_{\mu/\alpha}(y). \tag{7}$$

(Here $a_1,\dots, a_r$ range independently over all indices of the $v$'s.)

<u>Sketch of proof</u>. We work in the ring $R = \mathbb{Q}((v)) \otimes \Lambda(x) \otimes \Lambda(y)$, which should be regarded as consisting of formal power series of bounded degree, symmetric in the $x$'s and $y$'s separately, with coefficients in the field $\mathbb{Q}((v))$ (the quotient field of $\mathbb{Q}[[v]]$). Define a scalar product on $R$ by letting the elements $s_\alpha(x)s_\beta(y)$ form an orthonormal basis.

If $f \in R$ then let $D(f)$ denote the linear transformation which is adjoint to multiplication by $f$, i.e.,

$$\langle D(f)g,\, h\rangle = \langle g,\, fh\rangle \quad .$$

Note that $D(f+g) = D(f) + D(g)$. Let $P(v) = \prod_{k\geq 1}(1 - p_k(v))$. The right-hand of (7) is given by

$$P(v)\sum_\alpha D(s_\alpha(x)s_\alpha(y))\sum_{\lambda,\mu}s_\lambda * s_\mu(v)s_\lambda(x)s_\mu(y)$$

$$= P(v)\, D\Big(\prod_{i,j}(1-x_iy_j)^{-1}\Big)\prod_{i,j,k}(1-x_iy_jv_k)^{-1}\;.$$

Thus writing LHS for the left-hand side of (7), we need to show that for all $f \in R$,

$$\langle \mathrm{LHS},\, f\rangle = \langle P(v)\,\Pi(1-x_iy_jv_k)^{-1},\; f\,\Pi(1-x_iy_j)^{-1}\rangle.$$

It suffices to check this for all $f$ forming a $\mathbb{Q}((v))$-basis for $R$. Choose $f = p_\alpha(x)p_\beta(y)$, and the verification becomes a routine computation using standard symmetric function techniques. □

If we now compare Lemmas 3.1 and 3.2 and expand the right-hand side of (7) in terms of the $p_\lambda(v)$'s, we obtain:

3.3. LEMMA. We have

$$c_{\alpha\beta}(0;v) = P(v)^{-1}\, s_\alpha * s_\beta\Big(p_k \to \frac{p_k(v)}{1-p_k(v)}\Big) \quad . \tag{8}$$

The notation indicates that we are to expand $s_\alpha * s_\beta$ in terms of the $p_k$'s, as given explicitly by (5), and substitute $p_k(v)/(1-p_k(v))$ for $p_k$. □

In order to find a similar formula for $c_{\alpha\beta}(u;v)$, we first replace the variables $v$ in (8) by the two sets of variables $u$ and $v$. Now let $\omega_u$ denote the algebra automorphism described in [11, pp. 14-17, 26] acting on symmetric functions in $u$ (regard all other variables as scalars commuting with $\omega_u$). By standard properties of $\omega_u$,

$$\omega_u\left[\prod_k (1+u_k)^{-1}(1-v_k)^{-1}\right]\det\prod_k (1+u_k \text{ ad } X)^{-1}(1-v_k \text{ ad } X)^{-1}$$

$$= \left[\prod_k \frac{1-u_k}{1-v_k}\right] \det\prod_k \frac{1-u_k \text{ ad } X}{1-v_k \text{ ad } X} .$$

Hence from (3) we get

$$\omega_u \, c_{\alpha\beta}(0;-u,v) = c_{\alpha\beta}(u;v) .$$

On the other hand, from [11, (2.13)] there follows

$$\omega_u \, p_k(-u,v) = p_k(v) - p_k(u).$$

We deduce from Lemma 3.3 our main result:

3.4. THEOREM. We have

$$c_{\alpha\beta}(u;v) = \left[\prod_k (1+p_k(u)-p_k(v))^{-1}\right] s_\alpha * s_\beta\left(p_k \rightarrow \frac{-p_k(u)+p_k(v)}{1+p_k(u)-p_k(v)}\right) . \quad □$$

The above theorem is essentially implicit in the work of P. Hanlon. He computed maximal weight vectors for certain virtual representations of $SL(n,\mathbb{C})$, and it was apparent that his result implied an identity involving symmetric functions. The actual identity turned out to be a special case of Theorem 3.4, but there is no difficulty in obtaining all of Theorem 3.4 from Hanlon's technique. Earlier I had proved some special cases of Theorem 3.4, and the proof sketched here uses similar techniques.

4. APPLICATION TO THE Q-DYSON CONJECTURE. Recall that the q-Dyson conjecture corresponds to the substitution

$$u_i = \begin{cases} q^i, & 1 \le i \le \ell - 1 \\ 0, & i \ge \ell \end{cases}$$

$$v_i = 0,$$

$$\alpha = \beta = \phi .$$

Let us first state the result for arbitrary $\alpha,\beta$ .

4.1. COROLLARY. Let $\alpha,\beta \vdash m$. The coefficient of the character $s_{[\alpha,\beta]_n}$ in the expansion of the virtual character

$$\det(1-q \cdot \mathrm{ad}\, X)(1-q^2 \cdot \mathrm{ad}\, X) \cdots (1-q^{\ell-1} \cdot \mathrm{ad}\, X) \tag{9}$$

of $SL(n,\mathbb{C})$ approaches, as $n \to \infty$ , the value

$$\left[\prod_{k=1}^{\ell-1} (1-q^k)^{-1}\right] c_{\alpha\beta}(q,q^2,\cdots, q^{\ell-1}; 0)$$

$$= \left[\prod_{i\ge1} \prod_{j=1}^{\ell-1} (1-q^{\ell i+j})\right] s_\alpha * s_\beta\left(p_k \to \frac{-q^k(1-q^{(\ell-1)k})}{1-q^{\ell k}}\right) . \quad \square$$

If we let $\alpha = \beta = \phi$ (the void partition) above, then $s_\phi * s_\phi = s_\phi = 1$, so we obtain:

4.2. COROLLARY (q-Dyson conjecture for $a_i = \ell - 1$ and $n = \infty$). The coefficient of the trivial character $s_\phi$ in (9) approaches, as $n \to \infty$ , the value

$$\prod_{i\ge1} \prod_{j=1}^{\ell-1} (1-q^{\ell i+j}) . \quad \square$$

What can be said about the form of the generating function $c_{\alpha\beta}(q,q^2,\cdots, q^{\ell-1};0)$ appearing in Corollary 4.1? We will state a few results along these lines. Empirical evidence suggests that much stronger statements are possible, and that a fairly simple explicit formula may exist in many cases, if not in general.

For any partition $\lambda$, define the <u>hook-length</u> of $\lambda$ at $x = (i,j) \in \lambda$ to be

$$h(x) = h(i,j) = \lambda_i + \lambda'_j - i - j + 1 .$$

Here we identify $\lambda = (\lambda_1,\lambda_2,\cdots )$ with its Young diagram $\{(i,j)\colon 1 \le i \le \lambda'_1 = \ell(\lambda),\ 1 \le j \le \lambda_i\}$ . Following [11, p. 28], let

$$H_\lambda(q) = \prod_{x\in\lambda} (1-q^{h(x)}) ,$$

the "hook polynomial" of $\lambda$ .

The following lemma appears to be new. Its proof is based upon the combinatorial description due to Littlewood and Richardson [9, p.70] [11, Ex.5, p. 64] (equivalent to the "Murnagham-Nakayama formula") for computing the irreducible characters of the symmetric group $S_m$ . If $w \in S_m$ has cycle type $\rho = \rho(w)$, then we write $\chi^\lambda(\rho)$ for $\chi^\lambda(w)$.

4.3. LEMMA. Let $\lambda,\rho \vdash m$. Let $\chi^\lambda$ denote the irreducible character of $S_m$ corresponding to $\lambda$ . If $\chi^\lambda(\rho) \neq 0$, then $H_\lambda(q)$ is divisible by $\prod_{i=1}^{\ell(\rho)} (1 - q^{\rho_i})$ . □

4.4. PROPOSITION. Define a formal power series $D_{\alpha\beta}(q)$ (which depends on $\ell$) by

$$c_{\alpha\beta}(q,q^2,\cdots, q^{\ell-1};\, 0) = \left[\prod_{k\geq 1} \frac{1-q^k}{1-q^{k\ell}}\right] D_{\alpha\beta}(q) .$$

Then for some polynomial $L_{\alpha\beta}(q) \in \mathbb{Z}[q]$ (which depends on $\ell$), we have

$$D_{\alpha\beta}(q) = L_{\alpha\beta}(q)H_\alpha(q^\ell)^{-1} .$$

Proof. By Corollary 4.1 and (5), we have

$$D_{\alpha\beta}(q) = \frac{1}{m!} \sum_{w\in S_m} \chi^\alpha(w)\chi^\beta(w) \prod_{k\geq 1} \left(\frac{-q^k(1-q^{(\ell-1)k})}{1-q^{\ell k}}\right)^{m_k(w)} ,$$

where $m_k(w)$ parts of $\rho(w)$ are equal to $k$. By Lemma 4.3, every term of the above sum for which $\chi^\alpha(w) \neq 0$ is a rational function whose denominator divides $H_\alpha(q^\ell)$ . Hence the entire sum is a rational function whose denominator can be taken to be $H_\alpha(q^\ell)$, and it is easily seen that the numerator has integer coefficients. □

4.5. PROPOSITION. Let $\beta$ consist of the single part $m$. Then

$$L_{\alpha m}(q) = \prod_{(i,j)\in\alpha} (q^{i\ell} - q^{(j-1)\ell+1}) .$$

Equivalently, $L_{\alpha m}(q)$ is obtained by multiplying together for all $i$ the product of the first $\alpha_i$ terms from the $i$-th row of the array

$$\begin{array}{ccc} q^{\ell} - q & q^{\ell} - q^{\ell+1} & q^{\ell} - q^{2\ell+1} \ \cdots \\ q^{2\ell} - q & q^{2\ell} - q^{\ell+1} & q^{2\ell} - q^{2\ell+1} \ \cdots \\ q^{3\ell} - q & q^{3\ell} - q^{\ell+1} & q^{3\ell} - q^{2\ell+1} \ \cdots \\ \vdots & \vdots & \vdots \end{array}$$

The proof is essentially a consequence of Littlewood's work on "S-functions of special series", in particular, Theorem II on page 125 of [9].

5. GENERALIZED EXPONENTS. There is an additional specialization of $c_{\alpha\beta}(u;v)$ of independent interest. Let $\mathfrak{g} = s\ell(n,\mathbb{C})$. The adjoint action of $SL(n,\mathbb{C})$ extends to an action on the symmetric algebra $S(\mathfrak{g}) = \coprod_{k\geq 0} S^k(\mathfrak{g})$, where $S^k$ denotes the $k$-th symmetric power. It is well-known that the ring $J = S(\mathfrak{g})^{SL(n,\mathbb{C})}$ of invariants of this action is a polynomial ring in $n - 1$ variables $\theta_2,\cdots,\theta_n$, where $\theta_i$ is homogeneous of degree $i$. Namely, for $A \in \mathfrak{g}$, $\theta_i(A)$ is the coefficient of $t^{n-i}$ in the characteristic polynomial $\det(A-tI)$ of $A$.

By a theorem of Kostant [7, Thm. 0.2], we can write

$$S(\mathfrak{g}) = J \otimes H ,$$

where $H = \coprod H^k$ is a graded subspace of $S(\mathfrak{g})$ invariant under $SL(n,\mathbb{C})$. Let $H_\lambda$ denote the isotypic component of $H$ corresponding to $\lambda$, i.e., the sum of all subspaces of $H$ which afford the character $s_\lambda(x)$. We may then decompose $H_\lambda$ into irreducible subspaces $H_\lambda^i$,

$$H_\lambda = \coprod_i H_\lambda^i ,$$

where each $H_\lambda^i$ can be chosen to be homogeneous, i.e., to lie in $S^{d_i}(\mathfrak{g})$ for some $d_i$. The numbers $d_i$ are called the generalized exponents of $\lambda$. Define

$$G_\lambda(q) = \sum_i q^{d_i} ,$$

the generating function for the generalized exponents of $\lambda$. Kostant also shows in [7, Thm. 0.11] (when applied to $SL(n,\mathbb{C})$) that $G_\lambda(1)$ is equal to the dimension of the zero-weight space of the representation $\lambda$ and is therefore finite. Thus $G_\lambda(q)$ is a polynomial in $q$.

In terms of generating functions it is easy to see from the above discussion that

$$\det(1-q\cdot \text{ad } X)^{-1} = \frac{1}{(1-q^2)\cdots(1-q^n)}\sum_{\lambda} G_\lambda(q)s_\lambda(x_1,\cdots,x_n)$$

$$(\text{modulo } x_1\cdots x_n - 1). \tag{10}$$

Ranee Gupta conceived the idea of studying $G_{[\alpha,\beta]_n}(q)$ as $n \to \infty$, and showed that

$$G_{\alpha\beta}(q) := \lim_{n\to\infty} G_{[\alpha,\beta]_n}(q)$$

exists as a formal power series. She conjectured that $G_{\alpha\beta}(q)$ is a rational function $P_{\alpha\beta}(q)H_\alpha(q)^{-1}$, where $P_{\alpha\beta}(q)$ is a polynomial with nonnegative integer coefficients satisfying $P_{\alpha\beta}(1) = \chi^\beta(1)$. Later she and I conjectured on the basis of numerical evidence that $G_{\alpha\beta}(q) = s_\alpha * s_\beta(q,q^2,\cdots)$. We will indicate how all these conjectures follow immediately from our previous discussion, except for the nonnegativity of the coefficients of $P_{\alpha\beta}(q)$, which remains open.

Comparing (3) with (10), we see that

$$G_{\alpha\beta}(q) = \left[\prod_{k\geq 1}(1-q^k)\right] c_{\alpha\beta}(0;q) .$$

From Theorem 3.4 we deduce:

5.1. PROPOSITION. We have

$$G_{\alpha\beta}(q) = s_\alpha * s_\beta(q,q^2,\cdots) .$$

Additional properties of $G_{\alpha\beta}(q)$ follow from Proposition 5.1 in the same way Proposition 4.4 follows from Corollary 4.1. We merely state the results here.

5.2. PROPOSITION. (i) There is a polynomial $P_{\alpha\beta}(q) \in \mathbb{Z}[q]$ for which

$$G_{\alpha\beta}(q) = P_{\alpha\beta}(q)H_\alpha(q)^{-1} .$$

(ii) $P_{\alpha\beta}(1) = \chi^\beta(1)$, the number of standard Young tableaux of shape $\beta$.

(iii) $P_{\alpha'\beta'}(q) = P_{\alpha\beta}(q)$.

(iv) $q^{m+h(\alpha)}P_{\alpha\beta}(1/q) = P_{\alpha\beta'}(q)$, where $|\alpha| = |\beta| = m$ and $h(\alpha) = \sum_{x\in\alpha} h(x)$.

(v) $\deg P_{\alpha\beta}(q) \leq h(\alpha)$, and the coefficient of $q^{h(\alpha)}$ is the Kronecker delta $\delta_{\alpha\beta}$'.

(vi) $P_{\alpha\beta}(q)$ is divisible by $q^m$, and the coefficient of $q^m$ is $\delta_{\alpha\beta}$.

(vii) $P_{\beta\alpha}(q) = P_{\alpha\beta}(q)H_\beta(q)H_\alpha(q)^{-1}$.

(viii) Let $\beta$ consist of the single part $m$, and write $P_{\alpha m}(q)$ for $P_{\alpha\beta}(q)$. Then $P_{\alpha m}(q) = q^{m+n(\alpha)}$, where $n(\alpha) = \Sigma(i-1)\alpha_i = \sum\binom{\alpha_i'}{2}$.

Finally we state explicitly the conjecture mentioned above.

5.3. CONJECTURE (Gupta-Stanley). The coefficients of $P_{\alpha\beta}(q)$ are nonnegative.

Alain Lascoux has proved the above conjecture when $\beta$ is a "hook", i.e., a partition of the form $(m-k, 1^k)$ for some $0 \le k \le m-1$. He has shown that in this case $P_{\alpha\beta}(q)$ is the coefficient of $t^k$ in the product

$$q \prod_{\substack{(i,j)\in\alpha \\ (i,j)\neq(1,1)}} (q^i + tq^j) .$$

BIBLIOGRAPHY

1. G. E. Andrews, "Problems and prospects for basic hypergeometric functions", in Theory and Application of Special Functions, Academic Press, New York, 1975, pp. 191-224.

2. F. J. Dyson, "Statistical theory of energy levels of complex systems (I)", J. Math. Physics 3 (1962), 140-156.

3. I. J. Good, "Short proof of a conjecture of Dyson", J. Math. Physics 11 (1970), 1884.

4. J. Gunson, "Proof of a conjecture by Dyson in the statistical theory of energy levels", J. Math. Physics 3 (1962), 752-753.

5. E. C. Ihrig and M.E.H. Ismail, "The cohomology of homogeneous spaces related to combinatorial identities", preprint.

6. K. W. J. Kadell, "Andrews' q-Dyson conjecture: n = 4", preprint.

7. B. Kostant, "Lie group representations on polynomial rings", American J. Math. 85 (1963), 327-404.

8. D. E. Littlewood, "On invariant theory under restricted groups", Phil. Trans. Royal Soc. 239A (1944), 387-417.

9. D. E. Littlewood, The Theory of Group Characters, second ed., Oxford University Press, London, 1950.

10. D. E. Littlewood, "The Kronecker product of symmetric group representations", J. London Math. Soc. 31 (1956), 89-93.

11. I. G. Macdonald, Symmetric Functions and Hall Polynomials, Oxford University Press, London, 1979.

12. I. G. Macdonald, "Some conjectures for root systems", SIAM J. Math. Anal. 13 (1982), 988-1007.

13. I. Schur, "Über eine Klasse von Matrizen, die sich einer gegebenen Matrix zuördnen lassen", Dissertation, Berlin, 1901; Gesammelte Abhandlungen, Band I, Springer-Verlag, Berlin-Heidelberg-New York, 1973, pp. 1-72.

14. R. Stanley, "GL(n, $\mathbb{C}$) for combinatorialists", in Surveys in Combinatorics (E.K. Lloyd, ed.), London Math. Soc. Lecture Note Series #82, Cambridge University Press, Cambridge, 1983, pp. 187-199.

15. K. G. Wilson, "Proof of a conjecture by Dyson", J. Math. Physics 3 (1962), 1040-1043.

Late note. The entire q-Dyson conjecture has been proved by D. Zeilberger, A proof of Andrew's q-Dyson conjecture, Discrete Math., submitted.

DEPARTMENT OF MATHEMATICS
MASSACHUSETTS INSTITUTE OF TECHNOLOGY
CAMBRIDGE, MA 02139

Contemporary Mathematics
Volume 34, 1984

# SPHERICAL DESIGNS AND GROUP REPRESENTATIONS

Eiichi Bannai*

ABSTRACT. The nature of this paper is partly expository and partly original. We study finite subsets (i.e., spherical designs) of the unit sphere $S^{d-1}$ ($\subset \mathbb{R}^d$) using the representation theories of various groups, in particular of the orthogonal group $O(d)$ and the unitary group $U(\ell)$. Many of the results treated here concern spherical t-designs obtained as orbits of a finite subgroup $G$ of $O(d)$, and are based on my (more detailed) paper: Spherical t-designs which are orbits of finite groups, which is to be published elsewhere. As an example, we mention:

THEOREM 2. If $X_o = x_o^G$ (the orbit of $x_o$ by $G$) is a spherical t-design for some $x_o \in S^{d-1}$, then $X = x^G$ is a spherical $[\frac{t}{2}]$-design for any $x \in S^{d-1}$.

Also, we prove the following new result:

THEOREM 4. Let $G$ be a t-homogenous finite subgroup of $O(d)$, that is, $X = x^G$ is a spherical t-design for any $x \in S^{d-1}$. If $d \geq 3$, then $t \leq f(d)$ for a certain function $f(d)$ depending only on $d$.

Some informal discussions of further research problems and directions will also be given.

1. INTRODUCTION. This paper is a fairly precise reproduction of my talk at the conference. As the talk had, this paper has dual purposes of giving an introductory exposition and presenting some new results.

In Section 1 we give a very brief review of the representation theories of $O(d)$ and $U(\ell)$, which will be used in later sections. Section 2 contains a review of basic materials concerning spherical t-designs. This will include the updating of my previous expository paper [2] on similar subjects. Section 3 is based on [3]. (The full paper [3] will be published elsewhere.) Our main emphasis here is the study of spherical t-designs which are obtained

---

1980 AMS Subject Classification. Primary 05B, secondary 20C, 22E, 51M.
*Supported in part by NSF grant: MCS-8301826.

as orbits of a finite subgroup $G$ of $O(d)$. Here, the representation theories, in particular those of $O(d)$ and $U(\ell)$ play important roles. Section 4 is devoted to the proof of Theorem 4 (see Abstract), which is a new result and is not included in [3]. In Section 5, some miscellaneous remarks and questions will be discussed.

ACKNOWLEDGEMENT

The author thanks Tom Koornwinder (Amsterdam) for his help in proving Lemma 4.3.

1. SPHERICAL REPRESENTATIONS OF $O(d)$ AND $U(\ell)$. In this section we give a very brief review of spherical representations of the orthogonal group $O(d)$ and the unitary group $U(\ell)$, which will be used in later sections. No material in this section is new. We include this section for the convenience of those who are not familiar with this subject.

$O(d)$ denotes the real orthogonal group in $\mathbb{R}^d$. $O(d)$ acts naturally on the unit sphere $S^{d-1}$ $(\subset \mathbb{R}^d)$. Let $\mathrm{Hom}(i)$ be the space of homogeneous polynomials of degree $i$ in the variables $x_1, x_2, \ldots, x_d$. Then $\Delta = (\partial^2/\partial x_1^2) + \cdots + (\partial^2/\partial x_d^2)$ gives an onto homomorphism from $\mathrm{Hom}(i)$ to $\mathrm{Hom}(i-2)$. The kernel $\mathrm{Ker}\,\Delta \cap \mathrm{Hom}(i) = \mathrm{Harm}(i)$ denotes the space of homogeneous harmonic polynomials of degree $i$. Then, we have

$$\dim \mathrm{Harm}(i) = \dim \mathrm{Hom}(i) - \dim \mathrm{Hom}(i-2)$$

$$= \binom{d+i-1}{i} - \binom{d+i-3}{i-2}.$$

$O(d)$ acts on $\mathrm{Harm}(i)$ naturally by

$$(gf)(x) = f(g^{-1}x), \quad g \in O(d), \quad f \in \mathrm{Harm}(i).$$

This representation $\rho_i$ of $O(d)$ on $\mathrm{Harm}(i)$ is irreducible, and is called the $i^{th}$ spherical representation of $O(d)$ (on $S^{d-1}$). (Note that the regular representation of $O(d)$ on $S^{d-1}$ is decomposed into the direct sum of $\rho_i$'s $(i = 0, 1, 2, \ldots)$.)

Suppose that $d = 2\ell$ ($d$ even). Then the unitary group $U(\ell)$ (of matrix size $\ell$) is embedded in $O(d)$ naturally by the map

$$U = A + \sqrt{-1}\,B \longmapsto \begin{pmatrix} A & -B \\ B & A \end{pmatrix},$$

(with $A$, $B$ real matrices). Let $z_1, \ldots, z_\ell$ be the complex variables $z_i = x_i + \sqrt{-1}\,y_i$. $\mathrm{Hom}(i,j)$ denotes the space of all $\mathbb{C}$-coefficient homogeneous polynomials in $z_1, \ldots, z_\ell, \bar{z}_1, \ldots, \bar{z}_\ell$, of degree $i$ in $z_1, \ldots, z_\ell$ and of degree $j$ in $\bar{z}_1, \ldots, \bar{z}_\ell$. Then

$$\Delta = (\partial^2/(\partial z_1 \partial \bar{z}_1) + \dots + \partial^2/(\partial z_\ell \partial \bar{z}_\ell)),$$

where

$$\partial/\partial \bar{z}_i = \tfrac{1}{2}(\partial/\partial x_i + \sqrt{-1}\ \partial/\partial y_i)$$

and

$$\partial/\partial z_i = \tfrac{1}{2}(\partial/\partial x_i - \sqrt{-1}\ \partial/\partial y_i),$$

induces an <u>onto</u> homomorphism from $\mathrm{Hom}(i,j)$ to $\mathrm{Hom}(i-1,j-1)$. The kernel $\mathrm{Ker}\ \Delta \cap \mathrm{Hom}(i,j)$ is denoted by $\mathrm{Harm}(i,j)$. We have

$$\begin{aligned}\dim \mathrm{Harm}(i,j) &= \dim \mathrm{Hom}(i,j) - \dim \mathrm{Hom}(i-1,j-1)\\ &= \binom{\ell+i-1}{i}\binom{\ell+j-1}{j} - \binom{\ell+i-2}{i}\binom{\ell+j-2}{j}.\end{aligned}$$

$U(\ell)$ acts on $\mathrm{Harm}(i,j)$ naturally, and this representation $\rho_{i,j}$ of $U(\ell)$ on $\mathrm{Harm}(i,j)$ is irreducible for any $i$ and $j$. As $U(\ell)$ is embedded in $O(d)$ naturally if $d = 2\ell$, it would be interesting to know how $\mathrm{Harm}(i)$ is decomposed into $U(\ell)$-subspaces. The following fact is well known.

$$\mathrm{Harm}(i) = \bigoplus_{k=0}^{i} \mathrm{Harm}(k, i-k) \qquad (\text{as } U(\ell)\text{-spaces}).$$

We call the representation $\rho_{i,i}$ (of $U(\ell)$ on $\mathrm{Harm}(i,i)$) the $i^{\mathrm{th}}$ spherical representation of $U(\ell)$ (on $\mathbb{P}^{\ell-1}(\mathbb{C})$). (Note that the regular representation of $U(\ell)$ on $\mathbb{P}^{\ell-1}(\mathbb{C})$ is decomposed into the direct sum of $\rho_{i,i}$'s $(i = 0, 1, 2, \dots)$.)

The results described here are nothing but the very beginning of general theories called "spherical harmonics" and "complex spherical harmonics". It was Delsarte-Goethals-Seidel [8] who started to use these theories to study finite subsets of $S^{d-1}$ (and of $\mathbb{P}^{d-1}(\mathbb{R})$ and $\mathbb{P}^{\ell-1}(\mathbb{C})$). This approach is extended to other situations. It is well known that theories quite analogous to spherical harmonics and complex spherical harmonics hold for compact rank 1 symmetric spaces (i.e., compact 2-point homogeneous spaces, viz. sphere and projective spaces over $\mathbb{R}$, $\mathbb{C}$, $\mathbb{H}$ (quarternion) and Cayley's octanion, cf. e.g. Helgason [12].) Hoggar [13] (see also Neumaier [18]) studies finite subsets of these spaces in a very systematic way. We also remark that a study of finite subsets in noncompact rank 1 symmetric spaces has been carried out (although somewhat partially) along a similar line of study. (Cf. Bannai-Blokhuis-Delsarte-Seidel [5] for real hyperbolic space.)

A more systematic and thorough treatment (aimed at combinatorialists) of all the theories mentioned above is being planned in Part II of Bannai-Ito [6]. Roughly speaking, the concept of a rank 1 symmetric space in the continuous case corresponds to that of a (P and Q)-polynomial association scheme (with the group action) in the finite case. (Cf. [6, Part I, for the latter structure.) Both rank 1 symmetric spaces (in the continuous case) and (P and Q)-polynomial association schemes are very important frameworks on which coding theory and design theory etc. can be developed nicely. That is "Delsarte Theory" works there.

However, in the rest of this paper, we will devote ourselves to the case of "spherical harmonics" and "complex spherical harmonics", although many of the results hold in a more general setting, such as explained above.

2. SPHERICAL t-DESIGNS. The concept of spherical t-designs was defined by Delsarte-Goethals-Seidel [8] as a kind of concept analogous to that of combinatorial t-designs (i.e., t-$(v,k,\lambda)$ designs in the usual sense of block design theory).

DEFINITION 2.1 (Delsarte-Goethals-Seidel [8]). A finite nonempty subset $X$ of $S^{d-1}$ $(\subset \mathbb{R}^d)$ is called a spherical t-design in $S^{d-1}$ (or in $\mathbb{R}^d$) if (any) one of the following equivalent conditions (1), (2) and (3) holds:

(1) $$\frac{1}{|X|}\sum_{x\in X} f(x) = \frac{1}{|S^{d-1}|}\int_{S^{d-1}} f(x)dw$$

for all $f \in \mathrm{Hom}(i)$ with $i \leq t$ (i.e., for all polynomials $f(x) = f(x_1,\ldots,x_d)$ of degree $\leq t$), where $|S^{d-1}|$ denotes the area of the surface $S^{d-1}$ and the integral is over $S^{d-1}$.

(2) $$\frac{1}{|X|}\sum_{x\in X} f(x) = 0$$

for all $f \in \mathrm{Harm}(i)$ with $i = 1, 2, \ldots, t$.

(3) Every type of $k^{th}$ moment (e.g. $\frac{1}{|X|}\sum_{x\in X} x_1^{a_1} x_2^{a_2} \cdots x_d^{a_d}$ with $a_1 + a_2 + \ldots + a_d = k$) of $X$ is invariant when $X$ is transformed by any element $g$ of $O(d)$ for all $k \leq t$.

Here is a brief survey of what is known on spherical t-designs. (Cf. [8], [10], [11], [2], etc.)

THEOREM (Fisher type inequality, Delsarte-Goethals-Seidel [8]). If $X$ is a spherical t-design in $S^{d-1}$, then

(i) $|X| \geq \binom{d+s-1}{s} + \binom{d+s-2}{s-1}$ if $t = 2s$ (even)

and

(ii) $|X| \geq 2\binom{d+s-1}{s}$ if $t = 2s+1$ (odd).

(These bounds all come from spherical harmonics. The R.H.S. of (i) is $\sum_{i=0}^{s} \dim \mathrm{Harm}(s-i)$, and the R.H.S. of (ii) is $2 \cdot \sum_{i=0}^{[s/2]} \dim \mathrm{Harm}(s-2i)$. (Note that (i) is a special case of Theorem 3 in Section 3 by taking $G = 1$.)

REMARK 2.1. If the equality holds in one of the above inequalities (i) and (ii), then $X$ is called a tight spherical t-design. Tight spherical t-designs in $S^{d-1}$ $(d \geq 3)$ do not exist except for $t \leq 5$, $t = 7$, and $t = 11$ (with $d = 24$) (Bannai-Damerell, 1979/80). The tight 11-design with $d = 24$ is unique and consists of the 196,560 minimal vectors in the Leech lattice (Bannai-Sloane, 1981). The classification of tight spherical t-designs with $t \leq 5$ and $t = 7$ is still open.

REMARK 2.2. Many spherical t-designs were constructed as orbits of the action of a nice finite subgroup $G$ of $O(d)$ (e.g., real reflection groups, Conway's groups, etc.). However, the construction of such examples were limited to relatively small values of $t$ (for $d \geq 3$), and the general existence question of spherical t-designs in $S^{d-1}$ for any $t$ and any $d$ was open until recently. This was affirmatively solved by Paul Seymour and Tom Zaslavsky recently.

THE EXISTENCE THEOREM (Seymour-Zaslavsky [19]). For all $t$ and $d$, spherical t-designs in $S^{d-1}$ exist.

In view of this new development, the following questions seem to be interesting at the present.

QUESTIONS. (i) Explicitly construct spherical t-designs in $S^{d-1}$ for all $t$ and $d$. (The proof in [19] is an "existence proof".)

(ii) (Being less ambitious than (i)), find an explicit function $f(t,d)$ such that there exists a spherical t-design $X$ in $S^{d-1}$ with $|X| \leq f(t,d)$.

(iii) Classify (in some reasonable way if it is at all possible) those spherical t-designs which are rigid. (It seems that the rigid ones are particularly interesting, as they might represent certain stable physical states.)

(iv) Improve the lower bound for $|X|$ beyond the tight case, in particular for large $t$.

REMARK 2.3. Hong [14] may be interesting in connection with the above questions (iii) and (iv). For $d \geq 3$, it seems that if $n$ is small (i.e., close to the Fisher type lower bound), then spherical t-design $X$ with $|X| = n$ may not exist, but if it exists then it is likely to be rigid. While it seems that if $n$ is large, then spherical t-design $X$ with $|X| = n$ are quite abundant, but almost all of them are likely to be non-rigid.

## 3. SPHERICAL t-DESIGNS WHICH ARE ORBITS OF A FINITE SUBGROUP $G$ OF $O(d)$.

The following result is the start of this kind of study and allows us to construct many spherical t-designs explicitly.

THEOREM 1 ([1]). Let $G$ be a finite subgroup of $O(d)$, and let $\rho_i$ be the $i^{th}$ spherical representation of $O(d)$. If $\rho_i|_G$ remains (absolutely) irreducible for each $i = 0, 1, \ldots, s$, then for any $x \in S^{d-1}$, the set

$$X = x^G \quad (= \{gx \mid g \in G\} \subset S^{d-1})$$

is a spherical $2s$-design in $S^{d-1}$.

REMARK 3.1. Goethals-Seidel [10] generalized Theorem 1 in the following way: If $\rho_i|_G$ remains real irreducible for each $i = 0, 1, \ldots, s$, then the same conclusion holds. Also, Goethels-Seidel [11] showed that if we choose $x_o \in S^{d-1}$ nicely, then (under the same assumption as in Theorem 1) $X_o = x_o^G$ sometimes becomes a spherical $t$-design with $t$ larger than $2s$. However, the following theorem makes it clear that we cannot choose $x_o$ too nicely.

THEOREM 2 ([3]). Let $G$ be a finite subgroup of $O(d)$. If $X_o = x_o^G$ is a spherical $t$-design for some $x_o \in S^{d-1}$, then $X = x^G$ is a spherical $[\frac{t}{2}]$-design for any $x \in S^{d-1}$.

We introduce the following concept.

DEFINITION 3.1. A finite subgroup $G$ of $O(d)$ is called t-homogeneous if one of the following (equivalent) conditions holds.

(1) For any $x \in S^{d-1}$, $X = x^G$ is a spherical $t$-design.

(2) $(\rho_o, \rho_i)_G = \delta_{oi}$ for $i = 0, 1, \ldots, s$.

(3) $\dim \mathrm{Harm}(i)^G = 0$ for $i = 1, 2, \ldots, s$, where $\mathrm{Harm}(i)^G$ is the subspace of $\mathrm{Harm}(i)$ each of whose elements is fixed by the action of $G$. (Note that this concept of t-homogeneous subgroups of $O(d)$ can be defined for not necessarily finite subgroups $G$ of $O(d)$ by using (2) or (3).)

Using this terminology, Theorem 1 (together with Remark 3.1) is restated as follows:

THEOREM 1'. Let $G$ be a finite subgroup of $O(d)$. If $\rho_i|_G$ are real irreducible for $i = 0, 1, \ldots, s$, then $G$ is $2s$-homogeneous.

(It is claimed in Goethals-Seidel [10, Theorem 6.7] that the converse of Theorem 1' is also true. But this is not ture, as we show later.)

A PROOF OF THEOREM 2. Here we sketch a proof of Theorem 2 which goes through the following Theorem 3, which may be of independent interest.

THEOREM 3. Let $G$ be a finite subgroup of $O(d)$, and let $X$ be a spherical $t$-design in $S^{d-1}$ which is fixed as a whole by $G$. (The action of $G$ on $X$ need not be transitive.) Let $\pi$ be the permutation representation of $G$ on $X$. Then we have

$$\pi \supseteq (\rho_0 + \rho_1 + \cdots + \rho_{[t/2]})|_G .$$

SKETCH OF THE PROOF OF THEOREM 3. Let $H$ be the matrix of size $|X| \times \left(\binom{d + [\frac{t}{2}] - 1}{[\frac{t}{2}]} + \binom{d + [\frac{t}{2}] - 2}{[\frac{t}{2}] - 2}\right)$ whose rows are indexed by the elements $x \in X$ and whose columns are indexed by the basis elements $f$'s of $\bigoplus_{i=0}^{[t/2]} \mathrm{Harm}(i)$, and the $(x,f)$-entry is given by $f(x)$. Then we have

$$\pi(g) \cdot H = H \cdot (\rho_0 + \rho_1 + \cdots + \rho_{[t/2]})(g)$$

for all $g \in G$. It is known that $H$ has maximal possible rank, which is proved by using spherical harmonics (cf. Delsarte-Goethals-Seidel [8]). By Schur's lemma, this proves the assertion of Theorem 2. (Note that basically the same idea, which is a technique known in group theory, was also used by Stanley [20, §9] before.)

THEOREM 3 ⇒ THEOREM 2. Since $G$ is transitive on $X = x^G$, $\rho_0$ (the identity representation) appears in the permutation representation $\pi$ of $G$ on $X$ with multiplicity 1. Therefore, $\rho_0$ appears in $(\rho_0 + \cdots + \rho_{[t/2]})|_G$ with multiplicity one. So, we have $(\rho_0, \rho_i)_G = \delta_{0,i}$ for $i = 0, 1, \ldots, [t/2]$. That is $G$ is $[t/2]$-homogeneous by Definition 3.1.

Now, let us consider the question of whether one can construct t-designs of the form $X = x^G$ for large $t$ (with $d \geq 3$). According to Theorem 2, this question is equivalent to the question whether there exist t-homogeneous finite subgroups $G$ of $O(d)$ for large $t$. Unfortunately, the answer to this question is "essentially" negative. This is partially confirmed by the following.

THEOREM 4. If $G$ is a finite t-homogeneous subgroup of $O(d)$ with $d \geq 3$, then

$$t \leq f(d)$$

for a certain function $f(d)$ depending only on $d$.

CONJECTURE. I believe that $t$ is bounded by an absolute constant $t_0$ (which does not depend on $d$).

REMARK 3.2. If the converse of Theorem 1' <u>should</u> be true, then I could prove this conjecture by exploiting the classification of finite simple groups, although the details of my proof are very messy. However, the converse of Theorem 1' is not true as is shown in what follows. Anyway it would be very interesting to know to what extent the converse of Theorem 1' is true.

COUNTEREXAMPLES TO THE CONVERSE OF THEOREM 1'. Let $d = 2\ell$ (even). Then $U(\ell)$ is naturally embedded in $O(d)$, and $U(\ell)$ acts transitively on $S^{d-1}$. Actually, $U(\ell)$ is a t-homogeneous subgroup of $O(d)$ in the sense of Definition 3.1, (ii) or (iii). On the other hand,

$$\rho_2|_{U(\ell)} = \mathrm{Harm}(2,0) \oplus \mathrm{Harm}(1,1) \oplus \mathrm{Harm}(0,2),$$

and is not real irreducible. ($\mathrm{Harm}(2,0) \oplus \mathrm{Harm}(0,2)$ is real irreducible.) So, if $G$ is a large finite subgroup of $U(\ell)$, then $G$ may have properties similar to $U(\ell)$, say

$G$ is 4-homogeneous, but

$\rho_2|_G$ is not real irreducible.

We can find such explicit counterexamples. For example, take $G$ to be a 6-fold covering of $PSU_4(3)$ (with $\ell = 6$ and $d = 12$), or a 6-fold covering of Sz (with $\ell = 12$ and $d = 24$). (See [ 3 ] for the details.)

REMARK 3.3. It should be remarked that the concept of t-homogeneous groups and all the above theorems 1, 2, 3 and 4 have counterparts in finite permutation groups. This analogy is really deep, and should be pursued further. Some of the correspondence are described as follows.

$O(d)$ $\leftrightarrow$ $S_v$ (the symmetric group on a v-set).

$S^{d-1}$ $\leftrightarrow$ Johnson association scheme $J(v,k)$ (i.e., the set of k-element subsets of a v-set).

$\rho_i$ $\leftrightarrow$ $\chi_i$ =

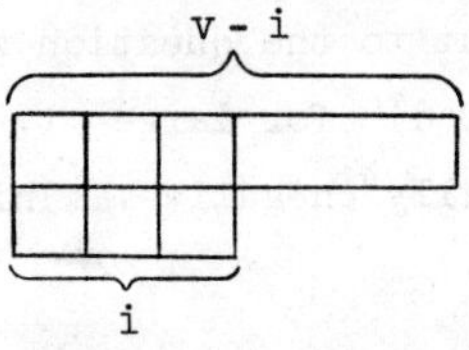


(the irreducible representation of $S_v$ corresponding to this Young diagram).

t-homogeneous $\leftrightarrow$ t-homogeneous permutation groups
($\Leftrightarrow$ $G$ acts transitively on t-element subsets)
($\Leftrightarrow$ all G-orbits of an element of $J(v,k)$ are combinatorial t-designs)
($\Leftrightarrow$ $(\chi_o, \chi_i) = \delta_{o,i}$ for $i = 0, 1, \ldots, t$).

? $\leftrightarrow$ t-transitive permutation groups.

It would be interesting if we could define the concept of t-transitive subgroups in $O(d)$ in a combinatorial context. (A representation theoretical definition of t-transitive groups is possible, cf. [3].) Anyway, "2s-transitive" subgroup $G$ of $O(d)$ should imply that the $\rho_i|_G$ are all irreducible for $i = 0, 1, \ldots, s$, and that the classification of t-transitive and t-homogeneous subgroups $G$ of $O(d)$ should be carried out as was done for permutation groups (although it heavily depends on the classification of finite simple groups). It would be very nice if one could prove that t-homogeneous ⇒ "t-transitive" for large $t$ for finite subgroups of $O(d)$, as was done for finite permutation groups by Livingstone and Wagner [17] for $t \geq 5$. If this should be proved, then the Conjecture (mentioned after Theorem 4) would be proved, as I can prove the nonexistence of 2s-transitive subgroups $G$ of $O(d)$ with $d \geq 3$, (i.e., $\rho_i|_G$ remain irreducible for $i = 0, 1, \ldots, s$) for large $s$ by using the classification of finite simple groups.

4. PROOF OF THEOREM 4. This section is devoted to the proof of Theorem 4. First we give the following well known lemma due originally to C. Jordan.

LEMMA 4.1. (Cf. [9, Theorem 5.7].) If $G$ is a finite linear group (over $\mathbb{C}$ or $\mathbb{R}$, say), then there exists an abelian normal subgroup $A$ of $G$ such that

$$|G : A| \leq f_1(d)$$

for a certain function $f_1(d)$ depending only on $d$.

The next lemma is also classically well known.

LEMMA 4.2. (Cf. [9, Theorem 4.2.A].) Let $G$ be an irreducible subgroup of $GL(V)$, where $V$ is a vector space (over $\mathbb{C}$ or $\mathbb{R}$, say). Let $N$ be a normal subgroup of $G$, $V_1,\ldots,V_n$ be the different homogeneous N-spaces of $V$ and let $W_1,\ldots,W_n$ be the corresponding (pairwise nonisomorphic) minimal N-spaces. (Note that $V_i$ is formed as the sum of all minimal N-spaces $W'$ which are N-isomorphic to $W_i$.) Then $V = \bigoplus_{i=1}^{n} V_i$, and each $x \in G$ permutes the set of $V_i$.

The author is indebted to Tom Koornwinder [16] for the following lemma.

LEMMA 4.3. Let $d = nm$, with $n \geq 1$ and $m \geq 2$ being integers. Let $H$ be the subgroup of $O(d)$ (acting naturally on $V = \mathbb{R}^d$) which fixes (as a whole) mutually orthogonal $m$ n-dimensional subspaces $V_1, V_2, \ldots, V_m$ of $V$. (That is, $H = O(n) \wr S_m$, the wreath product.) Then we have

$$(1) \qquad \begin{cases} \dim \operatorname{Hom}(2j+1)^H = 0 \\ \dim \operatorname{Hom}(2j)^H = p(j,m) , \end{cases}$$

where $p(j,m)$ is the number of partitions of $j$ in at most $m$ parts. So, we have

$$\dim \mathrm{Harm}(2j)^H = p(j,m) - p(j-1,m).$$

(2) In particular, we have

$$\dim \mathrm{Harm}(4)^H \geq 1.$$

Proof. (i) Any nonzero homogeneous polynomials of odd degree changes to its negative by $-I_d \in H$. Therefore, $\dim \mathrm{Hom}(2j+1)^H = 0$. $\mathrm{Hom}(2j)^H$ precisely consists of the symmetric polynomials in the variables $X_1 = x_1^2 + \dots + x_n^2$, $X_2 = x_{n+1}^2 + \dots + x_{2n}^2$, $\dots$, $X_m = x_{(m-1)n+1}^2 + \dots + x_{mn}^2$. Thus, we get the desired result.

(ii) This is immediately obtained from (i).

Proof of Theorem 4. In what follows we assume that $G$ is a $2s$-homogeneous $(s \geq 1)$ subgroup $O(d)$ acting naturally on the real vector space $V$. We want to show that $s$ is bounded by a certain function of $d$.

(4.1) $G$ is real irreducible.

Proof. If not, then for some $x_0 \in S^{d-1}$, $x_0^G$ is in a proper subspace of $V$. Take a line (through the origin) which is perpendicular to this proper subspace, and consider the second moment along this line. Clearly this second moment is 0. However, this contradicts the fact that the second moment of a spherical 2-design along a line is positive.

(4.2) By Lemma 4.1, $G$ has an abelian normal subgroup $A$ such that $|G:A| \leq f_1(d)$. Since $A$ is abelian, the minimal A-space of $V$ is either of dimension 1 or 2. We have the following possibilities.

(a) Some minimal A-spaces are not isomorphic.

(b) All minimal A-spaces are isomorphic and of dimension 1.

(c) All minimal A-spaces are isomorphic and of dimension 2.

(4.3) Suppose case (a) holds. Then, using Lemma 4.3, we obtain that $G$ is a subgroup of $O(n) \wr S_m$ for some integers $n \geq 1$ and $m \geq 2$. By Lemma 4.2, we have $\dim \mathrm{Harm}(4)^G \geq 1$. This implies that $G$ is not 4-homogeneous.

(4.4) Suppose case (b) holds. Then $|A| = 1$ or $2$. So $|x^G| \leq |G| \leq 2 \cdot f_1(d)$. If $G$ is $2s$-homogeneous, then $|G| \geq \binom{d+s-1}{s} + \binom{d+s-2}{s-1}$ by the Fisher type inequality. Since this lower bound increases as $s$ increases, combining with $|G| \leq 2 \cdot f_1(d)$, we have $2s \leq f_2(d)$ for a certain function $f_2(d)$ depending only on $d$.

(4.5) Suppose that case (c) holds. Then $A$ must be cyclic and $|A| \geq 3$. We then obtain, by making a suitable base change, that $N_{O(d)}(A)$ is isomorphic to (a subgroup of) the following subgroup

$$L = \left\langle \begin{pmatrix} A & -B \\ B & A \end{pmatrix}, \begin{pmatrix} I & 0 \\ 0 & -I \end{pmatrix} \middle| \begin{array}{l} A,\ B \text{ real } \ell \times \ell \text{ matrices, and} \\ A + B\sqrt{-1} \in U(\ell) \end{array} \right\rangle .$$

The group $L$ acts naturally on $\mathbb{P}^{\ell-1}(C)$, and $L$ also acts naturally on $\mathrm{Harm}(i,i)$. (Note that the matrix $\begin{pmatrix} I & 0 \\ 0 & -I \end{pmatrix}$ corresponds to the base change $z_i = x_i + \sqrt{-1}\, y_i \leftrightarrow \bar{z}_i = x_i - \sqrt{-1}\, y_i$.) Let us denote this representation of $L$ on $\mathrm{Harm}(i,i)$ by $\bar{\rho}_{i,i}$ (cf. the definition of $\rho_{i,i}$ in Section 1). As in Definition 3.1, we can prove the equivalence of the following three conditions for finite subgroups $G$ of $L$.

(1) For any $x \in \mathbb{P}^{\ell-1}(\mathbb{C})$, $X = x^G$ is an s-design in $\mathbb{P}^{\ell-1}(\mathbb{C})$.

(2) $(\bar{\rho}_o, \bar{\rho}_i)_G = \delta_{o,i}$ for $i = 0, 1, \ldots, s$.

(3) $\dim \mathrm{Harm}(i,i)^G = 0$ for $i = 0, 1, \ldots, s$.

It is known that if $X$ is an s-design in $\mathbb{P}^{\ell-1}(\mathbb{C})$, then

$$|X| \geq \binom{\ell + s - 1}{s}^2 ,$$

(cf. [8], [13], [18]). Now, we see that the subgroup $A$, which is in the center of $U(\ell) = \{A + B\sqrt{-1}\}$, acts naturally on $\mathbb{P}^{\ell-1}(\mathbb{C})$. Therefore, $|x^G| \leq f_1(d)$. If $G$ is a finite 2s-homogeneous subgroup of $O(d)$ and if $G \subset L$, then $\dim \mathrm{Harm}(i)^G = 0$ for all $i = 1, 2, \ldots, 2s$, and so $\dim \mathrm{Harm}(i,i)^G = 0$ for all $i = 1, 2, \ldots, s$ (cf. Section 1). That is $X = x^G$ must be an s-design in $\mathbb{P}^{\ell-1}(\mathbb{C})$ for all $x \in \mathbb{P}^{\ell-1}(\mathbb{C})$. Therefore, $|x^G| \geq \binom{\ell + s - 1}{s}^2$ for all $x \in \mathbb{P}^{\ell-1}(\mathbb{C})$. Combining this with $|x^G| \leq f_1(d)$, we obtain that $2s \leq f_3(d)$, where $f_3(d)$ is a certain function depending only on $d$. Thus, taking $f(d) = \max\{f_2(d), f_3(d)\}$, we complete the proof of Theorem 4.

5. MISCELLANEOUS REMARKS. In this section we mention several miscellaneous remarks and questions.

(1) For a finite subgroup $G$ of $U(\ell)$ ($\subset GL(\ell,\mathbb{C})$), the Molien series is defined by

$$(5.1) \qquad \Phi(\lambda) = \sum_{i=0}^{\infty} \dim \mathrm{Hom}(i)^G \lambda^i = \sum_{i=0}^{\infty} a_i \lambda^i .$$

While, the harmonic Molien series is defined by

$$(5.2) \qquad \Phi^h(\lambda) = \sum_{i=0}^{\infty} \dim \mathrm{Harm}(i)^G \lambda^i = \sum_{i=0}^{\infty} b_i \lambda^i \qquad (= \Phi(\lambda)/(1 - \lambda^2)).$$

(Notice that Theorem 4 implies that if $d \geq 3$, then

$$(5.3) \qquad b_1 = b_2 = \cdots = b_t = 0 \Rightarrow t \leq f(d).$$

Note that there exists no such similar bound for $t$ when $a_1 = a_2 = \dots = a_t = 0$ in (5.1). However, for a finite subgroup $G$ of $U(\ell)$, the complex Molien series is defined by

$$\Phi^{ch}(\lambda) = \sum_{i=0}^{\infty} \dim \mathrm{Harm}(i,i)^G \lambda^i = \sum_{i=0}^{\infty} c_i t^i . \tag{5.4}$$

Then, we have that (for $\ell \geq 2$),

$$c_1 = c_2 = \dots = c_t = 0 \Rightarrow t \leq f(\ell). \tag{5.5}$$

In [15], it is shown that most finite primitive linear groups have messy invariants. However, the reason comes from the possible existence of large centers. I believe that the "messiness" of invariants of $G$ should be discussed and measured for harmonic and complex harmonic Molien series. In this sense, the implications of (5.3) and (5.5) may be interesting.

(2) Thompson [21] obtained a result similar to Theorem 4. Is there any direct relation between the result in [21] and Theorem 4 (and the proof of Theorem 4)?

(3) A theory partly similar to spherical harmonics has been developed for (real) hyperbolic spaces (cf. [5]). As an application of this, it is shown that if $X$ is an s-distance subset in $\mathbb{R}^d$ or $\mathbb{H}^d$ (the real hyperbolic space of dimension $d$), then

$$|X| \leq \binom{d+s}{s},$$

(cf. [7], [4], [5]).

QUESTION. Can one define the concept of t-design for finite subsets in $\mathbb{R}^d$ or $\mathbb{H}^d$? (It would be nice if, and I believe that it should be that, s-distance subsets which attain the equality in the above inequality become 2s-designs.)

## REFERENCES

1. E. Bannai: On some spherical t-designs, J. of Combinatorial Theory (A), 26 (1979), 157-161.

2. E. Bannai: Orthogonal polynomials, algebraic combinatorics, and spherical t-designs, Proc. of Symp. in Pure Math. Vol. 37 (1980), 465-468.

3. E. Bannai: Spherical t-designs which are orbits of finite groups, (preprint).

4. E. Bannai, E. Bannai and D. Stanton: An upper bound for the cardinality of an s-distance subset in Euclidean space, II, to appear in Combinatorica (Hungary).

5. E. Bannai, A. Blokhuis, Ph. Delsarte and J. J. Seidel: An addition formula for hyperbolic space, to appear in J. of Combinatorial Theory (A).

6. E. Bannai and T. Ito: Algebraic Combinatorics, Part I: Association Schemes, to appear in the Benjamin/Cummings Lecture Note Series in Mathematics, Part II: Delsarte theory, codes and designs, in preparation.

7. A. Blokhuis: Few-distance sets, Ph.D. thesis, Tech. Univ. of Eindhoven, September, 1983.

8. Ph. Delsarte, J. M. Goethals and J. J. Seidel: Spherical codes and designs, Geometriae Dedicata 6 (1977), 263-288.

9. J. D. Dixon: The structure of linear groups, van Nostrand Reinhold Mathematical Series 37, London, 1971.

10. J. M. Goethals and J. J. Seidel: Spherical designs, Proc. of Symp. in Pure Math. Vol. 34 (1979), 255-272.

11. J. M. Goethals and J. J. Seidel: Cubature formulae, polytopes and spherical designs, in The Geometric Vein, Springer-Verlag, 1982, 203-218.

12. S. Helgason: Differential Geometry and Symmetric Spaces, Academic Press, N.Y., 1962 (a new edition, 1978).

13. S. G. Hoggar: t-designs in projective spaces, Europ. J. Combinatorics, 3 (1982), 233-254.

14. Y. Hong: On spherical t-designs in $\mathbb{R}^2$, Europ. J. of Combinatorics, 3 (1982), 255-258.

15. C. Huffman and N. J. A. Sloane: Most primitive groups have messy invariants, Advances in Math. 32 (1979), 118-127.

16. T. Koornwinder (personal communication).

17. D. Livingstone and A. Wagner: Transitivity of finite permutation groups on unordered sets, Math. Zeit. 90 (1965), 393-403.

18. A. Neumaier: Combinatorial configurations in terms of distances, (Tech. Univ. of Eindhoven, T.H. - Memo 81-09).

19. P. Seymour and T. Zaslavsky: Averaging sets: A generalization of mean values and spherical designs, to appear in Advances in Math.

20. R. P. Stanley: Some aspects of groups acting on finite posets, J. of Combinatorial Theory (A), 32 (1982), 132-161.

21. J. G. Thompson: Invariants of finite groups, J. of Algebra 69 (1981), 143-145.

DEPARTMENT OF MATHEMATICS
THE OHIO STATE UNIVERSITY
COLUMBUS, OHIO 43210

6. [illegible] association [illegible] de [illegible] Séries [illegible] in preparation.

7. [illegible] Techn. Report of Bielefeld [illegible]

8. Ph. Delsarte, J. M. Goethals and J. J. Seidel, Spherical codes and designs, Geometriae Dedicata 6 (1977) 363-388.

9. [illegible] Montréal [illegible] 1977.

10. J. M. Goethals and J. J. Seidel, Spherical designs, Proc. of Symp. in Pure Math. Vol. [illegible] (1979) [illegible]

11. J. M. Goethals and J. J. Seidel, Cubature formulae, polytopes and spherical designs, in The Geometric Vein, Springer-Verlag, 1982, 203-218.

12. S. Helgason, Differential Geometry and Symmetric Spaces, Academic Press, N.Y., [illegible]

13. [illegible] Combinatorics, 1980.

14. [illegible] J. [illegible] Combinatorics [illegible]

15. [illegible] maps [illegible] Advances in Math. [illegible] (1977) [illegible]

16. [illegible] measurement [illegible]

17. D. Livingstone and A. Wagner, Transitivity of finite permutation groups on unordered sets, Math. Z. 90 (1965) 393-403.

18. A. [illegible] in terms of [illegible] (1978) [illegible]

19. [illegible] a generalization of [illegible] designs, [illegible] Advances in Math.

20. [illegible] Some aspects of group actions on finite sets, J. of Combinatorial Theory (A) [illegible] (1979) [illegible]

21. J. G. Thompson, Invariants of finite groups, J. of Algebra 69 (1981) [illegible]

DEPARTMENT OF MATHEMATICS
THE OHIO STATE UNIVERSITY
COLUMBUS, OHIO 43210

Contemporary Mathematics
Volume 34, 1984

# ALGORITHMS FOR PLETHYSM

**Y. M. Chen [1], A. M. Garsia [2] and J. Remmel [2]**

**ABSTRACT:** The plethysm $P[Q]$ of two polynomials $P(x)$ and $Q(x)$ is essentially the polynomial obtained by substituting the monomials of $Q$ for the variables of $P$. This is one of the fundamental operations in the $\lambda$-ring of symmetric functions. It occurs in several areas including Particle Physics and Representation Theory. In this paper we present an Algorithm which yielded the plethysm $S_n[S_m]$ for values of $m$ and $n$ up to $mn=30$. Here $S_n$ denotes the classical homogeneous symmetric function of degree $n$. The algorithm permits us to produce tables of plethysm that go considerably further than any of the previously published work on the subject.

## Introduction

The term *plethysm* has been used with different meanings, we shall be concerned here with an operation on symmetric polynomials which has been introduced by D. E. Littlewood [14]. It can be defined as follows. We are given two symmetric polynomials $P$ and $Q$, the latter having integer coefficients. Assume first that the coefficients of $Q$ are positive. We can then write $Q$ in the form

$$Q(x) = \sum_{m \in M} m \quad .$$

where

$$M = \{m_1, m_2, \cdots, m_N\}$$

is an ordered multiset of monomials. For convenience we shall assume that the $m_i$ are in lexicographic order. For instance, if

$$Q(x_1, x_2) = x_1^2 + x_2^2 + 2x_1x_2$$

then

$$m_1 = x_1^2 ,$$
$$m_2 = x_1x_2 ,$$
$$m_3 = x_1x_2 ,$$
$$m_4 = x_2^2 .$$

1980 Mathematics Classification. 05A15, 05A19, 20C30, 20C35, 68-04, 68C05.
[1] work partially supported by a grant from the faculty research office of the Western Ill. Univ., Macomb Ill.
[2] work supported by NSF grant at the Univ. of Cal. San Diego.


0271-4132/84 $1.00 + $.25 per page

This given, we expand $P$ as a polynomial in $N$ variables, say

$$P = P(y_1, y_2, \cdots, y_N) \quad , \tag{I.1}$$

and set

$$P[Q] = P(m_1, m_2, \cdots, m_N)$$

To be precise, the expansion in I.1 can be obtained by first expressing $P$ as a polynomial in the elementary symmetric functions $a_1, a_2, \cdots$ and then making the replacements

$$a_k = \sum_{1 \le i_1 < i_2 < \cdots < i_k \le N} y_{i_1} y_{i_2} \cdots y_{i_k}$$

In previous literature the notation $P \circ Q$ [12] or worse yet $Q \otimes P$ [13] has been used for plethysm, we prefer our $P[Q]$ since we can at least remember which goes in and what stays out.

One of the fundamental problems of the theory of symmetric functions is to give an *efficient* combinatorial rule for the calculation of the coefficients $\pi^{\nu}_{\lambda\mu}$ in the expansion

$$S_{\lambda}[S_{\mu}] = \sum_{\nu} \pi^{\nu}_{\lambda\mu} S_{\nu} \tag{I.2}$$

where $S_{\lambda}, S_{\mu}, S_{\nu}$ denote Schur symmetric functions. By this we mean something analogous to the Littlewood-Richardson rule [11],[12] (LR rule in brief) giving the coefficients in the expansion

$$S_{\lambda} \cdot S_{\mu} = \sum_{\nu} g^{\nu}_{\lambda\mu} S_{\nu}$$

where the dot denotes ordinary multiplication.

Our main contribution here is an algorithm for carrying out large scale calculations of plethysms on the computer. Essentially we give a step by step procedure for reducing plethysm to ordinary multiplication. Thus, crudely speaking, we are eventually expressing the $\pi^{\nu}_{\lambda\mu}$ in terms of the $g^{\nu}_{\lambda\mu}$ .

Several such algorithms have been proposed in the literature. In particular, three different algorithms are due to Littlewood [14] and three further ones can be found in Todd [26] Robinson [22] and Foulkes [8]. Unfortunately, these algorithms are not easily adapted to the limitations of the computer. Some of them, especially those of Littlewood require human ingenuity and experimentation to carry through. Algorithms that require any form of pattern recognition, although may seem quite efficient for human consumption, usually turn out to be very difficult if not impossible to implement on an automatic device. Other less sophisticated but equally important limitations must also be taken into account. For instance, in our experience we have found memory more of a limiting factor than time. Thus efficiency had to be achieved by a deeper understanding of the nature of the calculations than at the expense of storage. In particular, we could not use Todd's algorithm for the simple reason that in calculations of the magnitude we have undertaken it requires an extravagant use of random access storage.

It may be worthwhile to point out the specific problems we encounter in producing a practical algorithm for plethysm. For example, in the expansion of

$$S_{11}[S_{224}] \qquad \text{I.3}$$

there are potentially $p(16) = 231$ terms, but in actual fact only 40 Schur functions do occur in the final expression. Since the number of partitions of $n$ grows quite rapidly with $n$, it is of paramount importance that the algorithm produces as few as possible *dead* partitions. By this we mean partitions that do not occur in the final expansion. We should note for instance that even a characterization of the 40 *live* partitions in I.3 is not necessarily helpful if each of the 231 partitions of 16 has to be individually tested before being discarded or accepted. Thus the algorithm, to be practical, must *construct* rather than *recognize* the live partitions.

In this connection, we should mention that most of the algorithms for plethysm, including ours, involve the calculations of Schur function expansions for expresssions of the form

$$P[x_1^p, x_2^p, \cdots, x_N^p] \qquad \text{I.4}$$

It is well known and easy to show that all Schur functions occurring in such expansions are indexed by partitions with empty $p$*-core*. The latter are partitions whose Ferrers' diagrams can be disassembled by removing hooks of length $p$. Now such a characterization is completely useless unless the algorithm is so devised that it will *automatically* only produce partitions with empty p-core.

Taking all this into account our efforts have been concentrated in putting together an algorithm which is as economical as possible with partitions, at least in two of the most crucial steps. Namely in calculating ordinary products and calculating plethysms of the form I.4.

More precisely, the range of our calculations derives mainly from our having developed efficient algorithms for obtaining direct expansions for multiple products of Schur functions

$$S_{\lambda_1} \cdot S_{\lambda_2} \cdots S_{\lambda_k} = \sum_{\nu} g^{\nu}_{\lambda_1,\lambda_2,\ldots,\lambda_k} S_\nu \qquad \text{I.5}$$

and for plethysms of the form

$$S_\lambda(x_1^p, x_2^p, \cdots, x_N^p) = \sum_{\lambda} c^{\mu}_{\lambda,p} S_\mu , \qquad \text{I.6}$$

To be sure the coefficients $g^{\nu}_{\lambda_1,\lambda_2,\ldots,\lambda_k}$ in the expansion I.5 may be calculated by a repetitive use of the LR rule, however this approach produces a large amount of dead partitions. Also, formulas for the the coefficients $c^{\lambda}_{n,r}$ in the expansion I.6 have been given by Littlewood [13]. In fact, these formulas are at the basis of Todd's algorithm. The problem is that in Todd's algorithm, the matrix of the $c^{\mu}_{\lambda,p}$ is, crudely speaking, constructed by *columns*, while in I.6 it is used by *rows* . This is a serious drawback when a single $S_\lambda$ is required.

Now it develops that recently, Remmel and Whitney [20] have obtained an algorithmic modification of the LR rule, (here an after referred to as the SD algorithm) which enables us to obtain the expansions in I.5 economically. Moreover, the SD algorithm combines beautifully with Littlewood's formulas to give an algorithm for calculating the matrix of the $c^{\mu}_{\lambda,p}$ by rows as needed in I.6. We shall refer to the latter as the SXP algorithm.

The whole package consisting of the SD and SXP algorithms and an appropriately designed data structure has been implemented by Egecioglu [6] and Egecioglu-Remmel [7]. This has yielded tables of plethysm which go considerably further than any that can be found in the literature. All the programs are available in PASCAL and designed to combine various operations with Schur functions including Plethysm in a UNIX environment. We shall have to refer the reader to [6] and [7] for further details on this matter.

In principle, we need only give an algorithm for expanding the plethysm $S_n[S_m]$ in terms of Schur functions. The reason for this is that the plethysm $P[Q]$ is linear and multiplicative in $P$ thus a use can be made of expansions in terms of homogeneous symmetric functions to reduce the outer polynomial to an $S_n$ . To reduce the inner polynomial to an $S_m$ we can work in a $\lambda$-ring setting and extend the definition of $P[Q]$ to include polynomials $Q$ with negative coefficients. This done, standard addition and multiplication formulas combined with an expansion in terms of homogeneous functions yield the desired reduction. A basic step in this aproach is the calculation of expansions of the form

$$S_n[x_1^p, x_2^p, \cdots, x_N^p] = \sum_{\mu} c^{\mu}_{n,p} S_\mu(x) \qquad \text{I.7}$$

In actual fact , considerable economy with partitions can be achieved by letting the inner polynomial be an arbitrary Schur function. And this as we shall see requires the direct calculation of the expansions in I.6.

Nevertheless when only the special expansion in I.7 is needed, a more efficient algorithm for the calculation of the $c^{\mu}_{n,p}$ than the SXP algorithm was given by Chen [2]. In this connection, we should mention that Todd [26] showed that the $c^{\mu}_{n,p}$ are always equal to $0$ , $1$ or $-1$ . Moreover, Todd has a *characterization* of the live partitions in I.7, with a rule for calculating the sign of $c^{\mu}_{n,p}$ . The latter rule was also given a combinatorial interpretation in terms of hooks by Duncan in [3].

We should also mention that one final reduction in our calculations is obtained by means of the recursion yielding $S_n$ in terms of $S_k$ $(k<n)$. The role of this reduction will be better understood when we work out an example.

Formulas giving special cases of plethysm have been given by Murnagham [19], Thrall [24], Todd [26] and Duncan [4]. However, the general case appears out of reach of the present methods.

Now it develops that the special expansions obtained Thrall [24] have a very illuminating explanation in our setting. Moreover, some non-trivial new properties of plethysm come to the surface. For instance, we can show that no non-trivial *hook* Schur functions do occur in the expansion of $S_n[S_m]$. Moreover, explicit expansions for the plethysms $S_3[S_m]$, $S_{12}[S_m]$ and $S_{1^3}[S_m]$ may also be easily obtained.

Our presentation is in four sections. In the first section we recall the standard identities and introduce some notation. In the second section we present our algorithm together with the SD, SXP and Chen algorithms.

For sake of completeness, in the third section we give proofs. We also give there a new and more transparent proof of Littlewood's formula for the $c^{\mu}_{\lambda,p}$ occurring in I.7. In the fourth section we prove Thrall's identities, give our expansions of $S_3[S_m]$, $S_{12}[S_m]$, $S_{1^3}[S_m]$ and prove the result concerning hook Schur functions. In the fifth and final section to illustrate the power of plethysm as a computational tool we give an application to representation theory.

We wish to express here our indebtedness towards A. Lascoux who introduced us all to the subject and suggested the problem in a series of very inspiring lectures and conversations.

## 1. Notation and basic identities

We shall denote partitions as vectors of integers with weakly increasing components. We shall use abreviations such as $(1^2\, 2^3\, 4\, 5^2)$ to denote the partition $(1,1,2,2,2,4,5,5)$. The partition conjugate to $I$ will be denoted by $\tilde{I}$. Thus $(1^2\, 2^3)^{\sim} = (3,5)$. The Schur function corresponding to the partition $I$ will be denoted by $S_I$. In particular $S_n$ and $S_{1^n}$ will respectively denote the *homogeneous* and *elementary symmetric functions of degree* $n$.

In accordance with $\lambda$-ring notation we shall set $\Lambda_I = S_{\tilde{I}}$. In particular $\Lambda_n$ and $\Lambda_{1^n}$ are alternate notations for $S_{1^n}$ and $S_n$ respectively. We shall also use the symbol $\psi_I$ to denote the power symmetric function corresponding to the partition $I$. More precisely, we set

$$\psi_p = x_1^p + x_2^p + \cdots + x_N^p$$

and if $I = (1^{q_1}, 2^{q_2}, \cdots, n^{q_n})$ then

$$\psi_I = \psi_1^{q_1}\psi_2^{q_2}\cdots\psi_n^{q_n} .$$

We shall use symbols such as $A$ , $B$ , $C$ , $E$ , $F$ , $G$ to denote ordered multisets of monomials and use the same symbols to denote the corresponding polynomials. That is

$$E = \sum_{m \in E} m \quad , \quad F = \sum_{m \in F} m \quad .$$

The letters $P$ , $Q$ , $R$ will be used to denote polynomials with integer coefficients some of which may be negative. For convenience we let $\Pi^+$ denote the set of polynomials with non negative integer coefficients.

The definition of plethysm is extended to include inner polynomials with arbitrary integer coefficients so as to make valid the general addition formula

$$S_n[P+Q] = \sum_{k=0}^{n} S_k[P]\, S_{n-k}[Q] \tag{1.1}$$

More precisely, for $E, F \in \Pi^+$ we set

$$S_n[E-F] = \prod_{m \in E} \frac{1}{1-tm} \prod_{m \in F} (1-tm) \Big|_{t^n} \tag{1.2}$$

and similarly

$$\Lambda_n[E-F] = \prod_{m \in E} (1+tm) \prod_{m \in F} \frac{1}{1+tm} \Big|_{t^n} \tag{1.3}$$

We can easily see that both definitions are consistent. That is, if $Q = E-F = E_1-F_1$ are two different decompositions of $Q$ as a difference of two polynomials in $\Pi^+$ then for instance 1.2 gives the same polynomial for both $E-F$ and $E_1-F_1$ .

The following identities are immediate

$$\begin{aligned} &a)\ S_o(Q) = \Lambda_o(Q) = 1 , \\ &b)\ S_1(Q) = \Lambda_1(Q) = Q , \\ &c)\ S_n(-Q) = (-1)^n \Lambda_n(Q) . \end{aligned} \tag{1.4}$$

For instance, with these conventions we see that

$$S_4[x_1+x_2+x_3-y_1-y_2] =$$

$$= S_4(x_1,x_2,x_3) - S_3(x_1,x_2,x_3)(y_1+y_2) + S_2(x_1,x_2,x_3)\, y_1 y_2 .$$

It is a classical result that the $S_n$ and the $\Lambda_n$ yield bases by setting, for $I = (1^{p_1}, 2^{p_2}, \cdots, n^{p_n})$ ,

$$S^I = S_1^{p_1} S_2^{p_2} \cdots S_n^{p_n}$$

and

$$\Lambda^I = \Lambda_1^{p_1} \Lambda_2^{p_2} \cdots \Lambda_n^{p_n}$$

Since we wish plethysm to be linear in the outer polynomial, we see then that using expansions in terms of the $S^I$ or the $\Lambda^I$ we can extend the definition of the plethysm $P[Q]$ to polynomials with arbitrary integer coefficients. Moreover, the multiplicative property of these bases makes it immediate that plethysm is itself multiplicative with respect to the outer polynomial.

Plethysm with the Schur functions $S_I$ , $\Lambda_I$ (for $I=(i_1,i_2, \ldots , i_s)$ ) may then be calculated by means of the determinantal formulas

$$a)\ S_I[Q] = \det\|S_{i_k+k-h}[Q]\|$$
$$b)\ \Lambda_I[Q] = \det\|\Lambda_{i_k+k-h}[Q]\| \qquad 1.5$$

In fact, we can proceed more generally and for any two partitions $I=(i_1, i_2, \ldots, i_s)$ $J=(j_1, j_2, \ldots, j_t)$ with $J \subseteq I$ set

$$a)\ S_{I/J}[Q] = \det\|S_{i_k+k-j_h-h}[Q]\|$$
$$b)\ \Lambda_{I/J}[Q] = \det\|\Lambda_{i_k+k-j_h-h}[Q]\| \qquad 1.6$$

The basic tools which enable us to reduce the calculation of $P[Q]$ to plethysms of the form $S_n[S_m]$ are the *addition* formula

$$S_{I/K}[P+Q] = \sum_{I \subseteq J \subseteq K} S_{I/J}[P]\, S_{J/K}[Q] \ , \qquad 1.7$$

and the *multiplication* formula

$$S_n[PQ] = \sum_{I \vdash n} S_I[P] S_I[Q] \qquad 1.8$$

The first of these can be easily derived from the addition formula in 1.1 and the determinantal formula in 1.5 a). The second is a bit more elaborate but follows with some manipulations using additivity and the classical Cauchy identity.

Finally we should observe that plethysm with the power symmetric function is commutative. That is

$$\psi_p[Q] = Q[\psi_p] = Q[x^p] \qquad 1.9$$

where for convenience we have set

$$x^p = (x_1^p, x_2^p, \cdots, x_n^p)$$

This simple fact has a most useful consequence which may be stated as follows

**Theorem 1.1**

*Let $P$ be a homogeneous symmetric polynomial of degree $n$ and let $\mathbf{Z}(y_1, y_2, \cdots, y_n)$ be the polynomial such that*

$$P(x) = \mathbf{Z}(\psi_1, \psi_2, \cdots, \psi_n)$$

*then*

$$P[Q] = \mathbf{Z}(Q(x), Q(x^2), \cdots, Q(x^n)) \tag{1.10}$$

**Proof**

Since plethysm is linear and multiplicative in the outer polynomial we need only show 1.10 for the case $P(x) = \psi_p$. However, in this case 1.10 reduces to 1.9.

It is known but unfortunately not that easy to show that the coefficients $\pi^{\nu}_{\lambda\mu}$ are non negative integers. It develops that 1.10 enables us to easily verify this fact in the cases of interest to us here.

Let $\mathbf{S}_n$ denote the symmetric group on $n$ symbols. For $\sigma \in \mathbf{S}_n$ let $\nu_i(\sigma)$ denote the number of cycles of length $i$ in $\sigma$. Set

$$I(\sigma) = (1^{\nu_1(\sigma)}, 2^{\nu_2(\sigma)}, \cdots, n^{\nu_n(\sigma)})$$

In other words, $I(\sigma)$ is the partition induced by the permutation $\sigma$. This given, if $\gamma$ is a real or complex valued function on $\mathbf{S}_n$ we set

$$F\gamma = \frac{1}{n!} \sum_{\sigma \in \mathbf{S}_n} \gamma(\sigma)\psi_{I(\sigma)}(x) \tag{1.11}$$

The map $\gamma \to F\gamma$ gives a vector space isomorphism of the space of class functions of $\mathbf{S}_n$ onto the space $H_n$ of homogeneous symmetric polynomials of degree $n$. We call $F\gamma$ the *Frobenius image* of $\gamma$ in $H_n$.

It is a fundamental result of Frobenius that $F$ sends the irreducible characters of $\mathbf{S}_n$ into the Schur functions. In fact, there is a labeling of the irreducible characters by partitions such that if $\chi^I$ is the irreducible character of $\mathbf{S}_n$ corresponding to the partition $I$ then

$$F\chi^I = S_I \tag{1.12}$$

Let now $\chi^A$ denote the character of some representation $A$ of $\mathbf{S}_n$, it is well known that the coefficients $c_I$ in the expansion

$$\chi^A = \sum_{I \vdash n} c_I \chi^I \tag{1.13}$$

are non negative integers. Indeed, they give the multiplicities of the corresponding irreducible representations in the decomposition of $A$ into irreducible components. Taking the Frobenius image of 1.13 we get

$$F\chi^A = \sum_{I \vdash n} c_I S_I$$

For convenience, let us say that a symmetric polynomial is *Schur positive* if its Schur function expansion has non negative integer coefficients.

The general result which implies that the coefficients $\pi^{\nu}_{\lambda\mu}$ are non negative is that the plethysm of two Schur positive polynomials is Schur positive.

From the above observations we see that proving that a certain homogeneous polynomial is Schur positive is equivalent to showing that it is the Frobenius image of the character of some representation. Now, the latter property is automatic for Polya enumerators. To be precise, recall that the Polya enumerator of the action of a subgroup $G$ of $\mathbf{S}_n$ on $\{1,2,...,n\}$ is given by the expression

$$P_G(x)=\mathbf{Z}_G(\psi_1,\psi_2,...,\psi_n) \qquad 1.14$$

where

$$\mathbf{Z}_G(Y)=\frac{1}{|G|}\sum_{\sigma\in G} y_1^{\nu_1(\sigma)}\, y_2^{\nu_2(\sigma)} \cdots y_n^{\nu_n(\sigma)} \qquad 1.15$$

is the *cycle index* polynomial of $G$. Now, it is quite easy to show that a Polya enumerator is none other than the Frobenius image of the character of the permutation representation corresponding to the action of $\mathbf{S}_n$ on the left cosets of $G$. Thus Polya enumerators are automatically Schur positive.

This given we are finally in a position to to state and prove the basic positivity result about plethysm which is implied by theorem 1.1.

**Theorem 1.2**

*Let $G$ and $H$ be two subgroups of $\mathbf{S}_n$ and $\mathbf{S}_m$ respectively. Then the plethysm*

$$P_G[P_H(x)] \qquad 1.16$$

*is also a Polya enumerator and therefore also Schur positive.*

**Proof**

Note first that expressions such as

$$\mathbf{Z}_G(Q(x),Q(x^2),\cdots,Q(x^n)) \qquad 1.17$$

do occur often in Polya theory. They are simply enumerators corresponding to a setting where the *colors* are themselves patterns and $Q(x)$ is the pattern inventory. Comparing with 1.10 we see that the expression in 1.17 is simply the plethysm $P_G[Q]$.

This given we define a Polya action on the squares of an $n \times m$ rectangle $\mathbf{R}$ in the following manner. Given a vector

$$\gamma = (g\ ; h_1,h_2,\cdots,h_n)$$

with $g \in G$ and each $h_i \in H$, then the action of $\gamma$ on **R** consists of permuting the squares within the $i^{th}$ row of **R** according to $h_i$ and, this done for each $i$, permuting the rows according to $g$.

We claim that the polynomial in 1.16 is precisely the Polya enumerator of this action. Indeed, another way of looking at the patterns resulting from this action is to think that the places are actually the rows of **R** and that the colors are the patterns resulting from the action of $H$ on a single row consisting of $m$ squares. By our initial observation the enumerator must be of the form 1.17 with $Q(x)$ the pattern inventory corresponding to the latter action. More precisely, the enumerator must be the polynomial

$$Z_G(P_H(x), P_H(x^2), \cdots, P_H(x^n)) .$$

However, by theorem 1.1, this is precisely the plethysm in 1.16. Thus our proof is complete.

**Remark 1.1**

Perhaps a word should be said about formula 1.9 in the general case when $Q$ is the difference of two multisets of monomials. First of all since we are defining plethysm through expansions in terms of the homogeneous symmetric functions, the point of departure for the definition of $\psi_p[Q]$ should be the classical Newton formula

$$\psi_p = (-1)^{p-1} \det \begin{bmatrix} S_1 & 1 & 0 & \cdots & 0 \\ 2S_2 & S_1 & 1 & \cdots & 0 \\ 3S_3 & S_2 & S_1 & \cdots & 0 \\ \cdot & \cdot & \cdot & \cdot & 0 \\ \cdot & \cdot & \cdot & \cdot & 1 \\ pS_p & S_{p-1} & S_{p-2} & \cdots & S_1 \end{bmatrix} \tag{1.18}$$

However, this is equivalent to the generating function definition

$$e^{\sum_{p\geq 1} \psi_p[Q] \frac{t^p}{p}} = \sum_{n\geq 0} t^n S_n[Q] \tag{1.19}$$

Now, for $Q = E - F$, 1.2 yields

$$e^{\sum_{p\geq 1} \psi_p[Q] \frac{t^p}{p}} = \prod_{m\in E} \frac{1}{(1-tm)} \prod_{m\in F} (1-tm) = e^{\sum_{p\geq 1} (\psi_p[E] - \psi_p[F]) \frac{t^p}{p}}$$

In other words our definition 1.18 reduces to the simple formula

$$\psi_p[Q] = \psi_p[E-F] = \psi_p[E] - \psi_p[F] . \tag{1.20}$$

Since

$$\psi_p[E] - \psi_p[F] = E(x^p) - F(x^p) = Q(x^p)\ ,$$

we see that 1.9 does hold in full generality.

**Remark 1.2**

It is well known that

$$S_n = \mathbf{Z}_{\mathbf{S}_n}(\psi_1, \psi_2, \ldots, \psi_n) \qquad 1.21$$

In other words, $S_n$ is the Polya enumerator of the action of $\mathbf{S}_n$ on $\{1,2,\ldots,n\}$. This fact combined with theorem 1.2 yields immediately that all plethysms $S_n[S_m]$ are Schur positive. Indeed, theorem 1.2 also gives them a very simple combinatorial interpretation.

For convenience, let us call a $WW[n,m]$ *pattern* a filling of the rows of the $n \times m$ rectangle with m-letter words in the alphabet $\{x_1, x_2, \ldots, x_N\}$ having the following properties

1) The words are lexicographically *weakly* increasing from top to bottom.

2) The letters in each word are *weakly* increasing.

Similarly we can define $SW[n,m]$, $WS[n,m]$ and $SS[n,m]$ patterns by replacing the words *weakly* by *strictly* in conditions 1) or 2) or both.

The arguments in the proof of theorem 1.2 yield that $S_n[S_m]$ is the inventory of $WW[n,m]$ patterns. The reader may also verify that $\Lambda_n[S_m]$, $S_n[\Lambda_m]$ and $\Lambda_n[\Lambda_m]$ are inventories of $SW[n,m]$, $WS[n,m]$ and $SS[n,m]$ patterns respectively.

## 2. The algorithms.

There are basically two approaches to the calculation of $S_I[S_J]$ which stem from the considerations of last section. One derives from theorem 1.1 and the other from formula 1.5. More precisely, when $I$ is a partition of $n$, formula 1.10 combined with Frobenius formula (1.12) yields

$$S_I[S_J] = \sum_{p_1+2p_2+\ldots+np_n = n} \frac{\chi^I_{1^{p_1}2^{p_2}\ldots n^{p_n}}}{p_1!p_2!\ldots p_n!} \left(\frac{S_J(x)}{1}\right)^{p_1}\left(\frac{S_J(x^2)}{2}\right)^{p_2} \cdots \left(\frac{S_J(x^n)}{n}\right)^{p_n} \qquad 2.1$$

where $\chi^I_J$ denotes the value of the character $\chi^I$ at the permutations of shape $J$. On the other hand formula 1.5 for $Q = S_J$ gives

$$S_I[S_J] = \det\| S_{i_k+k-h}[S_J] \| \qquad 2.2$$

For instance, when $I = \{1\ ,3\}$, 2.1 reduces to

$$S_{13}[S_J] = \frac{1}{24}\, \chi^{13}_{1111}(S_J(x))^4 + \frac{1}{4}\, \chi^{13}_{112}(S_J(x))^2 S_J(x^2) + \qquad 2.3$$

$$+\frac{1}{8}\chi_{22}^{13}(S_J(x^2))^2+\frac{1}{3}\chi_{13}^{13}S_J(x)S_J(x^3)+\frac{1}{4}\chi_4^{13}S_J(x^4)\ ,$$

while 2.2 yields

$$S_{13}[S_J]=\det\begin{bmatrix} S_1[S_J] & S_4[S_J] \\ 1 & S_3[S_J]\end{bmatrix} = S_J(x)S_3[S_J]-S_4[S_J]\ . \qquad 2.4$$

Formula 2.2, on the surface may appear less efficient than 2.1 since it only reduces the calculation to plethysms of the form $S_m[S_J]$ which in turn are given by the formula

$$S_n[S_J]=\sum_{p_1+2p_2+\ldots+np_n=n}\frac{1}{p_1!p_2!\ldots p_n!}\left(\frac{S_J(x)}{1}\right)^{p_1}\left(\frac{S_J(x^2)}{2}\right)^{p_2}\cdots\left(\frac{S_J(x^n)}{n}\right)^{p_n} \qquad 2.5$$

However, this relation is simply what we get from 1.19 for $Q=S_J$ that is

$$\sum_{n\geq 0}t^n S_n[S_J] = e^{\sum_{p\geq 1}S_J(x^p)\frac{t^p}{p}} \qquad 2.6$$

and this in turn yields the recursion

$$S_m[S_J]=\frac{1}{m}\sum_{p=1}^{m}S_J(x^p)\,S_{m-p}[S_J]\ . \qquad 2.7$$

Thus in the particular case $J=\{1\ ,3\}$ we obtain the identities

$$S_4[S_J]=\frac{1}{4}\{S_J(x)S_3[S_J]+S_J(x^2)S_2[S_J]+S_J(x^3)S_J(x)+S_J(x^4)\}$$

$$S_3[S_J]=\frac{1}{3}\{S_J(x)S_2[S_J]+S_J(x^2)S_J(x)+S_J(x^3)\} \qquad 2.8$$

$$S_2[S_J]=\frac{1}{2}\{S_J(x)S_J(x)+S_J(x^2)\}$$

and in ultimate analysis 2.4 and 2.8 are seen to compete favorably with 2.3. In general, formula 2.2 turns out to be more efficient than 2.1 when $I$ is a partitition with a small number of parts.

At any rate whichever method we choose we are reduced to multiplications of Schur functions and calculations of Schur function expansions for the plethysms $S_J(x^p)$. To carry out these calculations we use the SD and SXP algorithms respectively. These may be described as follows

*The SD Algorithm*

To cut down the cost of art work we shall, whenever possible, depict Ferrers diagrams by arrays of dots, each dot representing the center of the corresponding square. In this vein, tableaux will also be depicted without the squares and just place the entries in their appropriate positions. Keeping this in mind, given a skew Ferrers diagram $D$ with $n$ squares, the tabloid $D^{\bullet}$ obtained by filling the squares of $D$

with the integers $1,2,...,n$ from bottom to top and from right to left will be called the reverse labeling of $D$. For instance, if

$$D = \begin{array}{cccc} \cdot & & & \\ \cdot & \cdot & & \\ & & \cdot & \cdot \\ & & \cdot & \cdot \end{array} \qquad 2.9$$

then the reverse labeling of $D$ is

$$D^* = \begin{array}{cccc} 7 & & & \\ 6 & 5 & & \\ & & 4 & 3 \\ & & 2 & 1 \end{array} \qquad 2.10$$

This given, we shall say that a standard tableau $T$ is $D$-compatible if and only if for each pair $(i,i+1)$ and $(x,y)$ the following conditions are satisfied.

1) $i+1$ occurs in $T$ southeast of $i$ whenever $i+1$ is immediately to the left of $i$ in $D^*$.

2) $y$ occurs in $T$ northwest of $x$ whenever $y$ is immediately above $x$ in $D^*$.

It is good to visualize the conditions $i+1$ *southeast of* $i$ and $y$ *northwest of* $x$ respectively by the patterns

For instance for the skew diagram of 2.9 conditions 1) and 2) may be summarized by the patterns

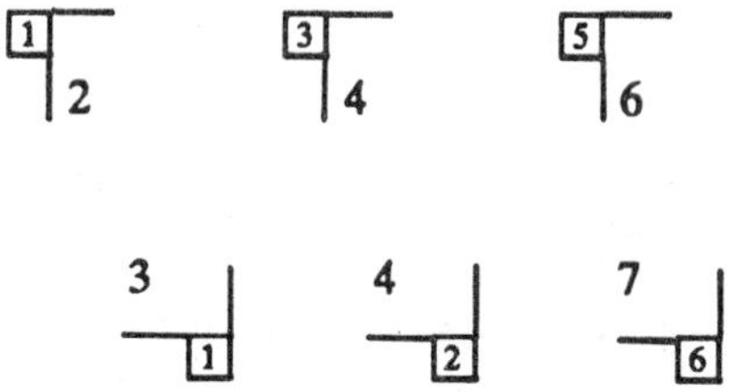


This given, the collection of $D$-compatible standard tableau may be constructed by adding squares labelled $1,2,...,n$ in succession, obeying each time the conditions 1) and 2). The reader may easily descern that in this manner we are brought to construct a tree whose internal nodes represent partially completed $D$-compatible tableaux. It is convenient to refer to this tree by the symbol $\mathbf{T}_D$ and the resulting class of $D$-compatible standard tableaux by the symbol $\mathbf{C}_D$. If in our calculations only tableaux of a given shape $J$ are needed, these may be obtained by a depth first transversal of the tree $\mathbf{T}_D(J)$ obtained by pruning away from $\mathbf{T}_D$ all the nodes representing tableaux not contained in $J$. It is easy to see that this pruning can be carried out by the standard backtracking procedure.

Figure 1 gives $\mathbf{T}_D$ for the diagram $D$ of 2.9. To illustrate how the above mentioned pruning can be carried out, we have also constructed in Fig. 2 the tree $\mathbf{T}_D(1344)$ for the diagram

2.11

to be precise, the tree given in Fig. 2, to save space, is only a modified version of $\mathbf{T}_D(1344)$. The modification consisting of combining *forced moves.* That is whenever, in the course of obeying the patterns coming from 1) and 2), we have a sequence of forced labelings, the resulting nodes of the tree are all combined into the last one.

Given $k$ Ferrers diagrams $I_1, I_2, \cdots I_k$ let us denote by $I_1 * I_2 * \cdots * I_k$ the skew diagram obtained by concatenating $I_1, I_2, \cdots I_k$ *corner to corner.* For instance, the concatenation of the diagrams

$I_1 =$  $I_2 =$  and $I_3 =$ 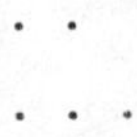

is precisely the diagram of 2.11.

With this convention, the observation that is at the basis of the SD algorithm may be stated as follows

**Theorem 2.1**

*The coefficient $c_J$ in the expansion*

$$S_{I_1} \cdot S_{I_2} \cdots S_{I_k} = \sum_J c_J S_J(x)$$

*is equal to the number of $I_1 * I_2 * \cdots * I_k$ -compatible standard tableau of shape $J$ . In other words we may write*

$$S_{I_1} \cdot S_{I_2} \cdots S_{I_k} = \sum_{T \in \mathbf{C}_{I_1 * I_2 * \cdots * I_k}} S_{J(T)}(x) \qquad 2.12$$

*where $J(T)$ , for a given tableau $T$ denotes the shape of $T$ .*

We can thus see that the procedure we have sketched above for constructing the tree $\mathbf{T}_D$ is simply an algorithm for calculating the product of Schur functions. In particular, we see that the tree in Fig. 1 gives the expansion

$$S_{12} \cdot S_{22} = S_{12^3} + S_{2^2 3} + S_{1^2 23} + S_{13^2} + S_{124} + S_{34} \ . \qquad 2.13$$

On the other hand the procedure for constructing the tree $\mathbf{T}_{I_1 \bullet I_2 \bullet \cdots \bullet I_k}(J)$ can be translated into an algorithm for calculating the coefficient of $S_J(x)$ in the product

$$S_{I_1} \cdot S_{I_2} \cdots S_{I_k}$$

In particular, the tree given in Fig. 2 yields that

$$S_{12} \cdot S_{22} \cdot S_{23} \, \big|_{S_{134^2}} = 6 \ .$$

To carry out all this efficiently on the computer we can proceed as follows. Each linear combination of Schur functions

$$\sum_I c_I \, S_I(x) \qquad 2.14$$

is represented as a dictionary.

We recall that a dictionary is a binary-tree-like structure whose nodes are records with four fields, the first field gives the key, the second field stores the information and the last two fields are pointers to left and right sons. In our case each node represents a term $c_I \, S_I$ , the key being the partition $I$ and the information being the coefficient $c_I$ . The records are constantly kept arranged so that when we read the dictionary in symmetric order the partitions come out in lexicographic order.

This given, as we read $\mathbf{T}_D$ in depth first order, each term $S_{J(T)}$ is recorded in the dictionary by locating the node with key $J(T)$ and increasing the corresponding coefficient by one. If $J(T)$ is not found, a new node is added in the appropriate location and the dictionary is updated to keep a suitable balance.

Since the frequencies with which the various shapes are produced in traversing $\mathbf{T}_D$ ,. are not known in advance (they are precisely the coefficients $c_I$ we are trying to determine) a most appropriate updating procedure to use here is one recently discovered by Tarjan and Sleator [23].

To carry out our calculations, we start with an empty dictionary, and record one by one the terms $S_{J(T)}$ coming out of $\mathbf{T}_D$ . When we are through traversing $\mathbf{T}_D$ , the resulting dictionary will be the expansion of the skew Schur function $S_D(x)$ .

*The SXP algorithm*

Let $J$ be a partition of $n$ and let us suppose we are given to calculate the expansion of $S_J(x^p)$ for some $p \geq 2$. The point of departure is the formula

$$S_J(x^p) = \sum_{|I_1|+\ldots+|I_p| = n} c_{I_1,\ldots,I_p} \, SS_{I_1,\ldots,I_p}(x) \qquad 2.16$$

where

a) the sum is to be carried out over all p-tuples of partitions $I_1,\ldots,I_p$ whose diagrams are contained in $J$ and whose sum of parts add up to $n$,

b) we have

$$c_{I_1,\ldots,I_p} = S_{I_1} S_{I_2} \cdots S_{I_p} \, |_{S_J} \qquad 2.17$$

and

c) the expression $SS_{I_1,\ldots,I_p}(x)$ denotes a certain signed Schur function indexed by a partition with empty p-core whose construction is best explaned through an example.

For instance, in the expansion of $S_{113}(x^3)$, since $n = 5$ and $p = 3$, one of the terms in 2.16 is that which corresponds to the triple of partitions

$$I_1 = \square \, , \; I_2 = \begin{smallmatrix}\square\\ \square\end{smallmatrix} \, , \; I_3 = \square\square$$

The SD algorithm based on the diagram

$$D^{\bullet} = \begin{array}{cccc} 5 & & & \\ & 4 & & \\ & 3 & & \\ & & 2 & 1 \end{array}$$

produces two D-compatible tableaux of shape 113, namely

$$\begin{array}{ccc} 5 & & \\ 4 & & \\ 1 & 2 & 3 \end{array} \qquad \textit{and} \qquad \begin{array}{ccc} 4 & & \\ 3 & & \\ 1 & 2 & 5 \end{array}$$

This gives that

$$c_{\square, \begin{smallmatrix}\square\\ \square\end{smallmatrix}, \square\square}(113) = 2 \qquad 2.18$$

to construct

$$SS_{\square, \begin{smallmatrix}\square\\ \square\end{smallmatrix}, \square\square}(x)$$

we proceed as follows. First of all we represent $\square$ , $\boxminus$ and $\square\square$ as partitions with an equal number of parts, that is we set

$$\square = (0,1) \ , \ \boxminus = (1,1) \ , \ \square\square = (0,2)$$

This given we construct the *circle diagram* given below

$$\begin{array}{ccc}
\square & \boxminus & \square\square \\
\bigcirc & \cdot & \bigcirc \\
 & \cdot\ \bigcirc\ \cdot & \\
\bigcirc\ \bigcirc & & \cdot \\
 & \cdot\ \cdot\ \bigcirc &
\end{array}$$

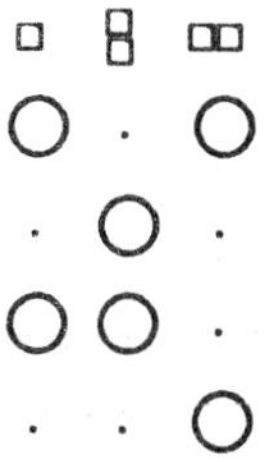

2.19

The precise rule for putting together a column of this diagram from a partition $(i_1, i_2, \cdots, i_m)$ is that the distance (in dots) between the $s^{th}$ and $(s+1)^{st}$ circle is given by the difference $i_{s+1} - i_s$ , or (which is the same) the $s^{th}$ circle is at distance $i_s + s - 1$ from the top. That is pictorially we have

$$(i_1, i_2, \cdots, i_m) \quad \rightarrow \quad
\begin{array}{l}
\cdot \\
\cdot \\
\bigcirc \leftarrow i_1 \\
\cdot \\
\cdot \\
\bigcirc \leftarrow i_2 + 1 \\
\cdot \\
\cdot \\
\cdot \\
\bigcirc \leftarrow i_m + m - 1
\end{array}$$

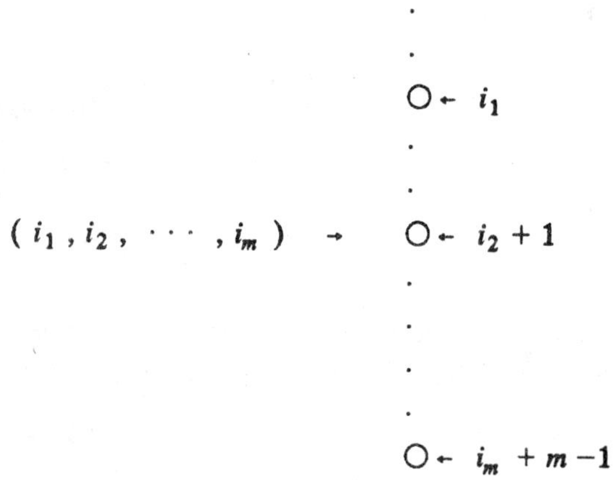

Accordingly, in the column labelled by $\boxminus = (1,1)$ the first circle is at distance 1 from the top and the second circle is at distance zero from the first. Proceeding in the same manner for the other two partitions we obtain the circle diagram given in 2.19. This done, we assign to the positions in the diagram (indicated by dots when not by circles) the labels 0,1,2,3,... successively from left to right and from top to bottom, and record the label only when it falls in one of the circles. This gives the labelled diagram

$$\begin{array}{ccc}
\square & \boxminus & \square\square \\
\textcircled{0} & \cdot & \textcircled{2} \\
 & \cdot\ \textcircled{4}\ \cdot & \\
\textcircled{6}\ \textcircled{7} & & \cdot \\
 & \cdot\ \cdot\ \textcircled{11} &
\end{array}$$

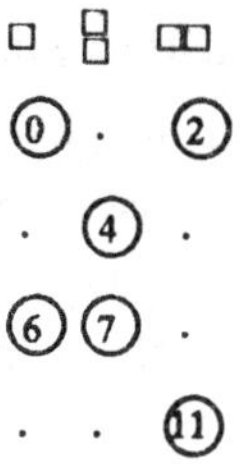

2.20

In the case of a general p-tuple $I_1, I_2, \ldots, I_p$ we obtain a circle diagram with $m$ circles in each column, where $m$ is the maximum number of non-zero parts appearing in any of the partitions $I_1, I_2, \ldots, I_p$.

Let

$$b_1 < b_2 < b_3 < \cdots < b_{mp}$$

be the labels placed on the circles and

$$q_{s,1} < q_{s,2} < \cdots < q_{s,m}$$

be the labels appearing in the column corresponding to the partition $I_s$. Finally, let $inv(I_1,\ldots,I_p)$ denote the number of inversions of the permutation

$$q_{1,1}\ q_{1,2}\ \cdots\ q_{1,m}\ q_{2,1}\ q_{2,2}\ \cdots\ q_{2,m}\ \cdots\ q_{p,1}\ q_{p,2}\ \cdots\ q_{p,m} \tag{2.21}$$

and let

$$I(I_1, \ldots, I_p) = (b_1, b_2 - 1, b_3 - 2, \cdots, b_{mp} - mp + 1).$$

This given, we set

$$SS_{I_1, \ldots, I_p} = (-1)^{\binom{m}{2}\binom{p}{2} + inv(I_1, \ldots, I_p)} S_{I(I_1,\ldots,I_p)}. \tag{2.22}$$

Going back to our particular example, we can easily see that to calculate the number of inversions of the permutation

$$0, 6, 4, 7, 2, 11$$

we need only count, for each circle in the diagram of 2.20, the number of circles that are northeast of it, and add all these counts. This gives

$$inv(\square, \begin{smallmatrix}\square\\ \square\end{smallmatrix}, \square\square) = 0 + 0 + 1 + 2 + 1 + 0 = 4$$

at the same time, we have

$$\binom{m}{2}\binom{p}{2} = \binom{2}{2}\binom{3}{2} = 3$$

and

$$I(I_1,\ldots,I_p) = (0-0, 2-1, 4-2, 6-3, 7-4, 11-5) = (0, 1, 2, 3, 3, 6)$$

So we finally obtain in this case

$$SS_{\square, \begin{smallmatrix}\square\\ \square\end{smallmatrix}, \square\square} = -S_{12336}(x)$$

It is worthwhile noting at this point that, in view of 2.17, the coefficient $c_{I_1,\ldots,I_p}$ does not depend on the order in which $I_1, I_2, \cdots, I_p$ are given. This means that in our particular example we can take advantage of the result in 2.18 and obtain 5 additional terms in the expansion of $S_{113}(x^3)$ by carrying out the above process for each of the remaining permutations of the triplet $(\square, \begin{smallmatrix}\square\\ \square\end{smallmatrix}, \square\square)$. In the table below we give the resulting diagrams and the corresponding partitions and signs.

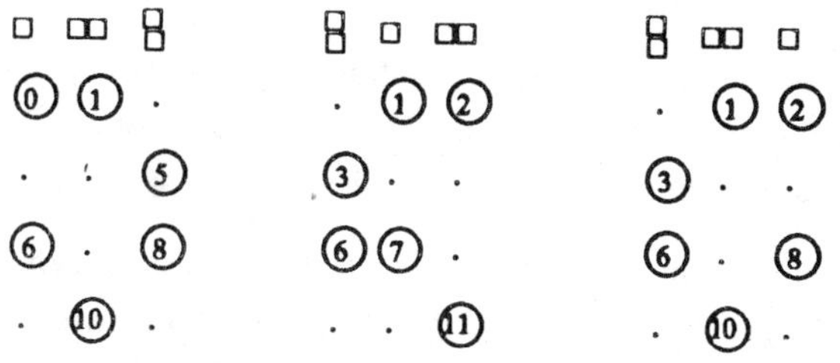

$-$ , $\{3^2 45\}$ $\quad$ $+$ , $\{1^3 3^2 6\}$ $\quad$ $-$ , $\{1^3 345\}$

$-$ , $\{124^3\}$ $\quad$ $+$ , $\{34^3\}$

We thus obtain that the contribution to the expansion of $S_{113}(x^3)$ coming from the triplet (□,⊟,□□) is the expression

$$-2(S_{123^26}(x) + S_{3^245}(x) - S_{1^33^26}(x) + S_{1^3345}(x) - S_{124^3}(x) + S_{34^3}(x)) \ .$$

Taking all of this into account, we can easily see that the SXP algorithm decomposes into successive applications of the following 3 basic steps. Namely, to calculate the expansion of $S_J(x^P)$ when $J$ is a partition of $n$, we proceed as follows

Step 1. We pick a p-tuple of partitions $I_1, I_2, \cdots, I_p$ satisfying the 3 conditions

a) The Ferrers' diagram of each $I_s$ is contained in that of $J$

b) $|I_1| \leq |I_1| \leq \cdots \leq |I_p|$

c) $|I_1| + |I_1| + \cdots + |I_p| = n$

This done we calculate the coefficient $c_{I_1,\ldots,I_p}$ by the SD algorithm. If this coefficient is not zero we proceed to the next step otherwise we repeat step 1.

Step 2. Pick a permutation $I_{\sigma_1}, I_{\sigma_2}, \cdots, I_{\sigma_p}$ of $I_1, I_2, \cdots, I_p$ and construct the labelled circle diagram whose $s^{th}$ column is indexed by $I_{\sigma_s}$.

Step 3. Calculate the partition $I(I_{\sigma_1}, I_{\sigma_2}, \cdots, I_{\sigma_p})$ and the corresponding sign.

Step 2. & 3. are to be repeated over all distinct permutations of $I_1, I_2, \cdots, I_p$. This done we go back to step 1 and repeat the process over all possible choices of $I_1, I_2, \cdots, I_p$ satisfying a) b) & c).

In the calculation of the plethysms $S_m[S_n]$ or, more generally, whenever the inner polynomial is $S_n(x)$, we are led to calculate plethysms of the form $S_n[x^p]$. For these the SXP algorithm simplifies considerably. We give a separate description of this important special case.

*The Chen algorithm*

We need here an additional notation for Schur functions which will also turn out to be convenient in the next section. It is well known that for any partition $I = \{i_1, i_2, \cdots, i_k\}$ we have

$$S_I(x_1, x_2, \cdots, x_k) = \det\|x_r^{i_s+s-1}\| \,/\, \det\|x_r^{s-1}\| \qquad (r,s=1,2,\ldots,k) \quad 2.23$$

This is the classical *bideterminantal formula* for Schur functions.

Now given $k$ distinct integers

$$a_1, a_2, \cdots, a_k$$

we set

$$[a_1, a_2, \cdots, a_k] = \det\|x_r^{a_s}\| \,/\, \det\|x_r^{s-1}\| . \qquad (r,s = 1,2,\ldots,k) \quad 2.24$$

We can easily see that if

$$b_1 < b_2 < \cdots < b_k$$

is the increasing rearrangemt of $a_1, a_2, \cdots, a_k$ then we have

$$[a_1, a_2, \cdots, a_k] = (-1)^{inv(\sigma)} S_{b_1, b_2-1, \cdots, b_k-k+1} \quad 2.25$$

where $\sigma$ is the permutation which rearranges the $a$'s into the $b$'s. In particular, for $I = (i_1, i_2, \cdots, i_k)$ we have of course

$$S_I = [i_1, i_2+1, \cdots, i_k+k-1]$$

This notation is convenient in that it expresses a Schur function indexed by $I$ in terms of the hook numbers of the first column of the Ferrers diagram of $I$.

This given, let us go back to our calculation of plethysms $S_n(x^p)$. To this end note that when $J$ consists of a single row condition a) above forces all the $I_s$ to be single rows as well. However, for such a choice of $I_1, I_2, \cdots, I_p$ the SD algorithm immediately delivers that

$$c_{I_1, I_2, \cdots, I_p} = 1 .$$

Thus no calculation of the coefficient is needed in this case and the whole algorithm for the expansion of $S_n(x^p)$ simply reduces to the construction of the partition $I(I_1, I_2, \cdots, I_p)$ and the corresponding sign. More precisely, we are reduced to a repetition of the following two steps

Step 1 We pick a composition of $n$, that is we pick $p$ non-negative integers $q_o, q_1, \cdots, q_{p-1}$ adding up to $n$.

Step 2 Calculate the p-tuple

$$a_o = pq_o, a_1 = pq_1+1, a_2 = pq_2+2, \cdots, a_{p-1} = pq_{p-1}+p-1, \qquad 2.26$$

determine the increasing rearrangement

$$b_1 < b_2 < \cdots < b_p$$

of the integers in 2.26 and the number of inversions of the permutation $\sigma$ which rearranges the $a$'s into the $b$'s

This done, we obtain

$$SS_{q_o, q_1, \ldots, q_{p-1}}(x) = (-1)^{inv(\sigma)} S_{b_o, b_1-1, \ldots, b_{p-1}-p+1} \qquad 2.27$$

In other words, in our new notation

$$SS_{q_o, q_1, \ldots, q_{p-1}}(x) = [a_o, a_1, \cdots a_{p-1}]$$

The expansion of $S_n(x^p)$ is thus obtained by repeating these steps over all compositions of $n$.

We can easily see then that the SXP algorithm in this case can be simply expressed by the following formula

$$S_n(x^p) = \sum_{q_o+q_1+\cdots+q_{p-1}=n} [pq_o, pq_1+1, pq_2+2, \cdots, pq_{p-1}+p-1] \qquad 2.28$$

It might be good to see for a moment how formula 2.27 comes out of the circle diagram. Observe that in the present case all our partitions $I_1, I_2, \cdots, I_p$ have a single row, (indeed $I_s = \{q_{s-1}\}$ ) and the circle diagram with columns indexed by

$$\{q_o\}, \{q_1\}, \cdots, \{q_{p-1}\}$$

will have a single circle in each column. Now, the labels of the squares in the $s^{th}$ column are (from top to bottom)

$$s, p+s, 2p+s, 3p+s, \cdots$$

Since the circle in the $s^{th}$ column has to appear at distance $q_s$ from the top, the label falling in it is precisely the integer

$$a_s = pq_s + s$$

we introduced in 2.26.

Thus we see that the permutation in 2.21 specializes to that in 2.26. This given, it is not difficult to conclude that formula 2.22 itself reduces to 2.27. This establishes our assertions.

It is good to see the significance of 2.28 by working out an example. For instance, to calculate the expansion of $S_2(x^3)$, we note that the compositions of 2 into 3 parts are

$$002 \ , \ 011 \ , \ 020 \ , \ 101 \ , \ 110 \ , \ 200$$

and these through formula 2.26 correspond to the permutations

$$018 \ , \ 045 \ , \ 072 \ , \ 315 \ , \ 342 \ , \ 612$$

which yield the following signs and partitions:

$$+, \{6\} \ ; \ +, \{3,3\} \ ; \ -, \{1,5\} \ ; \ -, \{1,2,3\} \ ; \ +, \{2,2,2\} \ ; \ +, \{1,1,4\}$$

Thus we derive that

$$S_2(x^3) = S_{[6]} + S_{[3,3]} - S_{[1,5]} - S_{[1,2,3]} + S_{[2,2,2]} + S_{[1,1,4]} \qquad 2.29$$

A rather colorful combinatorial algorithm for expanding $S_n(x^p)$ has been given by Chen in [2]. We will find it worthwhile to see here how it relates to formula 2.28.

To present Chen's algorithm we need further notation and we introduce it with an example. In the figure below we have depicted a Ferrers' diagram $F$ with an attached set of linked circles. We represent this configuration by the symbol $F\,Sl_{10}(4)$.

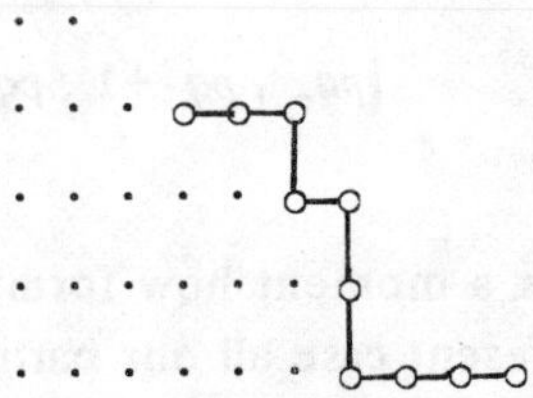

The symbol $Sl_{10}(4)$ refers to the configuration of linked circles with 10 giving the total number of circles in it and 4 the height of the first circle. We call $Sl_{10}(4)$ a 10*-slinky starting at* 4. The word *slinky* originating from the familiar toy capable of descending staircases. In the literature our slinkies are often referred to as *border strips* or worse yet *rim hooks.*

Given a Ferrers' diagram $F$ and an $Sl_p(i)$ the combined figure $F\,Sl_p(i)$ is constructed by finding the leftmost position adjacent to the boundary of $F$ at height $i$ then placing the first circle of the slinky in that position and the remaining $p-1$ circles following in a continuous band *down the staircase*. For instance if $F$ is the Ferrers' diagram corresponding to the partition $I = (1,3,5,6,9)$ then

$F\,Sl_9(5) \quad =$ 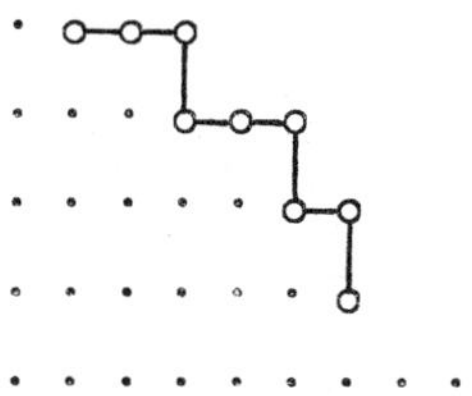

Clearly, $F$ $p$ and $i$ have to satisfy certain restrictions for the resulting figure to be a Ferrers' diagram. However, in the situations we shall deal with these restrictions will be automatically satisfied.

We should view the expression $F\,Sl_p(i)$ as a sort of multiplication, and our ultimate goal is to multiply several of these slinkies into a *signed* Ferrers' diagram. To this end we define the *sign* of a *placed* slinky to be $+1$ if the slinky extends over an **odd** number of rows and $-1$ otherwise. A product of slinkies is then given a sign which is the product of the signs of its factors. Finally, the resulting configuration will be used to represent a signed Schur function. An example should give the idea of what we have in mind here.

For instance, since

$Sl_5(4)Sl_5(4)Sl_5(3)Sl_5(2) \quad =$ 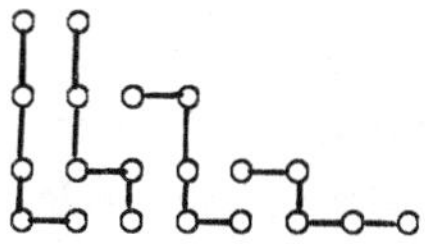

we shall use the symbol $Sl_5(4)Sl_5(4)Sl_5(3)Sl_5(2)$ to represent $-S_{[2,4,6,8]}$ .

This given Chen's algorithm may be summarized by the following formula:

$$S_m(x^p) = \sum_{p \geq i_1 \geq i_2 \geq \cdots \geq i_m \geq 1} Sl_p(i_1)\,Sl_p(i_2) \cdots Sl_p(i_m) \qquad 2.30$$

It may be good to compare this to 2.27 by working out again the example $S_2(x^3)$ . Now for $m=2$ and $p=3$ formula 2.30 reduces to

$$S_2(x^3) = Sl_3(3)Sl_3(3) + Sl_3(3)Sl_3(2) + Sl_3(3)Sl_3(1) +$$

$$+ Sl_3(2)Sl_3(2) + Sl_3(2)Sl_3(1) + Sl_3(1)Sl_3(1)$$

Replacing these expressions by the corresponding signed Ferrers' diagrams we get

$S_2(x^3) \quad =$ 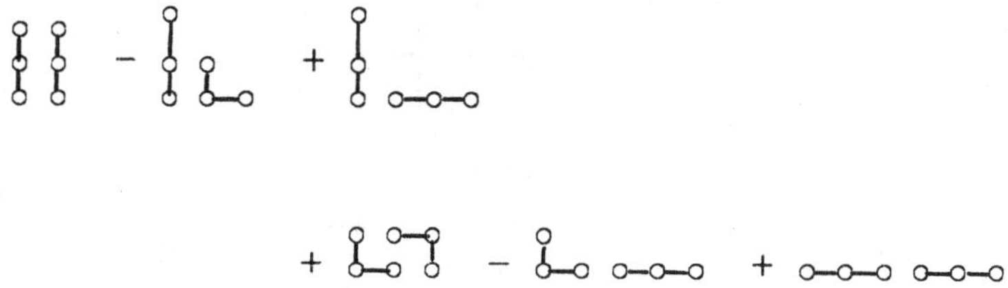

and this is to be interpreted to mean that

$$S_2(x^3) = S_{[2,2,2]} - S_{[1,2,3]} + S_{[1,1,4]} + S_{[3,3]} - S_{[1,5]} + S_{[6]}$$

We see that this is precisely what we had in 2.29.

## 3. Validity of the algorithms.

A derivation of the SD algorithm can be found in [20] and will not be repeated here. Thus in this section we shall only be concerned with the SXP and Chen algorithms.

Before we proceed with our arguments it will be good to see how *slinky constructions* are related to circle diagrams. This brings us to the theory of *p-cores* which is a rather interesting topic in its own merits. We shall however limit our treatment to that portion of the theory needed for our developments. The reader is referred to James-Kerber [11] for additional information.

In this context, it is convenient to identify partitions with their Ferrers' diagrams. This given, a diagram $I$ is said to be with *empty p-core* if it can be decomposed into a successive product of p-slinkies. Partitions with empty p-core are in correspondence with p-tuples of ordinary partitions by a remarkable construction of Robinson [21]. We shall see here that this correspondence arises very naturally in the course of proof of formula 2.16. Indeed, the circle diagrams are simply a convenient way of expressing the correspondence.

In dealing with a particular value of $p$ it is convenient to work with partitions with a number of parts which is a multiple of $p$. This can always be achieved, if necessary, by adding at most $p-1$ zero parts. For instance, in the case $p = 5$ we represent the partition $\{2^2 4 5^2 6\}$ by the diagram below (the four vertical unit segments representing zero parts.)

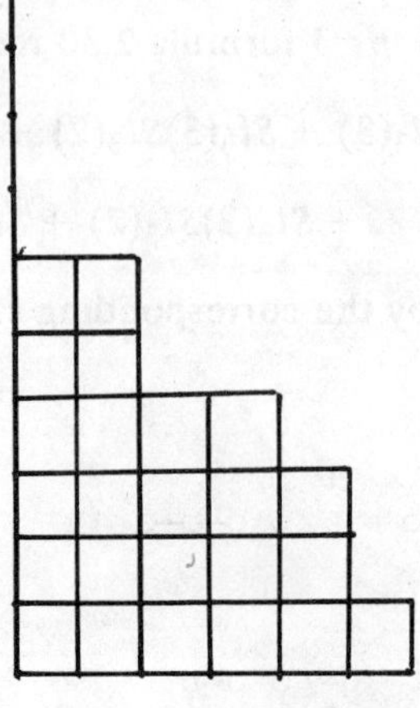

3.1

Given a partition $I = \{i_1, i_2, \cdots, i_k\}$, the vector

$$I^a = \{i_1, i_2+1, i_3+2, \cdots, i_k+k-1\}$$

is sometimes referred to as the *augmented partition*. We note that for our partition $I = \{0^4 2^2 4 5^2 6\}$ we have

$$I^a = \{0,1,2,3,6,7,10,12,13,15\} . \qquad 3.2$$

and these numbers can be interpreted as the row lengths of the diagram obtained by adding to the diagram of $I$ a suitable staircase. For this reason we shall refer to the components of $I^a$ as the *augmented row lengths* of $I$. In the figure below we have depicted the augmented diagram and filled its squares by their distances from the staircase.

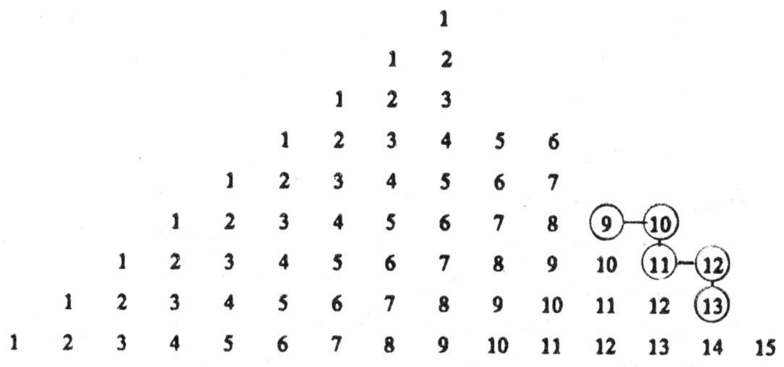

We have also outlined a 5-slinky there. Note that if we remove the slinky from the diagram then the rows with augmented lengths

$$13 \quad 12 \quad 10$$

are respectively replaced by rows with augmented lengths

$$12 \quad 10 \quad 8$$

Thus the net effect on the vector $I^a$ is simply the replacement of 13 by $13-5$. This is clearly true in general, and we may state that:

> removing from a partition $I$ with $mp$ parts a p-slinky whose bottom square is the last of a row of augmented length $i$ has the effect of or *is equivalent to* replacing $i$ by $i-p$ in $I^a$.

This may be quickly derived from the fact that such a p-slinky always contains the labels $i, i-1, \ldots, i-p+1$ in succession.

We see then that the removal of a p-slinky doesn't alter the (mod p) distribution of augmented lengths. The relation between slinky constructions and circle diagrams stems from this very simple observation. Indeed, let us represent the partition

$$I = \{i_1 , i_2 , \cdots , i_{mp}\}$$

by the p-column circle diagram obtained by placing our circles in the squares with labels

$$i_1 , i_2+1 , i_3+2 , \cdots , i_{mp} + mp-1 .$$

For $p = 5$ and $I = \{2^2 3\, 5^2 6\}$ this is the circle diagram given below

$$\begin{array}{ccccc} \textcircled{0} & \textcircled{1} & \textcircled{2} & \textcircled{3} & \cdot \\ \cdot & \textcircled{6} & \textcircled{7} & \cdot & \cdot \\ \textcircled{10} & \cdot & \textcircled{12} & \textcircled{13} & \cdot \\ \textcircled{15} & \cdot & \cdot & \cdot & \cdot \\ \cdot & \cdot & \cdot & \cdot & \cdot \end{array} \qquad 3.4$$

We can clearly see now the significance of this representation: the number of circles in a given column gives precisely the number of augmented row lengths in the corresponding equivalence class (mod p). Moreover, the removal of a slinky can be visualized as the motion of one of the circles one step up its column. That is circle "13" for the case illustrated in 3.3 and 3.4.

We can also immediately conclude from 3.4 that the diagram in 3.3 does not have empty 5-core. Indeed, the only remaining upward motions for the circles in 3.3 are "10" up one step followed by "15". This means that only 2 additional 5-slinkies may be removed from the diagram in 3.3.

In fact, the invariance of (mod p) distribution of augmented lengths (under removal of p-slinkies) implies that a partition has empty p-core if and only if in its p-column circle diagram each column has an equal number of circles. This is simply due to the fact that empty partitions correspond only to compacted circle diagrams with this property.

It is clear that many questions concerning slinky constructions may be easily answered by reinterpreting them in terms of circle diagrams. For the moment it is good to keep this parallelism in mind as we proceed with our arguments.

*Validity of the SXP algorithm.*

We need only verify 2.16. This formula is essentially due to Littlewood. A complete proof of it may be obtained by combining our remarks here with certain identities given in [13] (page 131) and some remarks made by Todd in [26]. Since this path is rather circuitous and not very transparent we give here a simple direct derivation. We shall also see from our arguments here how one might be naturally led to the discovery of the Robinson correspondence by working on the expansion of $S_I(x^p)$.

Our point of departure here as well as in [13] is the identity

$$\prod_{i,j} \frac{1}{1 - t^p x_i^p y_j^p} = \sum_I t^{p\,|I|}\, S_I(x^p)\, S_I(y^p)\,, \tag{3.5}$$

where $|I|$ denotes the sum of the parts of $I$. This is essentially the Cauchy identity with the variables replaced by their $p^{th}$ powers. Alternatively we may view it as a consequence of 1.8 with

$$P = \sum_i x_i^p = \psi_p(x)$$

$$Q = \sum_i y_i^p = \psi_p(y)$$

As in [13] we write the left hand side of 3.5 in the form

$$\prod_{i,j} \frac{1}{1-(t\,x_i y_j)^p} = \prod_{i,j} \frac{1}{1 - t\,x_i y_j}\; \frac{1}{1 - t\,x_i y_j \gamma} \cdots \frac{1}{1 - t\,x_i y_j \gamma^{p-1}} \tag{3.6}$$

where $\gamma$ is a primitive $p^{th}$ of unity. For instance we could take $\gamma = e^{2\pi i/p}$.

Note now that if we set

$$\begin{aligned} X &= x_1 + x_2 + \cdots + x_N\,, \\ Y &= y_1 + y_2 + \cdots + y_M\,, \\ Z &= z_1 + z_2 + \cdots + z_p\,, \end{aligned}$$

then the expression in 3.6 is none other than the sum of the series

$$\sum_{n \ge 0} t^n S_n(XYZ)$$

when we set

$$z_i = \gamma^{i-1} \qquad (i=1,2,\dots,p) \tag{3.7}$$

Making use of formula 1.8 with $P = X$ and $Q = YZ$ we deduce that

$$\prod_{i,j} \frac{1}{1 - t^p x_i^p y_j^p} = \sum_I t^{|I|}\, S_I(X)\, S_I(YZ) \tag{3.8}$$

with the understanding, of course, that the $z_i$'s are specialized as in 3.7, this will be tacitly assumed for the rest of this section.

Combining 3.5 and 3.8 and equating coefficients of $t^{np}$ we finally obtain the identity

$$\sum_{I \vdash n} S_I(x^p) S_I(y^p) = \sum_{I \vdash np} S_I(X) S_I(YZ) \tag{3.9}$$

It is not difficult to show, and it will be a byproduct of our proof, that each of the factors $S_I(YZ)$ is a polynomial in $y^p$. Since the Schur functions $S_I(y^p)$ are a basis for these polynomials we may conclude that for any $J \vdash n$ we have

$$S_J(x^p) = \sum_{I \vdash np} S_I(X)\, S_I(YZ)\, \Big|_{s_J(y^p)} \tag{3.10}$$

We aim to show that this is indeed formula 2.16. More precisely, we show that the only terms that survive in 3.10 are those corresponding to partitions $I$ produced by circle diagrams containing an equal number of circles in each column, and for those the coefficient and sign are precisely as given by 2.17 and 2.22. All of this follows very naturally upon evaluating the factors $S_I(YZ)$ by means of the determinantal formula 1.5 a). To see how it comes about we shall start with an example.

First of all it is good to observe that the expression

$$[a_1, a_2, \cdots, a_k]$$

defined in 2.24 may also be evaluated by means of the determinantal formula 1.5 a). Indeed, we simply place

$$S_{a_1}, S_{a_2}, \cdots, S_{a_k},$$

in the first row of the matrix and decrease the subscripts by one as we go down the columns. That is we have as well

$$[a_1, a_2, \cdots, a_k] = \det \| S_{a_j - i+1} \| . \tag{3.12}$$

Next note that from the identity

$$\sum_n t^n S_n(YZ) = \prod_i \frac{1}{1 - t^p y_i^p} .$$

we easily derive that

$$S_n(YZ) = \begin{cases} 0 \text{ if } n \neq 0 \, (mod p) \\ \\ S_m(y^p) \text{ if } n = mp . \end{cases} \tag{3.13}$$

This given, let us evaluate

$$S_{123^2 6}(YZ)$$

for $p=3$. According to one of our previous observations we write $\{123^2 6\}$ as a partition with a number of parts a multiple of 3, obtaining

$$\{0,1,2,3,3,6\}$$

then pass to the augmented partition

$$\{0,2,4,6,7,11\}$$

and get

$$S_{123^26}(YZ) = [0\,2\,4\,6\,7\,11] = \det\begin{bmatrix} S_o & S_2 & S_4 & S_6 & S_7 & S_{11} \\ 0 & S_1 & S_3 & S_5 & S_6 & S_{10} \\ 0 & S_o & S_2 & S_4 & S_5 & S_9 \\ 0 & 0 & S_1 & S_3 & S_4 & S_8 \\ 0 & 0 & S_o & S_2 & S_3 & S_7 \\ 0 & 0 & 0 & S_1 & S_2 & S_6 \end{bmatrix} . \qquad 3.14$$

Now, for $p=3$ 3.13 specializes to

$$S_n(YZ) = \begin{cases} 0 \text{ if } n \neq 0\ (mod 3) \\ \\ S_m(y^3) \text{ if } n = 3m\ . \end{cases} \qquad 3.15$$

In particular, this implies (as we had anticipated) that each $S_I(YZ)$ is a polynomial in $y^3$. But, more significantly for us here, each entry in 3.14 whose subscript is not divisible by 3 evaluates to zero. To visualize the consequences of this fact we have surrounded each entry with subscript divisible by 3 with a geometric figure (circle, triangle or square) obtaining:

$$S_{123^26}(YZ) = [0\,2\,4\,6\,7\,11] = \det\begin{bmatrix} \overset{\bigcirc}{S_o} & S_2 & S_4 & \overset{\bigcirc}{S_6} & S_7 & S_{11} \\ 0 & S_1 & \overset{\triangle}{S_3} & S_5 & \overset{\triangle}{S_6} & S_{10} \\ 0 & \overset{\square}{S_o} & S_2 & S_4 & S_5 & \overset{\square}{S_9} \\ \overset{\bigcirc}{0} & 0 & S_1 & \overset{\bigcirc}{S_3} & S_4 & S_8 \\ 0 & 0 & \overset{\triangle}{S_o} & S_2 & \overset{\triangle}{S_3} & S_7 \\ 0 & \overset{\square}{0} & 0 & S_1 & S_2 & \overset{\square}{S_6} \end{bmatrix} . \qquad 3.16$$

The rule in chosing our figures is as follows. Let us say that a column *belongs to class* $i$ if the subscript of the entry in the first row is congruent to $i$ $(mod\ p)$. With this convention, our rule may be easily stated. We use figure "i" for all terms in a column of class i. For our particular example in 3.15, we used circles for columns of class 0, triangles for columns of class 1 and squares for columns of class 2.

Now, replacing by zeros the terms not surrounded by figures we obtain

$$S_{123^26}(YZ) = \det\begin{bmatrix} \enclose{circle}{S_o} & 0 & 0 & \enclose{circle}{S_6} & 0 & 0 \\ 0 & 0 & \triangle{S_3} & 0 & \triangle{S_6} & 0 \\ 0 & \boxed{S_o} & 0 & 0 & 0 & \boxed{S_9} \\ \enclose{circle}{0} & 0 & 0 & \enclose{circle}{S_3} & 0 & 0 \\ 0 & 0 & \triangle{S_o} & 0 & \triangle{S_3} & 0 \\ 0 & \boxed{0} & 0 & 0 & 0 & \boxed{S_6} \end{bmatrix} .$$

The reader may discern that this matrix can be rearranged into block diagonal form. The reason for this is simple: if in a given row or column there is a circle then all other figures in that row or column are also circles, the same is true for triangles or squares. In other words in each row or column only one type of figure occurs.

Clearly, the same will be true in the general case. To be precise, we see that in a column of class $i$ the non zero terms are only in rows

$$i+1 \ , i+1+p \ , i+1+2p \ , \ \cdots \ , etc$$

This means that the $i^{th}$ figure can only occur in these rows. Thus, if the given partition has $mp$ parts, there will be exactly $m$ rows containing figure $i$ (for each $i=0 \ , 1 , \ldots , p-1$ ).

Let us now rearrange the rows and columns of our matrix so that the circles are contained in the first 2 rows or columns, triangles are contained in the next 2 rows or columns and squares in the last 2 rows or columns. This gives

$$S_{123^26}(YZ) = -\det\begin{bmatrix} \enclose{circle}{S_o} & \enclose{circle}{0} & 0 & 0 & 0 & 0 \\ \enclose{circle}{0} & \enclose{circle}{S_3} & 0 & 0 & 0 & 0 \\ 0 & 0 & \triangle{S_3} & \triangle{S_6} & 0 & 0 \\ 0 & 0 & \triangle{S_o} & \triangle{S_3} & 0 & 0 \\ 0 & 0 & 0 & 0 & \boxed{S_o} & \boxed{S_9} \\ 0 & 0 & 0 & 0 & \boxed{0} & \boxed{S_6} \end{bmatrix} .$$

and, using the second equation in 3.14, we finally deduce that

$$S_{123^26}(YZ) = -\det\begin{bmatrix} S_o(y^3) & S_2(y^3) \\ 0 & S_1(y^3) \end{bmatrix} \det\begin{bmatrix} S_1(y^3) & S_2(y^3) \\ S_o & S_1(y^3) \end{bmatrix} \det\begin{bmatrix} S_o(y^3) & S_3(y^3) \\ 0 & S_2(y^3) \end{bmatrix}$$

In other words we get

$$[0246711] = -[02](y^3)\cdot[12](y^3)\cdot[03](y^3) \ .$$

We see that the pairs $(0,2)$ , $(1,2)$ and $(0,3)$ are respectively

0) The augmented row lengths congruent to zero (mod 3) divided by 3,

1) The augmented row lengths congruent to 1 (mod 3) minus 1 divided by 3,

2) The augmented row lengths congruent to 2 (mod 3) minus 2 divided by 3.

What we have just discovered (in the notation of last section) is simply that

$$S_{I(\square\,,\,\boxminus\,,\,\square\square)}(YZ) = -S_{\square}(y^3)\,S_{\boxminus}(y^3)\,S_{\square\square}(y^3)$$

In general, if our partition $I$ has a p-column circle diagram with $m$ circles in each column, then the determinant giving $S_I(YZ)$ may be brought into block diagonal form with $p$ $m \times m$ blocks. Where the elements of the $i^{th}$ block come from the columns of class $i$ and the elements of the first row of the $i^{th}$ block have as subscripts the augmented row lengths congruent to $i$ minus $i$.

It is not difficult to conclude from these observations that, with a suitable choice of sign, we have

$$S_{I(I_1\,,\,I_2,\ldots,I_p)}(YZ) = \pm\, S_{I_1}(y^p)\,S_{I_2}(y^p)\cdots S_{I_p}(y^p)$$

Now the sign change is due to two factors: the row interchanges and the column interchanges. These are carried out as follows. To bring all the figure "1"'s in the first $m$ rows we need exactly

$$(p-1) + 2(p-1) + \cdots + (m-1)(p-1) = \binom{m}{2}(p-1)$$

interchanges. To bring all the figure "2"'s in the next $m$ rows we need

$$(p-2) + 2(p-2) + \cdots + (m-1)(p-2) = \binom{m}{2}(p-2)$$

interchanges,...etc. So we see that the contribution to the sign coming from the row interchanges is precisely

$$(-1)^{\binom{m}{2}\binom{p}{2}}$$

As for the column interchanges, we simply note that to bring all of the figure "1"'s in the first $m$ columns, all the figure "2"'s in the next $m$ columns,... etc, we need to carry out on the columns precisely the permutation which rearranges the augmented row lengths so that all those congruent to zero (mod p ) come first, then come those congruent to 1 , then those congruent to 2, ... etc. Thus the change of sign due to the column interchanges is (in the notation of last section) precisely

$$(-1)^{Inv(I_1\,,\,I_2\,,\ldots,\,I_p)}$$

We have thus obtained that

$$S_{I(I_1\,,I_2\,,\cdots\,,I_p)}(YZ) = (-1)^{\binom{m}{2}\binom{p}{2} + Inv(I_1\,,\,I_2\,,\ldots\,,\,I_p)}\, S_{I_1}(y^p)\,S_{I_2}(y^p)\cdots S_{I_p}(y^p)$$

to complete our argument we must examine the case when our partition $I$ does not have empty p-core. That is if the p-column circle diagram of $I$ does not have the same number of circles in each column. Since we can always arrange that our partition has $mp$ parts (for some $m$ ) we must conclude that in this case one of the columns must have less than $m$ circles. To see what happens then, it is best to look at an example. Let us suppose that $p=2$ and that $I$ is the partition

$$I = \{1245^2 6\} \quad . \quad I^a = \{1\ 3\ 6\ 8\ 9\ 11\} \ .$$

we have then

$$S_{1245^2 6}(YZ) = [\ 1\ \ 3\ \ 6\ \ 8\ \ 9\ \ 11\ ]$$

here we have circled the even augmented row lengths and squarred the odd ones. Without displaying the whole matrix we can see that all the circles will be in 3 rows and 2 columns. This implies that if we expand the determinant according to these 3 rows, all the $3 \times 3$ subdeterminants coming out of these rows will vanish, since they will have at most two non-zero columns, and the whole determinant will necessarily vanish as well.

In the general case there will be $m$ rows all of whose non-zero elements are contained in fewer than $m$ columns and the corresponding Schur function will necessarily vanish identically. We may thus conclude that the only terms occurring in 3.10 are those corresponding to partions $I$ of the form $I(I_1, I_2, \cdots, I_p)$

Our findings can thus be summarized by the formula

$$S_J(x^p) = \sum_{|I_1| + \ldots + |I_p| = n} (-1)^{\binom{m}{2}\binom{p}{2} + inv(I_1, I_2, \ldots, I_p)} S_{I(I_1, \ldots, I_p)}(x)\, S_{I_1}(y^p)\, S_{I_2}(y^p) \cdots S_{I_p}(y^p) \Big|_{s_J(y^p)} \qquad 3.17$$

and this is seen to be identical to 2.16.

**Remark 3.1**

The fact that $S_I(YZ)$ vanishes unless $I$ has empty p-core can be derived by a completely different path follows. We start by observing that for the power symmetric functions we have

$$\psi_n(YZ) = \psi_n(Y)\psi_n(Z) = \begin{cases} 0 \text{ if } n \neq 0 \pmod{p} \\ p\psi_{mp}(Y) \text{ if } n = mp \end{cases} .$$

This given, we deduce that the expansion of $S_I(YZ)$ in terms of power symmetric functions contains only terms whose factors are indexed by multiples of $p$ . Now, it follows from the Murnagham rule that products of such power symmetric functions expand into Schur functions that are indexed by partitions with empty p-core.

Keeping this fact in mind, we can see how our derivation of formula 2.16 would lead us to rediscover the Robinson correspondence. Indeed, on one hand we see that the *factorisation*

$$I \to (I_1, I_2, \cdots, I_p)$$

occurring for all $I$ leading to a non vanishing term $A_I(YZ)$, is simply suggested by the fact that the Schur function itself does factor in this manner. On the other hand the Murnagham rule gives that the only terms occurring in 2.16 are those coming from partitions with empty p-core. Thus the conclusion is inevitable: partitions which *factor* and empty p-core partitions must be one and the same thing!

*Validity of the Chen algorithm.*

We need only prove formula 2.30 and we shall derive it from 2.28. The reader is referred to [2] for a direct proof that is completely independent of the present developments.

Our program is to show that the Schur functions occurring in 2.30 and 2.28 are the same and their signs are in agreement.

Observe first that the partitions occurring in 2.30 are simply those which may be dismantled by pealing off p-slinkies which *end at the bottom row*. We can easily see that a circle diagram corresponds to such a partition if and only if it can be reduced to its compacted form by upward motions always involving the circle with highest label. However, we can easily see that this can be done if and only if no more than one circle per column needs to be moved. Now the circle diagrams producing 2.28 are precisely those having this property. This shows that the terms are the same in both sums.

We are left with having to check that the signs are the same. We proceed in full generality and show that the sign given in 2.22 is the same as the product of the signs of any set of p-slinkies which multiply into the partition $I(I_1, I_2, \cdots, I_p)$. We do this by evaluating the effect that the removal of a p-slinky has on the expression

$$(-1)^{\binom{m}{2}\binom{p}{2} + inv(I_1, \ldots, I_p)} \qquad 3.18$$

To do this we need only calculate the change in the number of inversions of the permutation 2.21. We recall that this number is obtained by counting the circles lying northeast of each circle in the diagram. Now clearly, as a result of the one step upwards motion of a circle "i", the count increases by the number of circles to the left of "i" and in the same row, and decreases by the number of circles to the right and in the next row. However, it is seen that these two counts add up precisely to one less than the number of rows occupied by the slinky that is *being removed*. In other words the sign changes precisely by the sign of that slinky. Now we see that for a compacted diagram representing the empty partition, the expression in 3.18 reduces to +1 and this is all that is needed to establish our "product of signs" assertion.

Thus the expressions in 2.30 and 2.28 are identical also in signs and our arguments are complete.

## 4. Explicit calculations.

In this section we shall use the algorithms to obtain explicit expansions for some plethysms of $S_n$ . However, before we can proceed with our presentation we need to recall the rule for expanding the product $S_I S_n$ for a given partition $I$ . This is usually referred to as *Pieri's rule* and can be easily derived from the SD algorithm. It may be stated as follows. Let us say that the skew diagram $I/J$ is *vertically thin* if no two of its squares are in the same column. This given we have

**Pieri's rule**

$$S_I S_n = \sum_{J \vdash |I|+n}^{\bullet} S_J$$

*Where the "*" is to indicate that the sum is to be restricted to the diagrams* $J$ *such that* $I/J$ *is vertically thin.*

In other words $S_I S_n$ is the sum of all Schur indexed by diagrams obtained by placing on top of $I$ $n$ squares no of them two in the same column.

This given, we start by deriving some formulas of Thrall [24]. Namely, the expansions

$$a)\quad S_2(S_n) = \sum_{0 \le 2a \le 2n} S_{2a\,,\,2n-2a}\,,$$

4.1

$$b)\quad S_{1^2}(S_n) = \sum_{0 \le 2a \le 2n-1} S_{2a+1\,,\,2n-2a-1}\,.$$

These are immediate consequences of the Chen algorithm. Indeed, since

$$a)\quad S_2 = \frac{1}{2}\,(\psi_1^2 + \psi_2)\,,$$

4.2

$$b)\quad S_{1^2} = \frac{1}{2}\,(\psi_1^2 + \psi_2)\,,$$

theorem 1.1 yields that

$$a)\quad S_2(S_n) = \frac{1}{2}\,(S_n^2 + S_n(x^2))\,,$$

4.3

$$b)\quad S_{1^2}(S_n) = \frac{1}{2}\,(S_n^2 - S_n(x^2))\,.$$

Now for $p=2$ formula 2.30 reduces to

$$S_n(x^2) = \sum_{i=0}^{n} (Sl(2))^i \, (Sl(1))^{n-i} \;, \qquad 4.4$$

and our definition gives

$$(Sl(2))^i \, (Sl(1))^{n-i} = (-1)^i \, S_{i,2n-i} \;.$$

On the other hand, Pieri's rule gives that

$$S_n^2 = \sum_{i=0}^{n} S_{i\,,\,2n-i}(x) \qquad 4.5$$

combining formulas 4.3 , 4.4 and 4.5 we get that

$$a) \quad S_2(S_n) = \sum_{i=0}^{n} \frac{1+(-1)^i}{2} \, S_{i\,,\,2n-i}$$

$$b) \quad S_{1^2}(S_n) = \sum_{i=0}^{n} \frac{1-(-1)^i}{2} \, S_{i\,,\,2n-i} \qquad 4.6$$

and this is another way of writing the formulas in 4.1.

In order of difficulty the next plethysms are

$$S_3(S_n) \quad , \quad S_{1^2}(S_n) \text{ and } S_{1^3}(S_n) \;.$$

An expansion of $S_3(S_n)$ is implicit in Thrall's work [24]. Here we shall derive explicit formulas for $S_{1^2}(S_n)$ and $S_{1^3}(S_n)$ as well. These may be stated as follows.

**Theorem 4.1**

If $b \leq n$

$$S_3(S_n)\,|_{S_{abc}} = \begin{cases} \lceil \frac{b-a+1}{6} \rceil & \text{if } b-a+1 \equiv 0,4 (mod\ 6) \\ \lceil \frac{b-a+1}{6} \rceil & \text{if } b-a+1 \equiv 1,3,5 (mod\ 6) \text{ and } a \text{ is even}, \\ \lfloor \frac{b-a+1}{6} \rfloor & \text{otherwise.} \end{cases} \qquad 4.7$$

$$S_{1^2}(S_n)\,|_{S_{abc}} = \lceil \frac{b-a}{3} \rceil \qquad 4.8$$

$$S_{1^3}(S_n)\,|_{S_{abc}} = \begin{cases} \lceil \frac{b-a+1}{6} \rceil & \text{if } b-a+1 \equiv 0,4 (mod\ 6) \\ \lceil \frac{b-a+1}{6} \rceil & \text{if } b-a+1 \equiv 1,3,5 (mod\ 6) \text{ and } a \text{ is odd}, \\ \lfloor \frac{b-a+1}{6} \rfloor & \text{otherwise.} \end{cases} \qquad 4.9$$

If $b > n$

$$S_3(S_n)\,|_{S_{abc}} = \begin{cases} \lceil \frac{c-b+1}{6} \rceil & \text{if } c-a+1 \equiv 0,4 (mod\ 6) \\ \lceil \frac{c-b+1}{6} \rceil & \text{if } c-a+1 \equiv 1,3,5 (mod\ 6) \text{ and } a+b-n \text{ is even}, \\ \lfloor \frac{c-b+1}{6} \rfloor & \text{otherwise.} \end{cases} \qquad 4.10$$

$$S_{12}(S_n)\,|_{S_{abc}} = \lceil \frac{c-b}{3} \rceil \qquad 4.11$$

$$S_{1^3}(S_n)\,|_{S_{abc}} = \begin{cases} \lceil \frac{c-b+1}{6} \rceil & \text{if } c-a+1 \equiv 0,4 (mod\ 6) \\ \lceil \frac{c-b+1}{6} \rceil & \text{if } c-a+1 \equiv 1,3,5 (mod\ 6) \text{ and } a+b-n \text{ is odd}, \\ \lfloor \frac{c-b+1}{6} \rfloor & \text{otherwise.} \end{cases} \qquad 4.12$$

Here we let $\lfloor x \rfloor$ and $\lceil x \rceil$ respectively denote the largest integer below and smallest integer above $x$. Note also that since $\{a,b,c\}$ must be a partition of $3n$, it should be understood here and in the following that $c = 3n-a-b$.

**Proof**

We start with the following three easily verified identities

$$S_3 = \frac{1}{6}(\psi_1^3 + 3\psi_1\psi_2 + 2\psi_3)$$

$$S_{12} = \frac{1}{3}(\psi_1^3 - \psi_3)$$

$$S_{1^3} = \frac{1}{6}(\psi_1^3 - 3\psi_1\psi_2 + 2\psi_3)$$

From theorem 1.1 we then derive that

$$S_3(S_n) = \frac{1}{6}(S_n^3 + 3S_nS_n(x^2) + 2S_n(x^3))$$

$$S_{12}(S_n) = \frac{1}{3}(S_n^3 - S_n(x^3)) \qquad 4.13$$

$$S_{1^3}(S_n) = \frac{1}{6}(S_n^3 - 3S_nS_n(x^2) + 2S_n(x^3))$$

Thus to obtain our desired expansions we need only calculate the coefficients of

$$S_n^3 \quad , \quad S_nS_n(x^2) \quad , \quad S_n(x^3) \; .$$

What remains to be proved can be derived with some arithmetic from the following three lemmas.

**Lemma 4.1**

$$S_n^3 \,\Big|_{s_{abc}} = \begin{cases} b-a+1 & \text{if } b \le n\ , \\ c-b+1 & \text{if } b > n\ . \end{cases} \qquad 4.14$$

**Proof**

This is a simple successive application of Pieri's rule. We can reason as follows. Imagine that we are to place n red, n blue, and n green squares in the diagram of $\{a,b,c\}$ according to the following rules. No two squares of the same color are to be on the same column. Moreover, all the red squares are in the bottom row contiguously starting from the left. The blue squares are to be placed some on the middle row (contiguously and starting from the left) and some on the bottom row following the red squares. (The resulting diagram at this point would be one occurring in the expansion of $S_n S_n$ .) Finally, the green squares are placed in the remaining squares to fill the diagram of $\{a,b,c\}$ . The desired coefficient counts the number of different ways of doing this. Indeed, formula 4.14 simply gives the number of different ways of placing the blue squares so that the resulting diagram does remain inside the diagram of $\{a,b,c\}$ and no two blue nor two green squares end up on top of each other.

**Lemma 4.2**

$$S_n(x^3) \,\Big|_{s_{abc}} = \begin{cases} 1 & \text{if } b-a \equiv 0 \pmod 3\ , \\ -1 & \text{if } b-a \equiv 1 \pmod 3\ , \\ 0 & \text{if } b-a \equiv 2 \pmod 3\ . \end{cases} \qquad 4.15$$

**Proof**

Formula 2.30 for $p=3$ gives

$$S_n(x^3) = \sum_{i+j+k=n} (SI(3))^i\,(SI(2))^j\,(SI(1))^k\ .$$

However, our definition gives that the slinky product

$$(SI(3))^i\,(SI(2))^j\,(SI(1))^k$$

for $j$ even is equal to

$$S_{i\,,\,i\,+\,3j/2\,,\,i\,+\,3j/2\,+\,3k}$$

and for $j$ odd is equal to

$$-S_{i\ ,\ i+3(j-1)/2+1\ ,\ i+3(j-1)/2+2+3k}$$

and this gives 4.15 as asserted.

We are left with the least obvious case, namely

**Lemma 4.3**

*When* $b \le n$

$$S_n S_n(x^2)\ |_{s_{abc}} = \begin{cases} 0 \text{ if } b-a+1 \text{ is even}, \\ 1 \text{ if } b-a+1 \text{ is odd and } a \text{ is even}, \\ -1 \text{ if } b-a+1 \text{ is odd and } a \text{ is odd} \end{cases}.$$

*and when* $b > n$

$$S_n S_n(x^2)\ |_{s_{abc}} = \begin{cases} 0 \text{ if } c-b+1 \text{ is even}, \\ 1 \text{ if } c-b+1 \text{ is odd and } a+b-n \text{ is even}, \\ -1 \text{ if } c-b+1 \text{ is odd and } a+b-n \text{ is odd} \end{cases}.$$

**Proof**

We have seen that

$$S_n(x^2) = \sum_{0 \le k \le n} (-1)^k\ S_{k\ ,\ 2n-k}$$

Thus,

$$S_n S_n(x^2)\ |_{s_{abc}} = \sum_{0 \le k \le n} (-1)^k\ S_{k\ ,\ 2n-k} S_n\ |_{s_{abc}}$$

Now it is easy to see from Pieri's rule that the terms corresponding to $k < a$ must all vanish in any case. But, for $b \le n$ non-zero terms can be found up to $k = b$ obtaining

$$S_n S_n(x^2)\ |_{s_{abc}} = \sum_{k=a}^{b} (-1)^k\ . \qquad 4.16$$

On the other hand for $b > n$ simple containement of the diagram of $\{k\ ,2n-k\}$ in $\{a\ ,b\ ,c\}$ yields that

$$2n-k \le 3n-a-b$$

Thus we must have $k \ge a+b-n$ for the correspoonding term to be different from zero (which is stronger than $k \ge a$ ). Furthermore, from Pieri's rule we also get that we must have $2n-k \ge b$ . (For otherwise the difference of $\{k\ ,2n-k\}$ and $\{a\ ,b\ ,c\}$ has two squares in the same column.) Putting all this together gives that

$$S_n S_n(x^2)\ |_{s_{abc}} = \sum_{k=a+b-n}^{2n-b} (-1)^k\ . \qquad 4.17$$

Formulas 4.16 and 4.17 are easily shown to imply our assertion.

This completes the proof of theorem 4.1.

Or next task is to derive the following result

**Theorem 4.2**

*In the expansion of* $S_n(S_m)$ *there are no hook Schur functions other than the trivial one* $S_{nm}$ .

**Proof**

Our point of departure is the special case $S_J = S_m$ of formula 2.7. Namely

$$S_n[S_m] = \frac{1}{n}\sum_{p=1}^{n} S_m(x^P)\, S_{n-p}[S_m] \ . \qquad 4.18$$

We proceed by induction on $n$ (the case $n=0$ being trivial). We shall use the symbol

$$Q \ |_{hooks}$$

to denote the portion of the Schur function expansion of $Q$ corresponding to hook Schur functions. This given, We assume, by induction, that for all $p$ in the interval $[1\ ,n]$ we have

$$S_{n-p}[S_m]\ |_{hooks} = S_{(n-p)m} \qquad 4.19$$

It is well known and easily derived from the SD algorithm that the coefficient $g_{I\ ,J}^{K}$ in the expansion of the product

$$S_I \cdot S_J = \sum_K g_{I\ ,J}^{K}\, S_K$$

is zero unless the diagrams of both $I$ and $J$ separately fit inside that of $K$ . In particular this implies that for a product $S_I \cdot S_J$ to contain a hook Schur function in its expansion, it is necessary that both $I$ and $J$ be hooks. This observation (combined with 4.18) implies that

$$S_n[S_m]\ |_{hooks} = \frac{1}{n}\sum_{p=1}^{n} S_{n-p}[S_m]\ |_{hooks}\ S_m(x^P)\ |_{hooks}\ \Big|_{hooks} \ . \qquad 4.20$$

On the other hand from Chen's algorithm it is easily derived that

$$S_m(x^P)\ |_{hooks} = \sum_{i=1}^{p-1} (-1)^i\, S_{1^i\ ,\ mp-i} \ . \qquad 4.21$$

Substituting 4.19 and 4.21 into 4.20 we get

$$S_n[S_m]\ |_{hooks} = \frac{1}{n}\sum_{p=1}^{n}\sum_{i=1}^{p-1}(-1)^i\, S_{nm-pm}\cdot S_{1^i\ ,\ mp-i}\ |_{hooks} \qquad 4.22$$

Now, Pieri's rule yields that for $p<n$

$$S_{nm-pm} \cdot S_{1^i, mp-i} \mid_{hooks} = S_{1^{i+1}, nm-i-1} + S_{1^i, nm-i} .$$

Substituting this into 4.22 (after isolating the $p=n$ term) gives then

$$S_n[S_m] \mid_{hooks} = \frac{1}{n}\Big\{ \sum_{i=1}^{n-1} (-1)^i S_{1^i, mn-i} +$$

$$+ \sum_{p=1}^{n-1} \sum_{i=1}^{p-1} (-1)^i \{ S_{1^{i+1}, nm-i-1} + S_{1^i, nm-i} \} \Big\}$$

and with some simple arithmetic we can see that this identity telescopes down to

$$S_n[S_m] \mid_{hooks} = S_{nm}$$

as desired. This completes our proof.

## 5. Calculations with plethysm.

It is not often realized that plethysm is a powerful computational tool in the study of representations. Since we have all the ingredients at hand here, we take the opportunity to illustrate by an example the uses that plethysm may be put to.

The problem was posed to one of the authors by R. Stanley. To introduce it we need some notation. Let $x_1, x_2, \ldots, x_m$ be variables with the commutativity relations

$$x_i \cdot x_j = -x_j \cdot x_i \qquad (i \neq j) \qquad 5.1$$

For a given subset

$$S = \{ i_1 < i_2 < \cdots < i_k \} \subseteq [1,m]$$

let us set

$$x(S) = x_{i_1} \cdot x_{i_2} \cdots x_{i_k} .$$

Given an $m \times m$ matrix $A = \| a_{ij} \|$ let us set

$$y_j(x) = x_1 a_{1j} + x_2 a_{2j} + \cdots + x_m a_{mj} \qquad (j=1,2,\ldots,m)$$

The map

$$x(S) \to \Lambda^k A x(S) = y_{i_1}(x) \cdot y_{i_2}(x) \cdots y_{i_k}(x)$$

defines through 5.1 and linearity a transformation of the $\binom{m}{k}$-dimensional vector space that is the linear span of the monomials

$$\{ x(S) : S \text{ of cardinality } k \}$$

It is not difficult to see that if

$$\sigma \to A(\sigma)$$

is a representation of $\mathbf{S}_n$ (or any other group for that matter) then so is the map

$$\sigma \to \Lambda^k A(\sigma)$$

R. Stanley asked for a Schur function proof of the following result:

**Theorem 5.1**

*If* $\sigma \to B(\sigma)$ *is the irreducible representation of* $\mathbf{S}_n$ *corresponding to the partition* $\{1, n-1\}$ *then the map*

$$\sigma \to \Lambda^k B(\sigma)$$

*is the irreducible representation corresponding to the partition* $\{1^k, n-k\}$

In view of formula 1.12, theorem 5.1 can be simply expressed by the identities

$$F\ trace\ \Lambda^k B = S_{1^k, n-k}(x) \qquad (k = 1, 2, \ldots, n-1) \tag{5.2}$$

Now an easy calculation yields that for any $m \times m$ matrix $A$ we have

$$\sum_{k=0}^{m} t^k\ trace\ \Lambda^k A = \det \| I + tA \| \tag{5.3}$$

Incidentally this identity may be viewed as a *conjugate* version of the Master Theorem, and as such it may be given a very lucid combinatorial proof.

At any rate, we can see then that the identities in 5.2 (using 1.11) can be translated into the single generating function identity

$$\frac{1}{n!} \sum_{\sigma \in S_n} \det \| I + tB(\sigma) \| \psi_{I(\sigma)} = \sum_{k=0}^{n-1} t^k S_{1^k, n-k} \tag{5.4}$$

Let $P(\sigma) = \| p_{ij}(\sigma) \|$ be the standard permutation representation of $\mathbf{S}_n$, that is

$$p_{ij}(\sigma) = \begin{cases} 1 & \text{if } i = \sigma_j , \\ 0 & otherwise . \end{cases}$$

It is well known that $P(\sigma)$ breaks up into two irreducible components: the trivial representation and our representation $B(\sigma)$. The first part of this statement is an immediate consequence of the identity

$$\frac{1}{n!} \sum_{\sigma \in S_n} \nu_1^2(\sigma) = 2 ,$$

($\nu_1(\sigma) =$ the number of fixed points of $\sigma$ ). It is also immediate that one of the components is the trivial representation. However, it is not entirely obvious that the other component is $B(\sigma)$. Nevertheless, if that is the case we necessarily have that

$$\det \| I + t P(\sigma) \| = (1+t) \det \| I + t B(\sigma) \| . \tag{5.5}$$

Using plethysm we shall show that the following identity holds:

$$\frac{1}{n!} \sum_{\sigma \in S_n} \frac{\det \| I + t P(\sigma) \|}{(1+t)} \psi_{I(\sigma)}(x) = \sum_{k=0}^{n-1} t^k S_{1^k, n-k}(x) . \tag{5.6}$$

This done, by equating the coefficients of $t$ we will deduce that $B(\sigma)$ is indeed the other irreducible component of $P(\sigma)$ , in particular this will give us 5.5. But then, of course, we can substitute 5.5 in 5.6 and derive formula 5.4 which is the desired result.

For technical reasons it is easier to work with $t$ replaced by $-t$ in 5.6. This given, note that for any permutation $\sigma$ we have

$$\det \| I - t P(\sigma) \| = \prod_i (1 - t^i)^{\nu_i(\sigma)}$$

replacing $t$ by $-t$ and using this identity the left hand side of 5.6 can be written in the form

$$\frac{1}{n!(1-t)} \sum_{\sigma \in S_n} \prod_i ((1-t^i)\psi_i)^{\nu_i(\sigma)} . \tag{5.7}$$

Note that, in view of 1.20, we have

$$\psi_i[1-t] = 1 - t^i$$

and thus

$$\psi_i[(1-t)X] = (1-t^i)\psi_i(x) .$$

This simple fact reveals that the expression in 5.7 is none other than the plethysm

$$S_n[(1-t)X]$$

divided by $(1-t)$ . We can thus use the multiplication formula 1.8 and obtain that

$$LHS(-t) = \frac{1}{(1-t)} \sum_{I \vdash n} S_I[1-t] S_I[X] . \tag{5.8}$$

Now the addition formula gives that

$$S_I[1-t] = \sum_{J \subseteq I} S_{J/I}(1) S_J[-t] \tag{5.9}$$

but $S_{J/I}(1)$ vanishes unless $J/I$ is a single row, and

$$S_J[-t] = (-1)^{|J|} \Lambda_J(t) ,$$

which gives that the other factor vanishes unless the diagram of $J$ consists of a single column.

Putting all this together we derive that the only terms surviving in 5.8 are those for which $I$ is a *hook*. But for $I = \{1^k, n-k\}$ formula 5.9 reduces to

$$S_I(1-t) = (-t)^k + (-t)^{k+1} .$$

So the final conclusion is that

$$LHS(-t) = \frac{1}{(1-t)} \sum_{k=0}^{n-1} ((-t)^k + (-t)^{k+1}) S_{1^k, n-k}(x)$$

which we can see is another way of writing the right hand side of 5.6 for $t$ replaced by $-t$. This completes our proof of theorem 5.1

## Illustrations

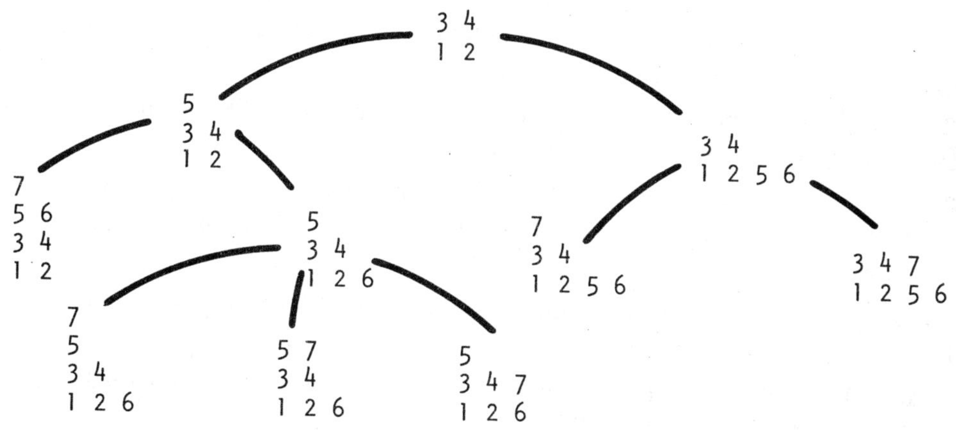


FIG 1

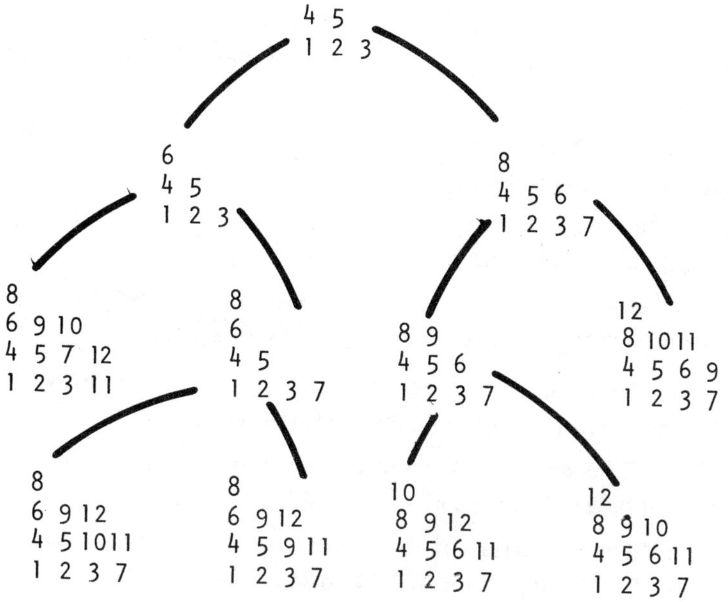


FIG 2

## Bibliography


[1] Butler, P.H., and King, R.C., "Branching rules for $U(N) \subseteq U(M)$ and the evaluation of outer plethysms", J. Math. Phys. 14(1973), 741-745; ZB 262, #20048.

[2] Chen, Y. M. "Combinatorial Algorithms for Plethysm", Doctoral thesis, UCSD(1982).

[3] Duncan, D.G., "Note on a formula by Todd", J. London Math. Soc. 27(1952a), 235-236; MR 13, p. 910.

[4] Duncan, D.G., "On D.E. Littlewood's algebra of S-functions", Can. J. Math. 4 (1952b),504-512; ZB 48,p. 11; MR 14, p.443.

[5] Duncan, D.G., "Note on the algebra of S-functions", Can. J. Math. 6 (1954),509-510; ZB 56, p. 17; MR 16, p.328.

[6] Egecioglu, O., "Computation of outer products of Schur functions", Computer Physics Comm. 28 (1982),183-187.

[7] Egecioglu, O., and Remmel J., "Symmetric and Antisymmetric Plethysms", Atomic Data and Nuclear Data Tables" (To appear).

[8] Foulkes, H.O., "Concomitants of the quintic and sextic up to degree four in coefficients of the ground form", J. London Math. Soc. 25 (1950), 205-209; ZB 37, p. 149.

[9] Foulkes, H.O., "The new multiplication of S-functions", J. London Math. Soc. 26 (1951), 132-139; ZB 42, p. 251; MR 12, p. 666.

[10] Foulkes, H.O., "Plethysm of S-functions", Philos. Trans. Roy. Soc. London Ser. A 246 (1954), 555-591; MR 15, p. 926.

[11] James G.D., and Kerber A., The Representation Theory of the Symmetric Group, Encyc. of Math. and its Appl. Add. Wes. 1981.

[12] Macdonald, I.G., Symmetric Functions and Hall Polynomials, Oxford Math. Monographs 1979.

[13] Littlewood, D.E., The Theory of Group Characters and Matrix Representations of Groups, Oxford University Press, 1940; MR 2, p.3.

[14] Littlewood, D.E., "Invariant theory, tensors and group characters", Philos. Trans. Roy. Soc. London Ser. A 239(1944), 305-365; MR 6, p.41.

[15] Littlewood, D.E., "Modular representations of symmetric groups", Proc. Roy Soc. London Ser. A 209(1951), 333-353;ZB 44, p.257; MR 14, p.243.

[16] Littlewood, D.E., "The characters and representations of imprimitive groups", Proc. London Math. Soc. (3) 6 (1956), 251-266;MR 17, p.1182.

[17] Murnaghan, F.D., "The analysis of representations of the linear group", Anais Acad. Brasil. Ci. 23(1951), 1-19; ZB 43, p.260; MR 13, p.204.

[18] Murnaghan, F.D., "The characters of the symmetric group", Anais Acad. Brasil. Ci. 23(19510, 141-154; ZB 45, p.157; MR 14, p.843.

[19] Murnaghan, F.D., "A generalization of Hermite's law of reciprocity", Anais Acad. Brasil Ci. 23(1951), 347-368; ZB 44, p.258.

[20] Remmel, J. and Whitney, R., "Multiplying Schur Functions", (to appear in J. of Algorithms).

[21] Robinson, G. de B., "On the representations of the symmetric group III", Amer. J. Math. 70 (1948), 277-294; ZB 36, p.155; MR 10, p. 678.

[22] Robinson, G. de B., "Induced representations and invariants", Canad. J. Math. 2 (1950), 334-343; ZB 39, p.20; MR 12, p. 74.

[23] Sleator, D. and Tarjan, R., "Self adjusting binary trees", (to appear).

[24] Thrall, R.M., "On symmetrized Kronecker powers and the structure of the free Lie ring" Amer. J. Math. 64 (1942a), 371-388; ZB 61, p. 42.

[25] Thrall, R.M., and Robinson, G. de B., "Supplement to a paper by G. de B. Robinson", Amer. J. Math. 73 (1951), 721-724; ZB 43, p.260; MR 13, p.205.

[26] Todd, J.A., "A note on the algebra of S-functions", Proc. Cambridge Philos. Soc. 45 (1949), 328-334.

[27] Todd, J.A., "Note on a paper by Robinson", Canad. J. Math. 2(1950), 331-333; ZB 39, p. 20.

DEPARTMENT OF MATHEMATICS
UNIVERSITY OF CALIFORNIA SAN DIEGO
LA JOLLA CALIFORNIA 92093

Contemporary Mathematics
Volume 34, 1984

# Combinatorial correspondences for Young tableaux, balanced tableaux, and maximal chains in the weak Bruhat order of $S_n$

Paul Edelman
University of Pennsylvania

Curtis Greene*
Haverford College

1. INTRODUCTION. R. Stanley conjectured in [8] that the number of maximal chains in the (weak) Bruhat order of $S_n$ is equal to the number $f$ of Young tableaux of "staircase" shape $\lambda = \{n-1,n-2,\ldots,2,1\}$. (The weak Bruhat order on $S_n$ is defined by letting $\sigma \leq \theta$ iff $\theta = \sigma\tau_1\tau_2\cdots\tau_k$, where each $\tau_i$ is a transposition of adjacent increasing elements). Stanley subsequently proved his conjecture [9], by algebraic methods, but his proof does not give a direct combinatorial correspondence and leaves a number of interesting combinatorial questions unanswered. The purpose of this note is to describe three bijections, which prove Stanley's conjecture and contain a wealth of information relating tableaux to chains in the weak Bruhat order. In the course of this work we were led to introduce and study a new class of tableaux (equinumerous with standard Young tableaux), called balanced tableaux. These tableaux have a rich structure, which has only begun to be explored. In what follows, we give a sketch of the main ideas, with an indication of the major results obtained to date. Complete proofs and other details will be published elsewhere [1].

2. BALANCED TABLEAUX. Let $\lambda = \{\lambda_1 \geq \lambda_2 \geq \cdots \geq \lambda_m\}$ be a partition of n, with conjugate partition $\lambda^* = \{\lambda^*_1 \geq \cdots \geq \lambda^*_m\}$. A tableau of shape $\lambda$ is a doubly indexed array T of integers $t_{ij}$, $1 \leq i \leq m$, $1 \leq j \leq \lambda_i$. For each cell $(i,j)$ define the hook $H_{ij}$ to be the multiset $\{t_{ik}\}_{k \geq j} \cup \{t_{kj}\}_{k>i}$. A tableaux T is said to be balanced if (1) its entries are a permutation of $1,2,\ldots,n$ and (2) each $t_{ij}$ is the $r_{ij}$th largest element of its hook $H_{ij}$, where $r_{ij} = \lambda^*_j - i + 1$. For example,

---

1980 A.M.S. Subject Classification: 05A15, 05A17, 20C30

* Research supported in part by N.S.F. Grant No. MCS 79-03209

5 7 4 8 9
6 10
2 1
3

is a balanced tableau of shape $\lambda = \{5,2,2,1\}$.

Our first result is that maximal chains in the weak Bruhat order of $S_n$ correspond bijectively to balanced tableaux of shape $\{n-1,n-2,\ldots,1\}$. The proof is not difficult.

THEOREM 1. Let $\Gamma$ denote a maximal chain in the weak Bruhat order of $S_n$. Let $B_\Gamma$ denote the tableau of shape $\{n-1,n-2,\ldots,2,1\}$ defined by setting $b_{(n+1-i)j} = k$ if the kth step in $\Gamma$ transposes $i > j$. Then $B_\Gamma$ is a balanced tableau, and the correspondence $\Gamma \longrightarrow B_\Gamma$ between maximal chains and balanced tableaux is bijective.

EXAMPLE 1. If $\Gamma$ is the chain in $S_4$ whose successive elements are (1234), (1324), (3124), (3142), (3412), (4312), (4321), then the corresponding balanced tableau $B_\Gamma$ is

4 3 5
2 1
6

Let $\lambda$ be an arbitrary partition, and let $b_\lambda$ denote the number of balanced tableaux of shape $\lambda$. Then Stanley's conjecture is equivalent to proving that $b_\lambda = f_\lambda$ when $\lambda$ is of staircase type, that is $\lambda = \{n-1,n-2,\ldots,1\}$ for some n. Our principal result is the following:

THEOREM 2. $b_\lambda = f_\lambda$ for any partition $\lambda$.

The proof is nontrivial. One might hope to show, for example, that $b_\lambda$ obeys the well-known recursion

$$f_\lambda = \sum f_{\lambda^-} \tag{1}$$

summed over all partitions $\lambda^-$ obtained by removing a cell from the Ferrers diagram of $\lambda$. This recursion is elementary, and underlies many basic properties of the $f_\lambda$'s. We can show that formula (1) holds for $b_\lambda$, but only as a consequence Theorem 2, not as a step in its proof.

While the original motivation for this work was to prove Stanley's conjecture by generalizing it to non-staircase shapes, the proof of Theorem 2 is ultimately based on an explicit bijection for staircase tableaux. The ideas rest heavily on an elegant theory of tableau transformations developed by M.P. Schützenberger ([5],[6],[7]), and on a new variant of the Robinson-

Schensted-Knuth insertion algorithm (see [4] for details). We describe these ideas briefly in the next section.

3. SCHÜTZENBERGER OPERATORS. For purposes of this discussion, define a <u>standard tableau</u> to be a tableau to be a tableau whose entries form a permutation of k+1,k+2,...,k+n, with rows and columns strictly increasing. When k=0, we call T a <u>standard Young tableau</u>. If T is standard, define a transformation $T \longrightarrow T^{\partial}$ (called "promotion*") as follows: if the largest entry in T occurs in position (i,j), i.e. $t_{ij} = k+n$, define

$$t_{ij}^{\partial} = \max\{t_{(i-1)j}, t_{i(j-1)}\} \tag{2}$$

with the convention that $t_{i0} = t_{0j} = 0$ for all $i,j > 0$. If $t_{(i-1)j} \geq t_{i(j-1)}$ replace i by i-1 (otherwise replace j by j-1), and iterate (2) in this manner until i = j = 1. Finally define $t_{11}^{\partial} = k$. This process results in a new tableau $T^{\partial}$ with entries k,k+1,...,k+n-1 whose rows and columns strictly increase. Let $T^{[i]}$ denote the result of applying $\partial^i$ to T. Thus $T^{[1]} = T^{\partial}$ and $T^{[0]} = T$. Now define operators P and S on standard tableaux T as follows*:

(a) $T^P = T^{[n]} + n$. In other words, $T^P$ is the result of promoting T n times, then adding n to each entry.

(b) $T^S$ is the tableau of shape $\lambda$ obtained by letting $t_{ij}^S = q$ if $t_{ij}^{[q]} \leq k$ but $t_{ij}^{[q-1]} > k$. In other words, $T^S$ records the times at which cells in T receive "new" labels, as $\partial$ is iterated.

EXAMPLE 2. As an illustration of the operators $\partial$, P, and S, consider the standard Young tableau

$$T = \begin{matrix} 1 & 3 & 4 \\ 2 & 6 & \\ 5 & & \end{matrix}$$

Then

$$T^{\partial} = \begin{matrix} 0 & 1 & 4 \\ 2 & 3 & \\ 5 & & \end{matrix} \qquad T^P = \begin{matrix} 1 & 2 & 5 \\ 3 & 6 & \\ 4 & & \end{matrix} \qquad T^S = \begin{matrix} 1 & 3 & 6 \\ 2 & 4 & \\ 5 & & \end{matrix}$$

A remarkable result of Schützenberger states that the map $T \longrightarrow T^S$ is an involution on standard tableaux: indeed an analogous result can be shown to hold more generally, when T is replaced by an arbitrary partially ordered set with a monotone labelling [7]. The situation with P is is more complicated.

---

*Our notation differs slightly from the definitions of $\partial$ found in [5], [6], and [7], and from that of S found in [4].

When T has rectangular shape one can deduce from results in [6] and [7] that $T^P = T$, although the derivation is not obvious. The following theorem, proved in [3], is somewhat more difficult. It underlies several of our major results, although it is not used explicitly in the proofs.

THEOREM 3. <u>If T is a standard tableau of staircase shape, then $T^P$ is the transpose of T. Thus the map $T \longrightarrow T^P$ is an involution on standard tableaux of staircase shape.</u>

4. THE STAIRCASE BIJECTION. Our second explicit bijection is obtained by studying in detail how the promotion operator $\partial$ acts on staircase tableaux. Let T be a standard Young tableau of shape $\lambda = \{n-1,n-2,\ldots,2,1\}$, with entries $1,2,\ldots,\binom{n}{2}$. Let $\Omega = \{X_1,X_2,\ldots,X_{n-1}\}$ be an alphabet of n-1 letters. Associate with T a word

$$W_T = W_1W_2\ldots W_{\binom{n}{2}}$$

with letters in $\Omega$ as follows: for $i = 0,1,\ldots,\binom{n}{2}-1$ define $W_{i+1} = X_j$ if the largest entry in $T^{[i]}$ occurs in cell (n-j,j). Thus the letters $X_j$ record the columns from which the largest entries in T "exit" as $\partial$ is iterated. For example, if T is defined as in Example 2, then $W_T = X_2X_1X_3X_2X_1X_3$.

THEOREM 4. <u>Let T be a standard Young tableau of (staircase) shape $\lambda = \{n-1,n-2,\ldots,2,1\}$, and let $W_T$ be defined as above. Then $W_T$ represents a maximal chain in the weak Bruhat order of $S_n$, if $X_i$ is identified with the transposition (i,i+1). Moreover the map $T \longrightarrow W_T$ defines a bijection between standard Young tableaux (of shape $\lambda$ ) and maximal chains in $S_n$</u> .

For example, if T is the tableau defined in Example 2, then $W_T$ represents the chain $\Gamma$ defined in Example 1.

It is not obvious that the correspondence defined by Theorem 4 is one-to-one, onto, or even well-defined for all staircase tableaux. To show the latter, one must prove that the adjacent elements transposed are always increasing. The proof of this fact relies on "Jeu-de-Taquin" arguments (see [6]), and leads to many other results concerning actions of tableaux and skew tableaux on permutations.

Combining Theorems 4 and 1, we obtain a bijection

$$T \longrightarrow W_T \longrightarrow B_{W_T} \tag{3}$$

from standard Young tableaux to balanced tableaux of the same shape. This bijection is valid for all staircase shapes, and provides a proof of Theorem 2 in this special case.

5. THE INVERSE MAPPING. The inverse to (3) is based on a variant of the Robinson-Schensted-Knuth insertion algorithm, and has a surprisingly simple form. Let $X_i$ denote the transposition $(i,i+1)$, and let

$$C = C_1C_2\dots C_{\binom{n}{2}}$$

be a word in letters $X_1,X_2,\dots,X_{n-1}$ which represents a maximal chain in the weak Bruhat order of $S_n$. We regard the alphabet $X_1,X_2,\dots,X_{n-1}$ as linearly ordered in the obvious way, and define a tableau $K_C$ of shape $\{n-1,n-2,\dots,2,1\}$ by successively inserting the letters $C_1,C_2,C_3,\dots$ according to the Robinson-Schensted-Knuth scheme (see [4] for definitions), with one exception: if the letter $X_i$ bumps $X_{i+1}$ from a row in which $X_i$ is already present, the result is $\dots X_iX_{i+1}\dots$ rather than $\dots X_iX_i\dots$ . This variation on the rules is derived from the standard algorithm by replacing certain of the so-called "elementary Knuth equivalence relations" (see [5]) by the "Coxeter relations $X_{i+1}X_iX_{i+1} = X_iX_{i+1}X_i$ which hold for adjacent transpositions in $S_n$.

As in the Robinson-Schensted-Knuth case, one can define a second tableau $L_C$, which records the order in which new cells appear in $K_C$. The construction of $K_C$ and $L_C$ is illustrated by the next example.

EXAMPLE 3. If C is the word $X_2X_1X_3X_2X_1X_3$ then

$$K_C = \begin{matrix} 1 & 2 & 3 \\ 2 & 3 & \\ 3 & & \end{matrix} \qquad\qquad L_C = \begin{matrix} 1 & 3 & 6 \\ 2 & 4 & \\ 5 & & \end{matrix}$$

THEOREM 5. <u>If C represents a maximal chain in $S_n$, then $K_C$ is row and column strict of (staircase) shape $\{n-1,n-2,\dots,1\}$, and always has the form</u>

$$\begin{matrix} 1 & 2 & 3 & \dots & n \\ 2 & 3 & \dots & n & \\ \cdot & & & & \\ \cdot & & & & \\ n-1 & n & & & \\ n & & & & \end{matrix}$$

THEOREM 6. <u>The map $C \longrightarrow L_C$ is a bijection from chains in the weak Bruhat order of $S_n$ to standard Young tableaux of staircase shape.</u>

Although Theorem 6 defines a bijection, the actual inverse of (3) is obtained by applying the operator S to $L_C$. For example, the reader can check that the tableaux appearing in Examples 2 and 3 satisfy the relations

$$W_T = C \qquad\qquad L_C^S = T$$

Thus the maps

$$\begin{aligned} T &\longrightarrow W_T \longrightarrow B_{W_T} \\ B &\longrightarrow C_B \longrightarrow L^S_{C_B} \end{aligned} \tag{4}$$

are inverses. Here, $C_B$ denotes the maximal chain (or word) associated with balanced staircase tableau B under the bijection given by Theorem 1.

6. BALANCED TABLEAUX OF ARBITRARY SHAPE. Finally, we sketch the proof of Theorem 2 for tableaux of arbitrary shape. A bijection from standard tableaux to balanced tableaux is obtained as follows: given a standard tableau T of arbitrary shape, first imbed T in a staircase tableau $\overline{T}$ in a canonical fashion (to be described later). Then apply the staircase bijection (4) to $\overline{T}$, obtaining a balanced staircase tableau $\overline{B}$. Finally, delete the extra cells from $\overline{B}$, obtaining a balanced tableau B of the original shape. The details are best explained by example.

EXAMPLE 4. Consider the Young tableau

$$T = \begin{matrix} 1 & 3 & 5 \\ 2 & & \\ 4 & & \\ 6 & & \end{matrix}$$

of shape $\{2,1,1,1\}$. First imbed T in a staircase tableau $\overline{T}$ of shape $\{4,3,2,1\}$ by adding the missing entries 7,8,9,10 from left to right in the first row, then the second row, then the third, and so forth. Thus

$$\overline{T} = \begin{matrix} 1 & 3 & 5 & 7 \\ 2 & 8 & 9 & \\ 4 & 10 & & \\ 6 & & & \end{matrix}$$

Next apply the staircase bijection, which associates with $\overline{T}$ the chain $W_{\overline{T}} = X_2X_3X_2X_4X_1X_3X_2X_3X_1X_4$, and the balanced tableau

$$\overline{B} = B_{W_{\overline{T}}} = \begin{matrix} 4 & 7 & 5 & 2 \\ 6 & 9 & 8 & \\ 3 & 10 & & \\ 1 & & & \end{matrix}$$

Now delete the entries 10,9,8,7 in this order, each time interchanging the columns directly above and to the right of the element deleted. It is easily seen that deleting elements in this way preserves the property of being balanced. The resulting tableau is

| | | | |
|---|---|---|---|
| | 4 | 2 | 5 |
| B = | 6 | | |
| | 3 | | |
| | 1 | | |

which (by definition) is the balanced tableau associated with T.

THEOREM 7. The correspondence described above is a bijection from standard Young tableaux to balanced tableaux of the same shape.

7. CONCLUDING REMARKS. We have not succeeded in finding a proof of Theorem 2 which is independent of (4), nor have we found any natural algebraic or geometric interpretation of balanced tableaux when the shape is not a staircase.

The bijections defined by Theorems 1, 4, and 6 contain enough information to count the number $c(\sigma)$ of maximal chains in $S_n$ with top element equal to a fixed permutation $\sigma$. Stanley conjectured that $c(\sigma) = \sum f_\lambda$, where the sum is over a certain multiset $M(\sigma)$ of shapes. His proof [9] gives this sum but does not preclude the possibility of negative coefficients. Our correspondence describes the multiset $M(\sigma)$ in a natural way, and shows that only positive coefficients occur. For further details, see [1].

## BIBLIOGRAPHY

1. P. Edelman, C. Greene, "Balanced tableaux", in preparation.

2. J.S. Frame, G. de B. Robinson and R.M. Thrall, "The hook graphs of the symmetric group," Canad. J. Math. 6 (1954), 316-324.

3. C. Greene, "Some remarks on Schützenberger's tableau operators," in preparation.

4. D.E. Knuth, The Art of Computer Programming, vol. 3, Addison-Wesley, 1973.

5. M.P. Schützenberger, "La correspondance de Robinson," in Combinatoire et Representation du Groupe Symetrique, D. Foata, ed., Springer Lect. Notes in Math. 579 (1977), 59-113.

6. M.P. Schützenberger, "Quelques remarques sur une construction de Schensted", Math. Scand. 13 (1963), 117-128.

7. M.P. Schützenberger, "Promotion des morphisms d'ensembles ordonnes", Discrete Math. 2 (1972), 73-94.

8. R.P. Stanley, "A Combinatorial conjecture concerning the symmetric group," preprint 1982

9. R.P. Stanley, "On the number of reduced decompositions of elements of Coxeter groups", European Jour. of Combinatorics, to appear.

DEPARTMENT OF MATHEMATICS
UNIVERSITY OF PENNSYLVANIA
PHILADELPHIA, PA 19104

DEPARTMENT OF MATHEMATICS
HAVERFORD COLLEGE
HAVERFORD, PA 19041

Contemporary Mathematics
Volume 34, 1984

# CONSTRUCTIONS ON RIM HOOK TABLEAUX

Dennis E. White[1]

1. INTRODUCTION. In [8] this author described a Schensted correspondence on rim hook tableaux. By interpreting the irreducible characters of $S_n$ as signed sets of rim hook tableaux, a bijection proving orthogonality was given.

If all the hooks and rim hooks were the same size, the correspondence simplified to an encoding of a wreath product group. Many of the properties of the ordinary Schensted correspondence had generalizations. These properties are described in detail by the author and D. Stanton in [7].

In this paper we will highlight the key results in these two papers.

2. BACKGROUND. A partition $\lambda$ of $n$ (written $\lambda \vdash n$ ) is a sequence of integers $\lambda = (\lambda_1, \lambda_2, \cdots, \lambda_\ell)$ with $\lambda_1 \geq \lambda_2 \geq \cdots \geq \lambda_\ell$ and $\lambda_1 + \lambda_2 + \cdots + \lambda_\ell = n$ . We say $\lambda$ has $\ell$ parts. We will frequently refer to a partition as a shape, particularly when we interpret it as successive rows of positions or cells.

We sometimes abbreviate the partition $\lambda$ with the notation $1^{j_1}2^{j_2}\ldots$ where $j_i$ is the number of parts of size $i$ . Sizes which do not appear are omitted and if $j_i = 1$, it is not written. Thus (5,2,2,2,1) can be written $12^35$ .

Suppose $\mu$ and $\lambda$ are partitions. If the shape of $\lambda$ is contained in the shape of $\mu$ , we denote by $\mu/\lambda$ the skew shape obtained by removing $\lambda$ from $\mu$ .

A composition $\rho$ of $n$ is a sequence of integers $(\rho_1, \cdots \rho_\ell)$, $\rho_i \geq 0$ and $\rho_1 + \cdots + \rho_\ell = n$ . Note that corresponding to every composition there is a unique partition, obtained by reordering the parts of $\rho$ .

Suppose $\lambda \vdash n$ and $\rho$ is a composition of $n$ . A tableau of shape $\lambda$ and content $\rho$ is the shape $\lambda$ with the cells filled with $\rho_1$ 1's, $\rho_2$ 2's, $\cdots$, $\rho_\ell$ $\ell$'s . We define skew tableaux similarly.

A k-hook (or hook) is a partition $1^ij$ where $i + j = k$ . If a hook has a single value in all of its cells, it is called a hook tableaux.

A rim hook (also called border strip) of $\lambda$ is a contiguous strip of

1980 Mathematical Subject Classification. 05A15, 20C30.
[1]Supported in part by National Science Foundation.

cells of width one on the border or outer rim of $\lambda$. A <u>rim hook tableau</u> of shape $\lambda$ and content $\rho$ may be defined recursively:

(a) hook tableaux are rim hook tableaux;

(b) P is a rim hook tableau with largest entry $\ell$ if all the cells in P containing $\ell$'s make up a rim hook and the removal of these cells from P leaves a rim hook tableau.

We may similarly define <u>skew rim hook tableaux</u>.

Any rim hook may be given the sign $(-1)^{\#\text{rows in rim hook} - 1}$. The sign of a (skew) rim hook tableau is the product of the signs of its rim hooks. Thus we have

$$P = \begin{array}{llll} 1 & 1 & 1 & 3 \\ 1 & 2 & 3 & 3 \\ 2 & 2 & 4 & \end{array} :$$

sign of rim hook of 1's is $-1$ ;
sign of rim hook of 2's is $-1$ ;
sign of rim hook of 3's is $-1$ ;
sign of rim hook of 4's is $+1$ ;
sign of P is $-1$ .

If all rim hooks in a rim hook tableau are the same size, say $k$, we say the tableau is a <u>k-rim hook tableau</u>.

Rim hook tableaux and their accompanying signs are significant in the representation theory of the symmetric group. The Murnaghan-Nakayama rule (see [2]) states that the irreducible character $\lambda$ of $S_n$ at class $\rho$ is given by summing the signs of all rim hook tableaux of shape $\lambda$ and content $\rho$. Furthermore, this sum is independent of the order of the parts of $\rho$. Thus, $\chi^{(234)}_{(12^24)} = 0$ because:

$$\begin{array}{l} 4\ 1\text{'s} \\ 2\ 2\text{'s} \\ 2\ 3\text{'s} \\ 1\ 4 \end{array} \left\{ \begin{array}{lllll} +: & \begin{array}{l}1111\\224\\33\end{array} & \begin{array}{l}1111\\234\\23\end{array} & \begin{array}{l}1122\\133\\14\end{array} & \begin{array}{l}1134\\123\\12\end{array} \\ -: & \begin{array}{l}1111\\233\\24\end{array} & \begin{array}{l}1114\\122\\33\end{array} & \begin{array}{l}1122\\134\\13\end{array} & \begin{array}{l}1133\\124\\12\end{array} \end{array} \right.$$

or because:

$$\begin{array}{l} 2\ 1\text{'s} \\ 1\ 2 \\ 2\ 3\text{'s} \\ 4\ 4\text{'s} \end{array} \left\{ \begin{array}{ll} + & \begin{array}{l}1233\\144\\44\end{array} \\ - & \begin{array}{l}1133\\244\\44\end{array} \end{array} \right.$$

or because there are no rim hook tableaux with shape (4,3,2) and 1 1, 2 2's, 2 3's and 4 4's.

One of the orthogonality formulas for the characters of $S_n$ is

$$\sum_{\lambda \vdash n} \chi^\lambda_\rho \chi^\lambda_\mu = \delta_{\rho\mu} \, 1^{j_1} j_1! \, 2^{j_2} j_2! \cdots \tag{1}$$

where $\rho = 1^{j_1}2^{j_2}\cdots \vdash n$ and $\mu \vdash n$.

In [8] we gave a combinatorial proof of (1) based on the Murnaghan-Nakayama rule. We defined a sign-reversing involution on the set of pairs of rim hook tableaux of the same shape. The sign of a pair is the product of the two signs. For example, the positive pair:

$$\begin{pmatrix} 1133 & & 1133 \\ -124 & , & -123 \\ 12 & & 22 \end{pmatrix}$$

is identified with the negative pair:

$$\begin{pmatrix} 11334 & & 11333 \\ -12 & , & +12 \\ 12 & & 22 \end{pmatrix} .$$

The special case of (1) where $\rho = \mu = 2^m$ says that pairs of domino tableaux ( 2-rim hook tableaux) of the same shape are equinumerous with the elements of the hyperoctahedral group: $2^m \cdot m!$ . Using the Schensted correspondence ([5], see also [3]), Lusztig actually gave an algorithm which produced a bijection in this case [4].

More generally, (1) says that pairs of k-rim hook tableaux of the same shape are equinumerous with elements of the wreath product of the symmetric group with the cyclic group, $S_m[Z_k]$: $k^m \cdot m!$ . The bijection which yields this identity is described in [7] and [8].

Using Robinson's "*-diagrams" (see [2]), it can be seen that k-rim hook tableaux are equinumerous with k-tuples of standard tableaux with different entries. In fact, shapes with empty k-core (see [2]) are equinumerous with k-tuples of partitions. Thus, the shape (7,6,5,4,3,2), which has empty 3-core, decomposes into ((3,2),(2,1),(1)). The 3-rim hook tableau:

```
1144499
125569
22566
3777
388
38
```

decomposes into $\begin{pmatrix} 269 & , & 14 & , & 5 \\ 37 & & 8 & & \end{pmatrix}$ . An alternative to the Robinson construction is given in [7].

In this paper we will discuss (1) the Schensted-like bijection between positive pairs of rim hook tableaux of the same shape and either negative pairs of rim hook tableaux of the same shape or permutations of hook tableaux, the number of which is $1^{j_1} j_i! 2^{j_2} j_2! \cdots$ ; and (2) the special case where all rim hooks have size $k$ . In this case, pairs of k-rim hook tableaux of the same shape correspond to k-hook permutations.

For proofs and further examples of these constructions, see [8].

We will also describe some of the properties of the Schensted correspondence (see [3], [5], [6]) which generalize to the rim hook version.

For further details, see [7].

3. K-RIM HOOK TABLEAUX. In this section we assume all hooks and rim hooks have size $k$. A <u>k-hook permutation</u> (in 2-line rotation) of length $m$ is an $m$-tuple of $k$-hooks together with a permutation (in 2-line notation - see [3]) of length $m$. Alternatively, we may insert the numbers of the permutation into the $k$-hooks to form hook tableaux. The $k$-hooks in each column will be the same. For example, $(13,13,1^4,4,1^2 2)$ and $\begin{pmatrix}13457\\31648\end{pmatrix}$ may be written as

$$\left(\begin{array}{ccccc}
\begin{array}{l}111\\1\end{array} & \begin{array}{l}333\\3\end{array} & \begin{array}{l}4\\4\\4\\4\end{array} & 5555 & \begin{array}{l}77\\7\\7\end{array}\\
\begin{array}{l}333\\3\end{array} & \begin{array}{l}111\\1\end{array} & \begin{array}{l}6\\6\\6\\6\end{array} & 4444 & \begin{array}{l}88\\8\\8\end{array}
\end{array}\right)$$

The general idea of the insertion algorithm is that a new $k$-rim hook tableau is constructed by inserting a $k$-hook into an old $k$-rim hook tableau. The new tableau will differ in shape from the old by a $k$-rim hook. Thus, a $k$-hook permutation may be encoded as a pair $(P,Q)$ of $k$-rim hook tableaux of the same shape. The content of $P$ will be the content of the second line of the permutation, while the content of $Q$ will be the content of the first line.

The insertion is accomplished by successively transferring rim hooks of larger content from the old tableau (called the <u>outside part</u>) to the new (called the <u>inside part</u>). The position of a rim hook in the inside part is determined by certain rules based upon how much the rim hook being transferred overlaps the inside part.

We describe the insertion process with an example. Suppose the hook $\begin{array}{l}33\\3\end{array}$ enters

$$\begin{array}{l}112677\\122678\\444688\\55999\\5\end{array}\ .$$

We begin by placing everything smaller than 3 into the inside part: $\begin{array}{l}112\\122\end{array}$. Next, we "position" the hook $\begin{array}{l}33\\3\end{array}$. This is done by sliding the hook down the first column three steps at a time until it no longer overlaps the inside part, then sliding it up three steps at a time along the outside of the inside part until it reaches a "legal" position. Thus, the next inside part is

$$\begin{array}{l}112\\122\\3\\3\\3\end{array}\ .$$

(If 333 had been the hook instead, the inside part would have been

112 .)
122
33
3

Now we successively transfer rim hooks from the outside part to the inside part. Since the inside part and the outside part overlap, we may not be able to place each rim hook in its same location. For example, when we attempt to move the 4's from the outside part to the inside part, we find that one of the 4's overlaps one of the 3's . When such overlapping occurs, but the entire rim hook doesn't overlap, we move the offending entries one step diagonally downward:

112
122
344 .
34
3

The 5's overlap entirely. In this case, they slide up three steps at a time along the outside of the inside part, until they reach a "legal" position:

11255
1225
344 .
34
3

Next, two 6's move down diagonals:

11255
12256
34466 .
34
3

Then two 7's move down diagonals:

112557
122567
344667 .
34
3

The 8's slide up:

11255788
1225678
344667 .
34
3

The 9's do not overlap the inside part, so we move them unchanged:

11255788
1225678
344667
34999 .
3

The deletion procedure reverses these steps. In fact, in the deletion algorithm, the rim hooks move from the inside part to the outside part. The new (smaller) tableau is build up starting as a skew tableau.

4. PROPERTIES. Many of the properties of the Schensted correspondence carry over to the rim hook Schensted correspondence. To begin with, we have an "evacuation" procedure or "jeu de taquin" (see [6]). In fact, this evacuation can be described for rim hooks of differing sizes. If the sizes differ, sign and direction reversals can occur. Using the Garsia-Milne involution principle [1], we obtain a direct combinatorial proof that the Murnaghan-Nakayama rule for calculating $\chi_\rho^\lambda$ does not depend upon the order of the parts of $\rho$. See [7] for details.

The idea behind the evacuation procedure is a switching mechanism which moves a rim hook of "holes" from within a tableau to its rim. For example, suppose * represents a hole and r is the content of the rim hook which must next move into the holes. Inside the tableau, we could have:

```
   *
 ***
 *rr
rrr
```

which becomes

```
   *
 rr*
 r**  .
rr*
```

Note that the * and r switch along diagonals, i.e., the r's fill holes by moving up diagonally.

Or, we might have:

```
 ***
 *
rr
r
r
```

which becomes

```
 rrr
 r
**
*    .
*
```

That is, if the holes and the r's are head-to-tail, they swap positions.

The properties of the Schensted correspondence and the evacuation procedure are proved using the idea of <u>orientation</u>. Each rim hook in a k-rim hook tableau is assigned an orientation: the <u>diagonal</u> <u>number</u> of its upperright cell (mod k). The diagonals are numbered with the main diagonal 0 and then 1,2,··· to the right and -1,-2,··· down. Thus, in

```
112
122
334 ,
344
```

for example, 4 has orientation 0 , 1 has orientation 1, and 2 and 3 have orientation 2.

The orientation of hooks and rim hooks is "preserved" by the Schensted correspondence and the evacuation procedure. Thus, evacuating all rim hooks except those of a given orientation amounts to erasing all rim hooks of other orientations and "pushing" all remaining values up along diagonals. The resulting tableau has all entries of one orientation. It is then easy to see that there is a natural bijection between k-rim hook tableaux all of whose hooks are of the same orientation and standard tableaux.

Thus, any k-rim hook tableau can be decomposed into a k-tuple of standard tableaux. This decomposition is exactly Robinson's *-diagram decomposition (see [7]). We call the $i^{th}$-component in the decomposition of P its <u>$i^{th}$-projection</u>, $\pi_{(i)}(P)$.

As an example, if

$$P = \begin{array}{l} 11337888 \\ 12347 \\ 22447 \\ 5559 \\ 6699 \\ 6 \end{array}$$

then $\pi_{(0)}(P) = \begin{array}{l} 23 \\ 69 \end{array}$

We can also decompose a k-hook permutation into a k-tuple of permutations. This decomposition we call $\Lambda_{(i)}$. If

$$\mathcal{H} = \left( \begin{array}{cccccc} 1 & & & 6 & & 9 \\ 1 & 44 & 55 & 6 & 777 & 9 \\ 1 & 4 & 5 & 6 & & 9 \\ \\ 8 & 22 & 66 & 3 & 777 & 4 \\ 8 & 2 & 6 & 3 & & 4 \\ 8 & & & 3 & & 4 \end{array} \right)$$

then $\Lambda_{(0)}(\mathcal{H}) = \begin{pmatrix} 1 & 6 & 9 \\ 8 & 3 & 4 \end{pmatrix}$

The central theorem which allows us to extend results about the Schensted and Schützenberger constructions to rim hook versions is this:

THEOREM. The following diagram commutes:

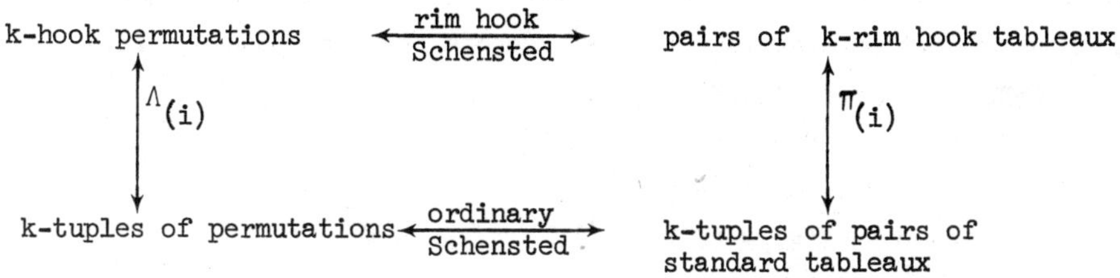


We give an example of the kinds of results which follow from this theorem. Suppose $\mathcal{H}$ is a k-rim hook tableau. E.g.,

$$\mathcal{H} = \begin{pmatrix} \begin{array}{l}111\\1\end{array} & \begin{array}{l}44\\4\\4\end{array} & 5555 & \begin{array}{l}66\\6\\6\end{array} \\ \\ \begin{array}{l}333\\3\end{array} & \begin{array}{l}11\\1\\1\end{array} & 8888 & \begin{array}{l}44\\4\\4\end{array} \end{pmatrix},$$

then Inv($\mathcal{H}$) is obtained by switching top and bottom rows and reordering:

$$\mathrm{Inv}(\mathcal{H}) = \begin{pmatrix} \begin{array}{l}11\\1\\1\end{array} & \begin{array}{l}333\\3\end{array} & \begin{array}{l}44\\4\\4\end{array} & 8888 \\ \\ \begin{array}{l}44\\4\\4\end{array} & \begin{array}{l}111\\1\end{array} & \begin{array}{l}66\\6\\6\end{array} & 5555 \end{pmatrix};$$

Rev($\mathcal{H}$) is obtained by subtracting the top line entries from $n + 1$, reordering and transposing: $(n = 9)$

$$\mathrm{Rev}(\mathcal{H}) = \begin{pmatrix} \begin{array}{l}444\\4\end{array} & \begin{array}{l}5\\5\\5\\5\end{array} & \begin{array}{l}666\\6\end{array} & \begin{array}{l}99\\9\\9\end{array} \\ \\ \begin{array}{l}444\\4\end{array} & \begin{array}{l}8\\8\\8\\8\end{array} & \begin{array}{l}111\\1\end{array} & \begin{array}{l}33\\3\\3\end{array} \end{pmatrix};$$

and Dual($\mathcal{H}$) is obtained by subtracting the bottom line from $n + 1$ and transposing:

$$\mathrm{Dual}(\mathcal{H}) = \begin{pmatrix} \begin{array}{l}11\\1\\1\end{array} & \begin{array}{l}444\\4\end{array} & \begin{array}{l}5\\5\\5\\5\end{array} & \begin{array}{l}666\\6\end{array} \\ \\ \begin{array}{l}77\\7\\7\end{array} & \begin{array}{l}999\\9\end{array} & \begin{array}{l}2\\2\\2\\2\end{array} & \begin{array}{l}666\\6\end{array} \end{pmatrix}.$$

Next, suppose $P$ is a k-rim hook tableau. In a manner exactly analogous to [6] (see also [3]) we define the evacuation tableau $\Sigma(P)$ by using the rim hook evacuation procedure.

Let Sch($\mathcal{H}$) denote the pair of k-rim hook tableaux constructed as described in Section 3 above. Let $P^T$ denote the transpose of $P$.

THEOREM. Suppose $\mathrm{Sch}(\mathcal{H}) = (P,Q)$. Then

$$\mathrm{Sch} \circ \mathrm{Inv}(\mathcal{H}) = (Q,P),$$
$$\mathrm{Sch} \circ \mathrm{Rev}(\mathcal{H}) = (P^T,\Sigma(Q)^T),$$
$$\mathrm{Sch} \circ \mathrm{Dual}(\mathcal{H}) = (\Sigma(P^T),Q^T),$$

(cf. [3]).

Other generalizations include fill-in procedures [6], the "Knuth relations" [3], and increasing and decreasing subsequences [5].

5. CANCELLATION. To produce a combinatorial proof of (1), we must answer these two questions:

A. What is being counted by $1^{j_1} j_1! 2^{j_2} j_2! \cdots k^{j_k} j_k!$ and how is it encoded as a pair of tableaux?

B. When decoding a pair of tableaux, how does cancellation occur?

The answer to A is not hard. Suppose $\rho = 1^{j_1} 2^{j_2} \cdots k^{j_k}$. Define a general hook permutation of type $\rho$ where the $j_k$ k-hooks appear first, followed by the $j_{k-1}$ (k-1)-hooks, etc. The entries in the k-hooks are the numbers $\{1, \cdots, j_k\}$, the entries in the (k-1)-hooks are the numbers $\{j_k+1, \cdots, j_k+j_{k-1}\}$, etc.

For example, such a permutation of <u>type</u> $2^3 3^3 4^2 5$ is:

$$\begin{pmatrix} 111 & 333 & 2222 & 44 & 66 & 555 & 88 & 9 & 7 \\ 1 & 3 & & 4 & 6 & & & 9 & 7 \\ 1 & & & & & & & & \end{pmatrix}.$$

(We are omitting the top line in the 2-line notation.)

For a given $\rho$, the number of such hook permutations of type $\rho$ is $1^{j_1} j_1! 2^{j_2} j_2! \cdots k^{j_k} j_k!$.

Now the insertion method of Section 3 can be used directly, since a hook of size $\ell$ will only "bump" hooks of the same size. The permutation above yields the tableau pair:

$$P = \begin{matrix} 1112333 \\ 12223 \\ 1444 \\ -5559 \\ 6669 \\ 77 \\ 88 \end{matrix} \qquad Q = \begin{matrix} 1112222 \\ 13333 \\ 1444 \\ -5569 \\ 5669 \\ 78 \\ 78 \end{matrix}$$

The answer to B is more difficult. To accomplish the necessary cancellation, the deletion algorithm must somehow "turn around" in the middle. When all rim hooks are of the same size, the deletion algorithm pushes a rim hook from the outside of the tableau to the inside. The new (smaller) tableau is built from the outside in, and at each stage the next smaller entry passes from the inside part to the outside part. To achieve cancellation, at some stage this process must reverse. I.e., instead of rim hooks passing from the inside part to the outside part, the reverse happens and the algorithm subsequently behaves like the <u>insertion</u> algorithm. When this direction reversal occurs, some kind of sign reversal also occurs.

In introducing rim hooks of varying size, we have added only one new feature to the insertion/deletion procedure described earlier. The portion of the inside part which overlaps with the outside part may be <u>strictly contained</u> in the rim hook which is next to pass from inside to outside (or vice versa). It therefore must be in these situations where we will find the direction/sign reversals.

Details of the exact nature of the sign and direction reversals may be found in [8]. We give here an example. Suppose

$$P = +\begin{array}{l}112222\\133455\\13445\\1\end{array} \qquad Q = -\begin{array}{l}111135\\123335\\22444\\2\end{array} .$$

We give, at each stage, the modifications being made to P: the inside part (with its sign and with the overlap with the outside part boxed); the outside part (with its sign and with the overlap with the inside part boxed); the sign of the overlapping rim hook; and the direction ("in" means rim hooks are moving from the inside part to the outside part; "out" means the opposite). Dots indicate no entry.

| Inside part | Outside part | sign of overlap | direction |
|---|---|---|---|
| 11222 $\boxed{2}$<br>+ 13345 $\boxed{5}$<br>13445<br>1 | + $\boxed{\begin{array}{c}\cdot\\\cdot\end{array}}$ | − | in |
| 1122 $\boxed{22}$<br>− 1334<br>1344<br>1 | $\boxed{5\ \cdot}$<br>+ 5<br>5 | + | in |
| 1122 $\boxed{22}$<br>+ 133<br>13<br>1 | $\boxed{5\ \cdot}$<br>− 45<br>445 | + | in |
| 1122 $\boxed{22}$<br>− 1<br>1<br>1 | $\boxed{5\ \cdot}$<br>+ 3345<br>3445 | + | in |
| 1 122<br>+ 1 $\boxed{22}$<br>1<br>1 | 5<br>+ $\boxed{33}$ 45<br>34 45 | + | out |
| 1122<br>+ 122<br>$\boxed{1}$<br>$\boxed{1}$ | 5<br>+ 45<br>$\boxed{3}$ 3445<br>$\boxed{3}$ | − | in |
| 11<br>− 1<br>$\boxed{1}$<br>$\boxed{1}$ | 225<br>− 2245<br>$\boxed{3}$ 3445<br>$\boxed{3}$ | − | in |
| 11 $\boxed{11}$<br>− 1 | $\boxed{22}$ 5<br>− 224 5<br>3344 5<br>3 | + | out |
| 1111<br>+ 122<br>$\boxed{22}$ | 5<br>+ 45<br>$\boxed{33}$ 445<br>3 | + | out |

```
Inside part    Outside part    sign of overlap    direction

  1111               5                               out
+ 122          -    45              —
  22              ..445
  3                .
 [3]              [.]
 [3]              [.]


  1111               5                               out
- 1224         +     5              —
  2244            ....5
  3                .
 [3]              [.]
 [3]              [.]


  11115                                              out
- 12245        +                    —
  22445
  3
 [3]              [.]
 [3]              [.]
```

Thus, the cancelling pair is

```
       11115           11113
       12245           12333
P =   -22445     Q =  -22444   .
       3               2
       3               5
       3               5
```

By arranging the content so that largest rim hooks have smallest entries and by choosing Q to be the rim hook tableau whose content is lexicographically first when read smallest part first (e.g. $12^2 4$ precedes $123^2$), we guarantee that either (a) a cancellation occurs or (b) a deletion occurs. In the former case, we are done; in the latter, we proceed recursively. We will finally end up with either a cancellation or a hook permutation, as discussed earlier.

For example, suppose

```
       11144           11156
P =   +12566     Q =  +13456   .
       225             234
       33              23
       3               2
```

Applying the deletion procedure produces

```
       1116            1115
P =   -1256      Q =  -1345
       225             234
       33              23
       3               2
```

with the hook $\begin{smallmatrix}4\\4\end{smallmatrix}$ removed.

Applying deletion again causes cancellation:

$$P = +\begin{array}{l}122\\126\\156\\15\\3\\3\\3\end{array} \qquad Q = -\begin{array}{l}111\\134\\234\\23\\2\\5\\5\end{array} .$$

Now insert $\begin{smallmatrix}4\\4\end{smallmatrix}$:

$$P = +\begin{array}{l}12244\\126\\156\\15\\3\\3\\3\end{array} \qquad Q = -\begin{array}{l}11166\\134\\234\\23\\2\\5\\5\end{array} ,$$

which cancels the original pair.

## BIBLIOGRAPHY


1. A. Garsia and S. Milne, "A Rogers-Ramanujan bijection", J. Combinatorial Theory Ser. A, 31 (1981), 289-339.

2. G. James and A. Kerber, The Representation Theory of the Symmetric Group, Addison-Wesley, Reading, Mass., 1981.

3. D. Knuth, The Art of Computer Programming, Vol. 3: Sorting and Searching, Addison-Wesley, Reading, Mass., 1973.

4. G. Lusztig, communicated by R. Stanley in seminar given at U.C.S.D., 1982.

5. C. Schensted, "Longest increasing and decreasing subsequences", Canad. J. Math., 13 (1961), 179-191.

6. M.-P. Schützenberger, "La correspondance de Robinson", in Combinatoire et Représentation du Groupe Symétrique, Strasbourg, 1976 (D. Foata, Ed.), 59-113, Lecture Notes in Mathematics No. 579, Springer-Verlag, Berlin, 1977.

7. D. Stanton and D. White, "A Schensted correspondence for rim hook tableaux", to appear.

8. D. White, "A bijection proving orthogonality of the characters of $S_n$", Adv. Math., to appear.



DEPARTMENT OF MATHEMATICS
UNIVERSITY OF MINNESOTA
MINNEAPOLIS, MINNESOTA 55455

Contemporary Mathematics
Volume 34, 1984

# ORDERINGS OF COXETER GROUPS

Anders Björner

1. INTRODUCTION. In this paper we shall report some discoveries about the general structure of two partial orderings of Coxeter groups. Both these orderings are naturally suggested by associated geometric structures; however, they can also be intrinsically defined and studied without external geometric reference. Our point of view is predominantly combinatorial, but some results have topological and algebraic ramifications.

Some familiarity with Coxeter groups will be assumed, such as what can be learned from the first few pages of Bourbaki [13]. Recall, in particular, that if $(W,S)$ is a Coxeter group, the <u>length</u> $\ell(w)$ of $w \in W$ is the least $k$ such that $w = s_1s_2\ldots s_k$, all $s_i \in S$. Such a minimal length expression $s_1s_2\ldots s_k$ for $w$ is called a <u>reduced decomposition</u>. Let $T = \{wsw^{-1} | w \in W, s \in S\}$. For $u,w \in W$ we shall say that $u$ precedes $w$ in the <u>strong ordering</u>, written $u \leq w$, if there exist $t_1,t_2,\ldots,t_k \in T$ such that $\ell(ut_1t_2\ldots t_i) = \ell(u) + i$, for $1 \leq i \leq k$, and $ut_1t_2\ldots t_k = w$. Similarly, we shall say that $u$ precedes $w$ in the <u>weak ordering</u>, written $u \preceq w$, if there exist $s_1,s_2,\ldots,s_k \in S$ such that $\ell(us_1s_2\ldots s_i) = \ell(u) + i$, for $1 \leq i \leq k$, and $us_1s_2\ldots s_k = w$. Clearly, $u \preceq w$ implies $u \leq w$.

For the purpose of general orientation, let us mention something about the geometric background to these orderings and at the same time make a few historical comments (without any claim of or even attempt at completeness).

The strong ordering (often called the "Bruhat ordering") was first considered in a geometric context, namely as describing the inclusion ordering of "Schubert" cells in Grassmannians, flag manifolds and other homogeneous spaces. In this form the strong ordering was first considered probably by Ehresmann (1934) [26], and later in more general settings by Chevalley and others. An important combinatorial characterization of the strong ordering in terms of the underlying Weyl group (the "subword property") was known to Chevalley in the 1950s (unpublished, [14]), cf. Steinberg [53, p.127], Borel and Tits [12,

1980 Mathematics Subject Classification. 05A99, 06F15, 14M15, 20F99, 22E46.

p. 267] and Demazure [19, p.75]. Since these beginnings, and much because of its intimate relationship with canonical cell decompositions of generalized flag manifolds and their Schubert varieties, the strong ordering of Coxeter groups has frequently figured in the vast literature on the geometry and representation theory of groups and algebras of Lie type, see e.g. Bernstein, Gelfand and Gelfand [2], Dixmier [59], Jantzen [33], Kazhdan and Lusztig [34,35], Lakshmibai, Musili and Seshadri [37,38] and Verma [56].

Motivated by representation theory, Verma conjectured (1968) [56] and later proved (1971) [57] a formula for the Möbius function of a general Coxeter group. This seems to have been the first step in the direction of a combinatorial study of the strong ordering as such. Further steps in this direction were later taken by Bernstein, Gelfand and Gelfand (1973) [ 2 ] and Deodhar (1977) [20]. Apart from these developments the strong ordering was independently discovered for the symmetric groups by Savage (1957) [46, 47], and for the symmetry groups of regular polytopes by Harper (1977) [29]. Some recent combinatorial investigations of the strong ordering appear in the work of Edelman [23], Lascoux and Schützenberger [39], Proctor [42, 43] and Stanley [50].

The weak ordering seems previously to have been explicitly considered mainly for the symmetric groups, sometimes under the name of "permutohedron". There are some papers in the statistical literature of the 1960s which treat this ordering of permutations, notably Lehmann [41], Savage [47] and Yanagimoto and Okamoto [58]. Other sources are Guilbaud and Rosenstiehl [28] and Berge [ 1 ]. From a very general point of view the weak ordering of a Coxeter group (W,S) is of course nothing other than the Cayley graph of the pair (W,S) directed away from the identity. Apart possibly from such generalities, there seems to be hardly any explicit mention of the weak ordering of general Coxeter groups in the literature. However, quite recently the weak ordering has figured in the work of Edelman [24], Edelman and Greene [25], Lascoux and Schützenberger [ 40 ] and Stanley [52]. For the most part the results still concern mainly the symmetric groups.

No mention has yet been made of the work of Tits, and to our knowledge, with the possible exception of [12], no explicit mention of these orderings is to be found in his writings. Nevertheless, both the weak and the strong orderings of general Coxeter groups can be found implicitly present in his 1961 notes [54], and the geometric point of view of [54, 55] provides a convenient and natural setting for their visualization. For instance, the weak ordering is easily seen to be a special case of Tits' notion of convexity: $u \preceq w$ if and only if $C_u$ lies in the convex hull of $C_e$ and $C_w$, where $C_w$ denotes the chamber corresponding to $w \in W$ in the Coxeter complex of (W,S). This is the geometric version of the weak ordering referred to in the beginning.

Figures 1, 2 and 3 depict the orderings of the Coxeter groups of type $A_3$ (i.e., the symmetric group $S_4$), $I_2(p)$ (i.e., the dihedral group of order $2p$ , $p \geq 2$) and $\widetilde{A}_2$ (i.e., the symmetry group of the tesselation of the plane by equilateral triangles), respectively. The last group is infinite and Figure 3 shows only the elements of length $\leq 3$. In all three figures the covering relations in the weak ordering are drawn with solid lines and the additional covering relations in the strong ordering are drawn with dotted lines.

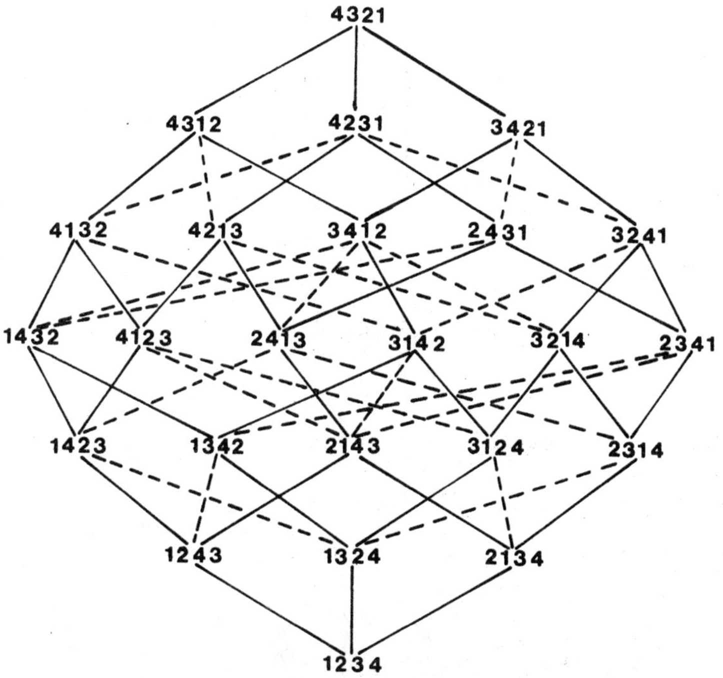


Figure 1

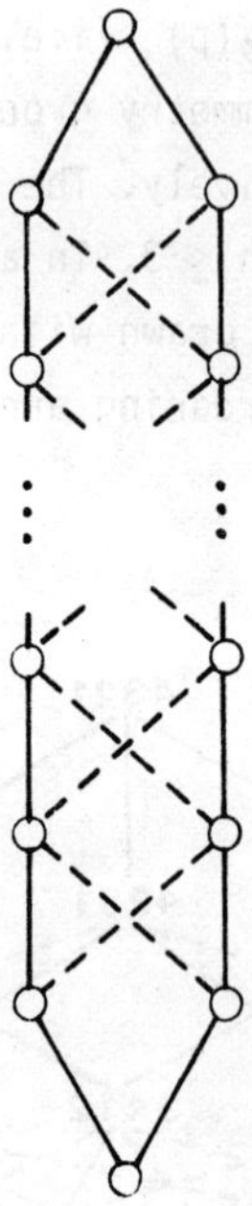

Figure 2

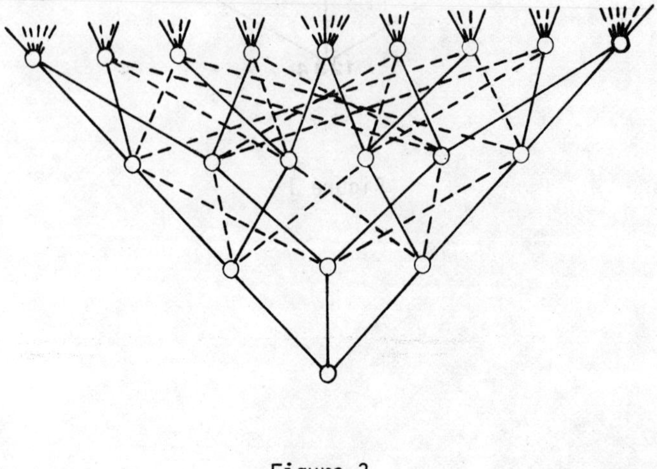

Figure 3

2. THE STRONG ORDERING. The results which are presented in this section were for the most part obtained jointly with Michelle Wachs [ 4, 9 , 11 ]. To be able to state them at the right level of generality we shall consider certain subsets $\mathcal{D}_I^K$ of Coxeter groups.

Let (W,S) be a Coxeter group. For $J \subseteq S$ let $\mathcal{D}_J = \{w \in W | ws < w \Leftrightarrow s \in J,$ for all $s \in S\}$ , and for $I \subseteq K \subseteq S$ let $\mathcal{D}_I^K = \bigcup_{I \subseteq J \subseteq K} \mathcal{D}_J$ .

The sets $\mathcal{D}_J$ , which we call <u>descent classes</u>, have several interesting properties. Let us mention a few. We assume in (1) - (3) that W is finite.

(1) The numbers card $\mathcal{D}_J$ have geometric meaning as the ranks of the top (and unique non-vanishing) homology groups of the type-selected subcomplexes of the Coxeter complex associated with (W,S) , and sums $\Sigma q^{\ell(w)}$ over $w \in \mathcal{D}_J$ play a similar role for buildings associated with groups over GF(q) [ 5 ].

(2) Each descent class $\mathcal{D}_J$ is a disjoint union of Kazhdan-Lusztig left cells [34], and the representation of W afforded by the homology module of the J-type-selected Coxeter complex is similarly a sum of the corresponding Kazhdan-Lusztig representations, cf. [ 5 , Section 6].

(3) The characteristic functions of descent classes form the basis of a subalgebra of the group algebra of a finite Coxeter group [48].

(4) Let $W_J$ denote the parabolic subgroup generated by $J \subseteq S$ . It is well-known that each left coset of $W_J$ has a unique member of minimal length, and this system of distinguished coset representatives coincides with $\mathcal{D}_\emptyset^{S-J}$ .

Let $I \subseteq K \subseteq S$ , and suppose that $u,w \in \mathcal{D}_I^K$ and $u < w$ . We shall write $[u,w]_I^K = \{z \in \mathcal{D}_I^K \mid u \leq z \leq w\}$ and $(u,w)_I^K = \{z \in \mathcal{D}_I^K \mid u < z < w\}$ for the closed and open intervals, respectively, in the strong ordering of $\mathcal{D}_I^K$ . Notice that $I' \subseteq I \subseteq K \subseteq K' \subseteq S$ implies that $[u,w]_I^K \subseteq [u,w]_{I'}^{K'}$ . We shall call the closed interval $[u,w]_I^K$ <u>full</u> if $[u,w]_I^K = [u,w]_\emptyset^S$ , and similarly for open intervals. Here are some basic facts about the structure of $\mathcal{D}_I^K$ as an ordered set.

PROPOSITION 1. (i) <u>$\mathcal{D}_I^K \neq \emptyset$ if and only if $W_I$ is finite. Suppose in the following parts that this is the case.</u>

(ii) <u>$\mathcal{D}_I^K$ has a least element.</u>

(iii) <u>$\mathcal{D}_I^K$ is a directed set</u>, i.e. <u>for every pair $u,v \in \mathcal{D}_I^K$ there exists $w \in \mathcal{D}_I^K$ such that $u,v \leq w$ .</u>

(iv) <u>Every interval $[u,w]_I^K$ is finite, and all maximal chains $u = u_0 < u_1 < \ldots < u_p = w$ in $[u,w]_I^K$ have the same length $p = \ell(w) - \ell(u)$ .</u>

(v) <u>Intervals such that $\ell(w) - \ell(u) = 2$ have the structure</u>

$[u,w]_I^K \cong$ 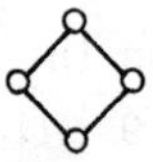 , if $[u,w]_I^K$ *is full*,  , *otherwise*.

The least element of $\mathcal{D}_\emptyset^K$ is the identity element $e \in W$ (i.e., $w \geq e$ for all $w \in \mathcal{D}_\emptyset^K$). It is a consequence of part (iii) that if $\mathcal{D}_I^K$ is finite, then $\mathcal{D}_I^K$ has a greatest element. In particular, if $W$ is finite then $W$ has a greatest element $w_0$. For the general case, if $W_I$ is a finite parabolic subgroup we shall write $w_0(I)$ for its greatest element. This $w_0(I)$ is then also the least element of $\mathcal{D}_I^K$ for all $I \subseteq K \subseteq S$.

Intervals of length 2 in the strong ordering are completely described by part (v) of Proposition 1. We will now seek a description of the structure of longer intervals. The general result will be stated in combinatorial terms but is perhaps best understood in topological language (cf. Theorems 2 and 4 below).

Let $(W,S)$ be a Coxeter group and $I \subseteq K \subseteq S$. Suppose that $u \leq w$, $u,w \in \mathcal{D}_I^K$, and $\ell(w) - \ell(u) = p$. Fix some reduced decomposition $w = s_1 s_2 \dots s_q$, and let $m\colon u = u_p < u_{p-1} < \dots < u_0 = w$ be a maximal chain in $[u,w]_I^K$. There is a unique way of erasing $p$ letters in the word $s_1 s_2 \dots s_q$, one at a time, so that after $j$ letters are erased the remaining word gives a reduced decomposition of $u_j$, for $0 \leq j \leq p$. Suppose that the j:th letter thus to be erased is $s_{\lambda_j}$, and let $\lambda(m) = (\lambda_1, \lambda_2, \dots, \lambda_p)$. Keeping $s_1 s_2 \dots s_q$ fixed and letting $m$ vary we get distinct p-tuples $\lambda(m)$ for all maximal chains $m$ in the interval $[u,w]_I^K$. It can be shown that it is possible to choose the reduced decomposition $w = s_1 s_2 \dots s_q$ in such a way that: (i) there exists a unique maximal chain $m_0$ in $[u,w]_I^K$ whose p-tuple $\lambda(m_0)$ is increasing $\lambda_1 < \lambda_2 < \dots < \lambda_p$, and (ii) if $m$ is any other maximal chain in $[u,w]_I^K$ then $\lambda(m)$ comes later than $\lambda(m_0)$ in the lexicographic ordering of $\mathbb{N}^p$. These are the essential features of the combinatorial notion of *dual CL-shellability* ("chainwise lexicographic shellability"). We refer to [8] or [10] for a complete definition and discussion of this concept. The following result from [11] is the key technical result of this section.

THEOREM 1. *Every interval* $[u,w]_I^K$ *in the strong ordering of* $\mathcal{D}_I^K$ *is dual CL-shellable.*

For the case $I = \emptyset$, $K = S$ there is an equivalent topological reformulation of Theorem 1 to which we now turn.

Let us say that a finite poset (i.e., partially ordered set) is *thin* if all its closed intervals of length two have precisely four elements. For any topological cell complex $K$ let $F(K)$, the *face poset* of $K$, denote the poset of closed cells of $K$ ordered by inclusion.

Suppose that $P$ is a finite poset of length $d+2$ with least element $\hat{0}$ and greatest element $\hat{1}$. It can be shown [ 4 ] that $P$ is dual CL-shellable and thin if and only if $P - \{\hat{0},\hat{1}\} \cong F(K)$ for some shellable regular CW decomposition $K$ of the d-dimensional sphere. For the precise meaning of the topological terms, see [ 4 ] and the references cited there. Thus, since the strong ordering of $W$ is known to be thin (Proposition 1(v)), the exact topological content of Theorem 1 for the case $K-I = S$ is the following, cf. [ 4 , Section 5].

THEOREM 2. *Suppose that* $u < w$ *and* $\ell(w) - \ell(u) = d+2 \geq 2$ , $u,w \in W$. *Then there exists a unique (up to cellular equivalence) regular CW complex* $K$ *such that* $F(K)$ *is order-isomorphic with the open interval* $(u,w)^S_\emptyset$ . *Furthermore, this complex* $K$ *is shellable and homeomorphic with the d-sphere.*

This result is illustrated in Figure 4, which shows the interval $(1234, 3241)^S_\emptyset$ in the strong ordering of $S_4$ and the corresponding regular CW complex $K$ . The 2-cell corresponding to $\theta = 2341$ fills the outer region of the compactified plane.

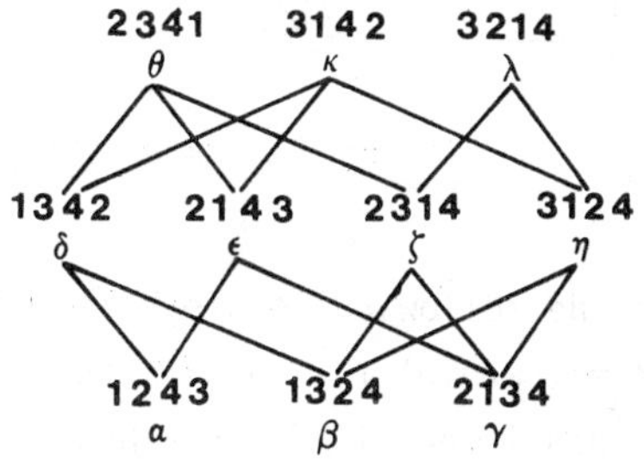


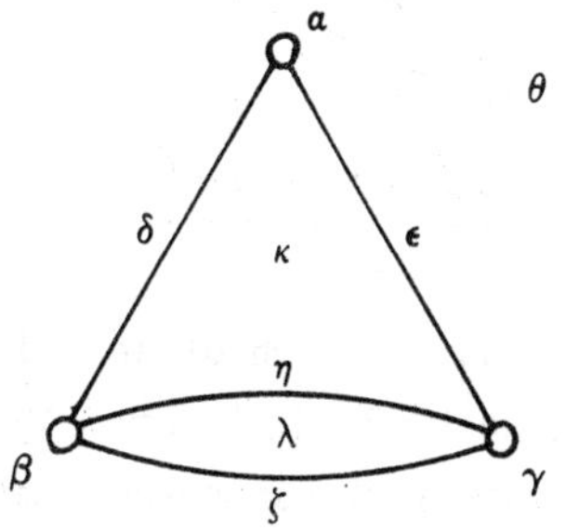


Figure 4

Variation of the same basic technique permits topological interpretation of the strong ordering of the full group $W$ as follows. Here the group elements will be associated with topological cells in such a way that $w \in W$ corresponds to a $(\ell(w)-1)$-dimensional cell.

THEOREM 3. *Let* $(W,S)$ *be a Coxeter group. Then there exists a unique (up to cellular equivalence) regular CW complex* $K$ *such that* $F(K) \cong W - \{e\}$ . *Furthermore,* (i) *if* $W$ *is finite,* $K$ *is homeomorphic with the* $(\ell(w_0)-1)$-*cell,* (ii) *if* $W$ *is infinite,* $K$ *is infinite-dimensional and contractible.*

When $K-I \neq S$ the strong ordering of $\mathcal{D}^K_I$ is not necessarily thin, and therefore a cellular interpretation of Theorem 1 similar to Theorem 2 is not always possible. However, a weaker simplicial formulation is always available, which in the situation of Theorem 2 would correspond to first barycentric subdivision.

For any poset $P$ let $\Delta(P)$ denote the abstract simplicial complex whose simplices are the finite chains $x_0 < x_1 < \dots < x_d$ in $P$, and let $\Delta(P)$ depending on context also denote its geometric realization.

THEOREM 4. *Suppose that* $u < w$ *and* $\ell(w) - \ell(u) = d+2 \geq 2$, *and that* $u,w \in \mathcal{D}_I^K$. *Then* $\Delta\big((u,w)_I^K\big)$ *is homeomorphic with*

(i) *the d-sphere* , *if* $[u,w]_I^K$ *is full*,

(ii) *the d-cell* , *otherwise*.

For example, consider the symmetric group $S_5$ with Coxeter generators $\{(12), (23), (34), (45)\}$ and let $J = \{(23), (45)\}$. Figure 5 shows the open interval $P = (13254, 35142)_J^J$ and the corresponding simplicial complex $\Delta(P)$.

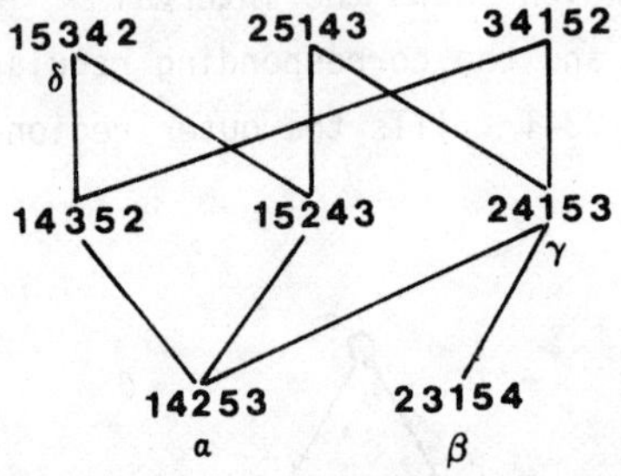


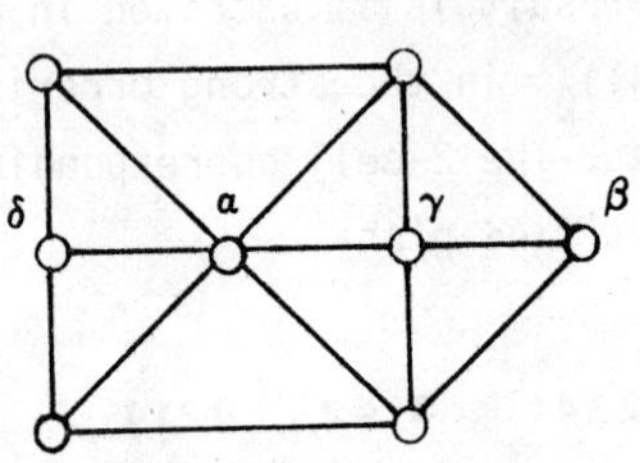


Figure 5

The Möbius function of an ordered set has the following topological interpretation (due to P. Hall, cf. [45]): for $u < w$, $\mu(u,w) + 1$ equals the Euler characteristic of the space determined by the open interval $(u,w)$. Hence, the following formula for the Möbius function $\mu_I^K(u,w)$ computed on $\mathcal{D}_I^K$ is a direct consequence of Theorem 4. The $I = \emptyset$ case is due to Verma [57] and Deodhar [20].

COROLLARY 1. *Suppose that* $u \leq w$ *and* $u,w \in \mathcal{D}_I^K$. *Then*

$$\mu_I^K(u,w) = \begin{cases} (-1)^{\ell(w)-\ell(u)}, & \text{if } [u,w]_I^K \text{ is full}, \\ 0, & \text{otherwise}. \end{cases}$$

The combinatorial structure of the strong ordering of $\mathcal{D}_I^K$ revealed by Theorem 1 has a ring-theoretic aspect, which is of some consequence for the algebraic geometry of generalized Schubert varieties. A detailed discussion of these topics would best be held in the theoretical setting of "Cohen-Macaulay posets" and "Hodge algebras", for which we refer to the surveys [8,16,18,51] and the references cited there. Let us here merely indicate the fashion in which ring-theoretic information is deduced from Theorem 1.

Suppose that $(W,S)$ is a Coxeter group and $I \subseteq K \subseteq S$. Let $u,w \in \mathcal{D}_I^K$ and $u \le w$. Suppose that $[u,w]_I^K = \{v_1,v_2,\dots,v_n\}$ and let $R$ be a commutative ring with unit. Define $R\Big[[u,w]_I^K\Big] = R[v_1,v_2,\dots,v_n]/I$, where $I$ is the ideal in the polynomial ring $R[v_1,\dots,v_n]$ generated by all monomials $v_iv_j$ such that $v_i \nleq v_j$ and $v_j \nleq v_i$. Suppose that $\ell(w) - \ell(u) = p$, and for $0 \le i \le p$ let $\theta_i = \Sigma\, v$, the sum extending over all $v \in [u,w]_I^K$ such that $\ell(v) = \ell(u)+i$. The forms $\theta_i$ are algebraically independent, so $R[\theta] = R[\theta_0,\theta_1,\dots,\theta_p]$ is a polynomial subring of $R\Big[[u,w]_I^K\Big]$. Let $m\colon u = u_p < u_{p-1} < \dots < u_0 = w$ be a maximal chain in $[u,w]_I^K$. Given a reduced decomposition of $w$, we construct the p-tuple $\lambda(m) = (\lambda_1,\lambda_2,\dots,\lambda_p)$ as was described in connection with Theorem 1. Let $D_\lambda(m) = \{i \mid \lambda_i > \lambda_{i+1}\} \subseteq \{1,2,\dots,p-1\}$, and if $D_\lambda(m) = \{i_1,i_2,\dots,i_j\}$ with $i_1 < i_2 < \dots < i_j$ then let $\eta_m = u_{i_1}u_{i_2}\cdots u_{i_j} \in R\Big[[u,w]_I^K\Big]$. The basic fact about the structure of the ring $R\Big[[u,w]_I^K\Big]$ is the following. Here $M$ will denote the set of all maximal chains in the interval $[u,w]_I^K$.

THEOREM 5. <u>The ring</u> $R\Big[[u,w]_I^K\Big]$ <u>is a free and finitely generated</u> $R[\theta]$-<u>module, and certain (for</u> $I=\emptyset$, <u>all) reduced decompositions of</u> $w$ <u>determine via the stated algorithm an explicit basis</u> $\{\eta_m \mid m \in M\}$ <u>for</u> $R\Big[[u,w]_I^K\Big]$ <u>over</u> $R[\theta]$.

Theorem 5 can be deduced, following ideas of Garsia [27] and Kind and Kleinschmidt [36], directly from Theorem 1. Also, concise and elementary proofs for this and most other consequences of shellability (i.e., Theorem 1) mentioned in this section can be found in [ 5 , Section 1].

Give $A = R\Big[[u,w]_I^K\Big]$ the standard grading which is induced by giving all indeterminates $v_i$ degree one and let $R$ be a field. One immediate consequence of Theorem 5 is then that the Hilbert series $F(A,z) = \sum_{i\ge0} (\dim_R A_i)z^i$ of the graded algebra $A = \bigoplus_{i\ge0} A_i$ takes the form $F(A,z) = (1-z)^{-p-1} \sum_{m\in M} z^{\operatorname{card} D_\lambda(m)}$.

To exemplify these results, let us consider the ring $R[S_n] = R\Big[[e,w_0]_\emptyset^S\Big]$ of the symmetric group $S_n$ for $n = 3$ and $4$. For $n = 3$ let $v_1 = 321$, $v_2 = 231$, $v_3 = 312$, $v_4 = 213$, $v_5 = 132$ and $v_6 = 123$. Then $R[S_3] = R[v_1,\dots,v_6]/(v_2v_3,\ v_4v_5)$ and the reduced decomposition $w_0 = v_1 = v_4v_5v_4$ yields by Theorem 5 that $R[S_3] = \bigoplus_{i=1}^{4} \eta_i\, R[v_1,\ v_2+v_3,\ v_4+v_5,\ v_6]$ where $\eta_1 = 1$, $\eta_2 = v_5$, $\eta_3 = v_2$ and $\eta_4 = v_2v_4$. Furthermore, $F(R[S_3],z) = (1-z)^{-4}(1+2z+z^2)$. Similarly, $R[S_4] = R[v_1,v_2,\dots,v_{24}]/I$ where the ideal $I$ is generated by 87 relations $v_iv_j$ corresponding to the incomparable pairs of permutations. We get $R[S_4] = \bigoplus_{i=1}^{168} \eta_i R[\theta_0,\theta_1,\ldots,\theta_6]$ where $\theta_j$ is the sum of all permutations having $j$ inversions and where the $\eta_i$'s are constructed from the 168 maximal chains in the strong ordering of $S_4$. One also computes $F(R[S_4],z) = (1-z)^{-7}(1+17z+66z^2+66z^3+17z^4+z^5)$.

From now on let $R$ be a field or the ring of rational integers. Theorem 5 is a more precise way of saying that $R\left[[u,w]_I^K\right]$ is a Cohen-Macaulay ring. Furthermore, if $[u,w]_I^K$ is thin then by general results (cf. Stanley [51, p.75]) the ring is even Gorenstein (i.e., Cohen-Macaulay of type one). See [51] for details concerning these ring-theoretic concepts. Thus we deduce:

COROLLARY 2. <u>The ring</u> $R\left[[u,w]_I^K\right]$ <u>is</u>

(i) <u>Gorenstein, if</u> $[u,w]_I^K$ <u>is full,</u>

(ii) <u>Cohen-Macaulay, in general</u>.

In this connection, let us remark that the Gorenstein property of the ring is not equivalent to full-ness of the interval. For instance, all intervals $[u,w]_I^K$ of length 2, whether full or not, yield Gorenstein rings.

Two recent developments, the "standard monomial theory" of Lakshmibai, Musili and Seshadri [38] and the theory of "Hodge algebras" of DeConcini, Eisenbud and Procesi [16], have revealed that some of the rings $R\left[[e,w]_\emptyset^K\right]$ of finite Weyl groups are closely related to homogeneous coordinate rings of generalized Schubert varieties $X(w)$ in spaces $G/P$ of corresponding semi-simple algebraic groups $G$. This relationship permits the transfer of some desirable properties such as Cohen-Macaulayness from the former ring to the latter. By such methods DeConcini, Huneke and Lakshmibai [17, 32] have established the Cohen-Macaulayness of certain classes of Schubert varieties. We refer to the cited sources for further information.

## 3. THE WEAK ORDERING.

This section will discuss the general structure of the weak ordering of a Coxeter group. Proofs and further details can be found in [7].

Let $(W,S)$ be a Coxeter group. At first glance one sees that the weak ordering of $W$ has some simple features in common with the strong ordering: (i) $e \preceq w$ for all $w \in W$, (ii) if $u \prec w$ then all maximal chains $u = u_0 \prec u_1 \prec \ldots \prec u_p = w$ have length $p = \ell(w) - \ell(u)$, (iii) closed intervals of length 2 have either 3 or 4 elements, and (iv) if $W$ is finite then $w \preceq w_0$ for all $w \in W$. One also sees several differences. For instance, the weak ordering of an infinite group is not directed. On the whole, as we shall find, the two orderings differ in many fundamental ways.

In this section we shall write $[u,w] = \{z \in W \mid u \preceq z \preceq w\}$ and $(u,w) = \{z \in W \mid u \prec z \prec w\}$ for the intervals in the weak ordering of $W$. Otherwise, all previously introduced notation remains unchanged.

The structure of intervals in the weak ordering can be determined up to homotopy type. Recall the notation $\Delta(\cdot)$ introduced in connection with Theorem 4.

THEOREM 6. *Suppose that* $u \prec w$ *and* $\ell(w) - \ell(u) \geq 2$, $u,w \in W$. *Then* $\Delta((u,w))$ *is*

(i) *homotopic with the* (card $J$ -2)-*sphere*, if $u^{-1}w = w_0(J)$ *for some* $J \subseteq S$,

(ii) *contractible, otherwise.*

Figure 6 shows the open interval $P = (1234, 3241)$ in $S_4$ and the corresponding simplicial complex $\Delta(P)$.

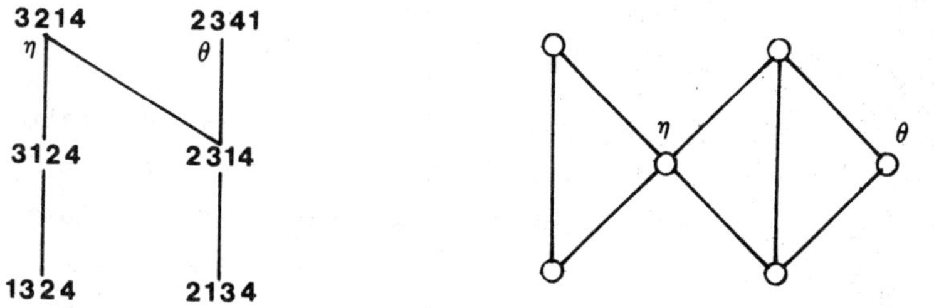


Figure 6

The nature of the Möbius function $\mu(u,w)$ of the weak ordering can immediately be deduced.

COROLLARY 3. *If* $u \preceq w$, *then*

$$\mu(u,w) = \begin{cases} (-1)^{\mathrm{card}\,J}, & \text{if } u^{-1}w = w_0(J) \text{ for some } J \subseteq S, \\ 0, & \text{otherwise.} \end{cases}$$

We now turn to the global structure of the weak ordering. First, from the topological point of view the following can be said (to be compared with Theorem 3). Here $\Gamma(W,S)$ will denote the abstract simplicial complex on the vertex set $S$ whose simplices are the nonempty subsets $J \subseteq S$ for which $W_J$ is finite.

THEOREM 7. $\Delta(W-\{e\})$ *and* $\Gamma(W,S)$ *are of the same homotopy type.*

Some experimentation with Coxeter diagrams and their spherical subdiagrams will show that the homotopy type of $\Delta(W - \{e\})$ for infinite Coxeter groups $W$ can vary greatly. For instance, if $W_n$ is the Coxeter group whose Coxeter diagram is the complete graph $K_n$ on $n$ vertices, then $\Delta(W_n - \{e\}) \simeq K_n$. One can also deduce that $\Delta(W - \{e\})$ is homotopic with the $(\operatorname{card} S - 2)$-sphere if and only if $(W,S)$ is of Euclidean or compact hyperbolic type.

Recall that an ordered set $L$ is said to be a complete meet-semilattice if every nonempty subset has a greatest lower bound, and a lattice if every pair of elements $x,y \in L$ has a greatest lower bound $x \wedge y$ and a least upper bound $x \vee y$ (called their meet and join, respectively). A self-map $\varphi: L \to L$ of a finite lattice $L$ with least element $\hat{0}$ and greatest element $\hat{1}$ is said to be an orthocomplementation if (i) $x \wedge \varphi x = \hat{0}$, $x \vee \varphi x = \hat{1}$ and $\varphi^2 x = x$ for all $x \in L$, and (ii) $x \leq y$ implies $\varphi x \geq \varphi y$ for all $x,y \in L$. See Birkhoff [3] for general information about lattices.

THEOREM 8. *Let $(W,S)$ be a Coxeter group.*

(i) *The weak ordering of $W$ is a complete meet-semilattice.*

(ii) *If $W$ is finite, the weak ordering is a lattice and $w \mapsto ww_0$ an orthocomplementation map.*

For a geometric perspective on part (ii) it is illuminating to picture the lattice diagram of $W$ (such as the solid line diagrams in Figures 1 and 2) represented as the chamber-adjacency graph on the spherical Coxeter complex of $(W,S)$. The map $w \mapsto ww_0$ will then coincide with the antipodal map of this sphere. For symmetric groups (the permutohedron) the lattice property is due to Guilbaud and Rosenstiehl [28] and Yanagimoto and Okamoto [58].

One consequence of Theorem 8 is that Coxeter groups can be faithfully represented as simple algebraic systems in a nonstandard way (with some irrelevant exceptions). Recall that a semilattice $(L, \wedge)$ is a ground set $L$ together with a binary operation $\wedge$ which is commutative, associative and idempotent. This algebraic notion is easily seen to be equivalent to the combinatorial notion of an ordered set where every pair of elements has a meet, cf. [3, p.10]. Now, let $(W,S)$ be a Coxeter group, and assume that the products $ss'$ have finite order for all $s,s' \in S$. The set $W$ together with the meet operation of the weak ordering determine a semilattice $(W, \wedge)$. Conversely, the semilattice $(W, \wedge)$ determines $S$ and the group structure of $W$. For this converse one argues that the semilattice determines the weak ordering of $W$, and that this in turn determines the group multiplication. It is for this last part that one needs the special requirement on $(W,S)$, which ensures that any automorphism $\varphi$ of the weak ordering of $W$ which leaves $S$ pointwise fixed must be the identity.

The lattice property of a finite Coxeter group $(W,S)$ is to some extent related to a geometric structure which can be introduced on the set $T = \{wsw^{-1} \mid w \in W,\ s \in S\}$ of reflections. We will give a general and abstract description of this connection, but we remark that for root systems everything can be concretely phrased in terms of convexity properties of rays and cones in Euclidean space.

Let $(W,S)$ be a Coxeter group. For every $w \in W$ let $T_w = \{t \in T \mid tw < w\}$. These sets are closely related to the weak ordering by the following observation.

PROPOSITION 2. *For all* $u,w \in W$:

$$u \preceq w \quad \text{if and only if} \quad T_u \subseteq T_w .$$

We shall seek to characterize those subsets of $T$ which are of the type $T_w$ for some $w \in W$. To this end, for subsets $A \subseteq T$ let $\varphi(A) = \{w \in W \mid t \notin T_w \text{ for all } t \in A\}$. Similarly, for $B \subseteq W$ let $\psi(B) = \{t \in T \mid t \notin T_w \text{ for all } w \in B\}$. Then by standard reasoning [ 3, p.123] the maps $\varphi$ and $\psi$ set up a Galois correspondence $2^T \rightleftarrows 2^W$, and $\bar{A} = \psi\varphi(A)$ and $\bar{B} = \varphi\psi(B)$ are a dual pair of closure operators on $2^T$ and $2^W$ respectively. The latter operator is the pointed convex hull closure on $W$, which sends $B \subseteq W$ to the convex hull (in the sense of Tits [55]) of $B \cup \{e\}$. The former operator also has the general nature of a convex hull closure. More precisely it satisfies the characteristic antiexchange condition [22]: If $A \subseteq T$, $x,y \in T$, $x \neq y$, $x \notin \bar{A}$ and $x \in \overline{A \cup \{y\}}$, then $y \notin \overline{A \cup \{x\}}$. For this reason we call a subset $A \subseteq T$ *convex* if $\bar{A} = A$ and *biconvex* if both $A$ and $T-A$ are convex.

PROPOSITION 3. *If* $W$ *is finite and* $A \subseteq T$ *then the following are equivalent:*

(i) $A = T_w$ *for some* $w \in W$,

(ii) $A$ *is biconvex*.

It is a property of our closure operator that the convex hull of a union of two biconvex sets is again biconvex. Hence we deduce the following relationship with the lattice join operation.

THEOREM 9. *For all* $u,w \in W$: $T_{u \vee w} = \overline{T_u \cup T_w}$.

4. ADDITIONAL COMMENTS. To complement the earlier sections we shall end with a brief review of various known order-theoretic properties of Coxeter groups and some remarks about recent work in this area. Our purpose here is mainly just to compile the information, and in most cases we must refer to the cited sources for further details. Throughout $(W,S)$ will denote a Coxeter group, and for simplicity we assume everywhere that $S$ is finite.

(4.1) Suppose that there is a partition $S = I \cup J$ such that every element of $I$ commutes with every element of $J$ (i.e., $(W,S)$ is reducible). Then $W$ decomposes as a direct product $W = W_I \times W_J$ not only in the group-theoretic sense [13, p.22] but also in the order-theoretic sense [3, p.8]. For the strong ordering this was pointed out by Stanley [50], and it is trivially true also for the weak ordering.

(4.2) If $\varphi: P \to P$ is a self-map of an ordered set $P$ we shall say that $\varphi$ is an automorphism [resp., antiautomorphism] if $\varphi$ is bijective and $x \leq y \Leftrightarrow \varphi x \leq \varphi y$ [resp., $\varphi x \geq \varphi y$] for all $x,y \in P$. $W$ enjoys the following symmetry properties as an ordered set: (i) Every automorphism of the Coxeter diagram induces automorphisms of both orderings of $W$, (ii) $w \mapsto w^{-1}$ is an automorphism of the strong ordering of $W$, and (iii) if $W$ is finite then $w \mapsto w_0 w w_0$ is an automorphism and $w \mapsto w w_0$ and $w \to w_0 w$ are antiautomorphisms of both orderings of $W$. Furthermore, (iv) if $W$ is finite and $J \subseteq S$ then $w \mapsto w_0 w w_0 (S-J)$ is an antiautomorphism of the strong ordering of $\mathcal{D}^J_\emptyset$, as was observed by Proctor [50, p.181]. The strong ordering of $\mathcal{D}^K_I$, for $\emptyset \neq I \subseteq K \subseteq S$, may in general lack non-trivial symmetries. On the other hand, for $I = \emptyset$ there may exist other symmetries than those mentioned above.

(4.3) Call a subset $A$ of an ordered set an antichain if no two distinct elements of $A$ are comparable. Every antichain in the strong ordering of $W$ is finite. Equivalently, every linear extension of the strong ordering is well-ordered. This is implied by a theorem of Higman [30] that in every infinite collection of words over a finite alphabet there must be a pair of distinct words such that one is a subword of the other. Examples show that the weak ordering of some groups $W$ can admit infinite antichains.

(4.4) For subsets $B \subseteq W$ let $B(t) = \sum_{w \in B} t^{\ell(w)}$. Given $I \subseteq K \subseteq S$ consider the rank-generating function $\mathcal{D}^K_I(t)$. The argument outlined in [13,p.45] immediately yields

$$(*)\quad \mathcal{D}^K_I(t) = \sum_{K-I \subseteq J \subseteq K} (-1)^{\operatorname{card}(K-J)} \mathcal{D}^J_\emptyset(t) \quad \text{and} \quad \mathcal{D}^J_\emptyset(t) = \frac{W(t)}{W_{S-J}(t)} .$$

Now, suppose that $W$ is finite. Then $(*)$ together with the known expressions $W_I(t) = \Pi(1+t+t^2+\dots t^{e_i})$ in terms of the exponents $e_i$ of $W_I$

(cf.[50, p.171]) give an explicit polynomial expression for $\mathcal{D}_I^K(t)$ . For the case of symmetric groups the polynomials $\mathcal{D}_J(t)$ have been given on determinantal form by Stanley [49, p.347]. For $I = K = S$ the formula (*) takes the form $t^{\ell(w_0)} = \sum_{J \subseteq S} (-1)^{\mathrm{card}(S-J)} \mathcal{D}_\emptyset^J(t)$ . This can be shown to be equivalent to the $w = w_0$ case of the following relation:

$$(**) \quad t^{\ell(w)} = \sum_{J \subseteq \mathcal{D}(w)} (-1)^{\mathrm{card}\, J} [e, ww_0(J)](t) \ , \quad \text{for all} \quad w \in W \ ,$$

which is the Möbius inversion formula implied by Corollary 3. Here $\mathcal{D}(w) = \{s \in S \mid ws < w\}$ , and $[\cdot,\cdot]$ denotes closed intervals in the weak ordering.

Suppose again that $W$ is finite, and let $\mathcal{D}_I^K(t) = t^p(c_0+c_1t+\ldots+c_rt^r)$ . $\mathcal{D}_I^K$ is said to be rank-symmetric if $c_i = c_{r-i}$ for all $i$ , and rank-unimodal if $c_0 \leq c_1 \leq \ldots \leq c_j \geq c_{j+1} \geq \ldots \geq c_r$ for some $j$ . The sets $\mathcal{D}_\emptyset^K$ are rank-symmetric because of the antiautomorphism (4.2 (iv)), and were shown to be rank-unimodal in essentially all cases by Stanley [50]. When $I \neq \emptyset$ the sets $\mathcal{D}_I^K$ may fail to be rank-symmetric, and whether they are rank-unimodal is at present undecided.

(4.5) Suppose that $W$ is finite and preserve the notation of the preceding paragraph. The strong ordering of $\mathcal{D}_I^K$ is said to have the Sperner property if for all $1 \leq n \leq r+1$ the largest union of $n$ antichains has at most $\max\{c_{i_1} + \ldots + c_{i_n} \mid 0 \leq i_1 < i_2 < \ldots < i_n \leq r\}$ elements. Stanley [50] showed that in essentially all cases the strong ordering of $\mathcal{D}_\emptyset^K$ has the Sperner property. Whether this remains true for general $\mathcal{D}_I^K$ is open.

It is also open whether the weak ordering of $W$ has the Sperner property or possibly even the stronger symmetric chain decomposition property (meaning that $W$ can be partitioned into mutually disjoint chains $u_i \prec u_{i+1} \prec \ldots \prec u_{\ell(w_0)-i}$ such that $\ell(u_j) = j$ for $i \leq j \leq \ell(w_0) - i$). When $W$ is of classical type the strong ordering of $W$ is known to admit symmetric chain decompositions [50, p.182].

(4.6) Suppose that $W$ is finite, irreducible and nontrivial. Then the strong ordering of $W$ is not a lattice. However, for $I \subseteq K \subseteq S$ the strong ordering of the subset $\mathcal{D}_I^K$ can sometimes be a lattice. For finite Weyl groups Proctor [42] has determined that $\mathcal{D}_\emptyset^K$ , $K \neq \emptyset$ , is a lattice if and only if $K = \{s\}$ where $s$ corresponds to a miniscule weight or $W$ is of type $G_2$ . Furthermore, in all these cases $\mathcal{D}_\emptyset^K$ actually is a distributive lattice and its poset of join-irreducibles has interesting combinatorial properties. Examples show that $\mathcal{D}_I^K$ can have the lattice property also when $I \neq \emptyset$ and $K \neq S$ , but this has not been explored.

(4.7) If $u < w$ and $\ell(w) - \ell(u) = 3$ then the combinatorial structure of the open interval $(u,w)$ in the strong ordering of $W$ is precisely determined:

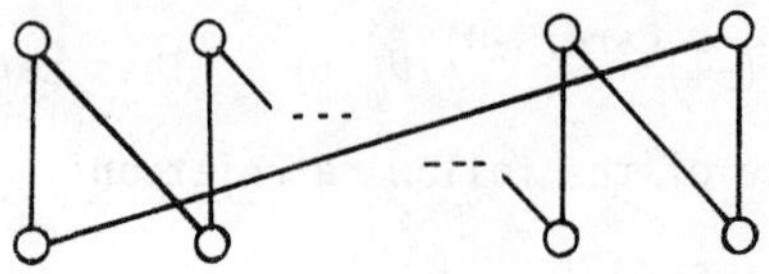

This is most clearly seen from Theorem 2 but follows already from Proposition 1. Here $\mathrm{card}(u,w) = 2n$ and $n \geq 2$. Such intervals of arbitrarily large width $n$ may occur in infinite groups, but according to Jantzen [33, p.177] in the finite Weyl groups these intervals satisfy $n \leq 4$. In view of Theorem 2 one is then led to ask also for $d \geq 2$: Which combinatorial types of regular CW decompositions of the d-sphere do occur as intervals in the strong ordering of a finite Coxeter group ? In particular, is for each $d$ only a finite number of such types possible ? Concerning the first question we note that the range of possibilities is narrowed already by the requirement that a permissible d-spherical CW complex $K$ must admit a dual regular CW complex $K^*$ such that $K$ and $K^*$ are shellable and their face posets $F(K)$ and $F(K^*)$ are anti-isomorphic. Unfortunately, there is an infinity of combinatorially distinct 2-spherical regular CW complexes which satisfy this requirement and the Jantzen condition. Nevertheless, the second question may well have an affirmative answer.

(4.8) In [34] Kazhdan and Lusztig construct for each interval $[u,w]$ in the strong ordering of $W$ a certain polynomial $P_{u,w}$ with integer coefficients. These polynomials play a fundamental role in the construction of certain representations of $W$ and its associated Hecke algebra and are also related to questions concerning Schubert varieties and Verma modules. The recursive definition of the polynomial $P_{u,w}$ does not establish any clear relationship to the structure of the interval $[u,w]$ itself. For finite and affine Weyl groups a cohomological interpretation has been given [35] which shows that $P_{u,w}$ has non-negative coefficients. It remains a challenging and important problem to find combinatorial and more easily computable interpretations of the Kazhdan-Lusztig polynomials. For the case of symmetric groups some progress has been made by Lascoux and Schützenberger [39]. See also Springer [60].

(4.9) Suppose that $J$ is an interval in either the strong or the weak ordering of $W$. The problem of computing <u>the number of maximal chains</u> in $J$ has considerable interest. Any interval $[u,w]$ in the weak ordering is translated by $u^{-1}$ to a lower interval, $[u,w] \cong [e, u^{-1}w]$, so for the weak ordering it suffices to determine the number of maximal chains in intervals of type $[e,w]$ or equivalently the number of reduced decompositions of $w$. For the symmetric groups a complete solution to this problem in terms of standard Young tableaux has been found by Stanley [52], who also offers some conjectures for the hyperoctahedral groups. Alternative solutions for the case of symmetric groups have been found by Edelman and Greene [25] and Lascoux and Schützenberger (unpublished). The situation for counting maximal chains in the strong ordering appears to be more difficult. Good solutions exist mainly for the case of subsets $\mathcal{D}_{\emptyset}^{\{s\}}$ in symmetric groups, where maximal chains in a given interval can easily be identified with standard tableaux of a given skew shape and therefore classical formulas from representation theory apply. Incidentally, in this situation the strong ordering of $\mathcal{D} = \mathcal{D}_{\emptyset}^{\{s\}}$ is isomorphic with the weak ordering of $\mathcal{D}^{-1}$, which in turn is a closed interval in the weak ordering of $W$. Therefore, counting chains for the strong ordering is in this case equivalent to counting chains for the weak ordering. The number of maximal chains of intervals in the strong ordering of $\mathcal{D}_{\emptyset}^{K}$ can sometimes have direct relevance for the Schubert calculus, see Hiller [31, p.176] and Lascoux and Schützenberger [40].

(4.10) The weak ordering can be viewed from the following general perspective. Let $(X,d)$ be a metric space and $x_0 \in X$. Define a relation on the set $X$ by $x \le y$ if and only if $d(x_0,x) + d(x,y) = d(x_0,y)$, $x,y \in X$. This makes $X$ into a partially ordered set with least element $x_0$. Now, if we set $X = W$, $d(u,w) = \ell(u^{-1}w)$ and $x_0 = e$ the construction specializes to the weak ordering of the Coxeter group $W$. Let us mention two generalizations of the weak ordering that come about in this fashion.

First, let $X$ be the set of connected regions in $\mathbb{R}^n$ which form the complement of the union of a finite set $H$ of hyperplanes, and let the distance $d(R_1,R_2)$ between two such regions be the number of hyperplanes in $H$ which separate $R_1$ from $R_2$. The resulting finite posets have been studied by Edelman [24], and were later generalized to the setting of oriented matroids by Cordovil [15].

Second, let $X$ be the set of chambers of a weak building $\Delta$ and let the distance between two chambers be the length of a minimal connecting gallery. The induced ordering of the chambers was considered in [5]. The most interesting case is when $\Delta$ is the building of a group $G$ with $(B,N)$-pair [55]. Then the chambers can be identified with the left cosets modulo $B$, and taking $x_0 = B$ the construction specializes to the weak ordering of the coset space $G/B$ as defined in [5, Section 3]. By formal analogy there is also a strong ordering of $G/B$ [ibid.].

The topological properties of the weak ordering (Theorems 6 and 7) survive generalization in some cases, cf. [15,24], but it seems unclear to what extent the lattice property does. The weak ordering of symmetric groups was adapted to infinite permutations by Rosenberg [44] and shown to remain a semilattice in that setting.

Another general context which includes the weak ordering of finite Coxeter groups as a special case is that of exchange languages and their "posets of flats", see [6].

(4.11) Let us end by mentioning a problem which is naturally suggested in connection with the lexicographic shellability of the strong ordering. Let Cov(W) denote the set of intervals of length one (or, "coverings") in the strong ordering of W. A map $\lambda$: Cov(W) $\to$ T is defined by $\lambda(u < v) = vu^{-1}$. If [u,w] is an interval of length p then the maximal chains m: $w = u_0 > u_1 > \dots > u_p = u$ produce distinct p-tuples $\lambda(m) = (\lambda_1, \lambda_2, \dots, \lambda_p) \in T^p$ where $\lambda_i = \lambda(u_i < u_{i-1})$. The question is: Can the set T be given a linear ordering in such a way that for every interval [u,w] in the strong ordering of W: (i) there exists a unique maximal chain $m_0$ in [u,w] such that $\lambda(m_0)$ satisfies $\lambda_1 < \lambda_2 < \dots < \lambda_p$, and (ii) if m is any other maximal chain in [u,w] then $\lambda(m)$ comes later than $\lambda(m_0)$ in the induced lexicographic ordering of $T^p$? An affirmative answer would provide a new proof and give a more global geometric content to Theorem 1 for the K-I = S case (and would from the combinatorial point of view also constitute a minor sharpening, replacing "CL-shellable" by "EL-shellable", in the terminology of [8,10]). The work of Edelman [23] shows that an affirmative answer can be given for the symmetric groups.

## BIBLIOGRAPHY

1. C. Berge, Principles of Combinatorics, Academic Press, New York, 1971.

2. I.N. Bernstein, I.M. Gelfand and S.I. Gelfand, Schubert cells and the cohomology of the spaces G/P, Russian Math. Surv. 28 (1973), 1-26. [Reprinted in: "Representation Theory", London Math. Soc. Lect. Notes No. 69, Cambridge Univ. Press, Cambridge, 1982.]

3. G. Birkhoff, Lattice Theory (3rd. ed.), Amer. Math. Soc. Colloq. Publ. No. 25, Amer. Math. Soc., Providence, R.I., 1967.

4. A. Björner, Posets, regular CW complexes and Bruhat order, European J. Combinatorics, (to appear).

5. A. Björner, Some combinatorial and algebraic properties of Coxeter complexes and Tits buildings, Advances in Math., (to appear).

6. A. Björner, On matroids, groups and exchange languages, in "Proceedings of Colloquium on Matroid Theory, Szeged 1982", (ed. L. Lovász), Coll. Math. Soc. János Bolyai, North-Holland, (to appear).

7. A. Björner, The weak ordering of a Coxeter group, (to appear).

8. A. Björner, A.M. Garsia and R.P. Stanley, An introduction to Cohen-Macaulay partially ordered sets, in "Ordered Sets" (ed. I. Rival), Reidel, Dordrecht, 1982, pp. 583-615.

9. A. Björner and M. Wachs, Bruhat order of Coxeter groups and shellability, Advances in Math. 43 (1982), 87-100.

10. A. Björner and M. Wachs, On lexicographically shellable posets, Trans. Amer. Math. Soc. 277 (1983), 323-341.

11. A. Björner and M. Wachs, Descent classes in Coxeter groups under the Bruhat ordering, (to appear).

12. A. Borel and J. Tits, Compléments à l'article "Groupes réductifs", I.H.E.S. Publ. Math. 41 (1972), 253-276.

13. N. Bourbaki, Groupes et algèbres de Lie, Chap. 4, 5 et 6, Éléments de mathématique, Fasc. XXXIV, Hermann, Paris, 1968.

14. C. Chevalley, Sur les décompositions cellulaires des espaces G/B, unpublished manuscript, circa 1958.

15. R. Cordovil, A combinatorial perspective on the non-Radon partitions, preprint, Lisbon, 1983.

16. C. De Concini, D. Eisenbud and C. Procesi, Hodge Algebras, Astérisque, Vol. 91, Soc. Math. France, 1982.

17. C. De Concini and V. Lakshmibai, Arithmetic Cohen-Macaulayness and arithmetic normality for Schubert varieties, Amer. J. Math. 103 (1981), 835-850.

18. C. De Concini and C. Procesi, Hodge Algebras: A survey, Astérisque, Vol. 87-88, Soc. Math. France, 1981, pp. 79-83.

19. M. Demazure, Désingularisation des variétés de Schubert généralisées, Ann. Scient. Éc. Norm. Sup. 7 (1974), 53-88.

20. V.V. Deodhar, Some characterizations of Bruhat ordering on a Coxeter group and determination of the relative Möbius function, Invent. Math. 39 (1977), 187-198.

21. V.V. Deodhar, On Bruhat ordering and weight-lattice ordering for a Weyl group, Indagat. Math. 40 (1978), 423-435.

22. P.H. Edelman, Meet-distributive lattices and the anti-exchange closure, Algebra Universalis 10 (1980), 290-299.

23. P.H. Edelman, The Bruhat order of the symmetric group is lexicographically shellable, Proc. Amer. Math. Soc. 82 (1981), 355-358.

24. P.H. Edelman, A partial order on the regions of $R^n$ dissected by hyperplanes, preprint, Univ. of Penn., 1982.

25. P.H. Edelman and C. Greene, Combinatorial correspondences for Young tableaux, balanced tableaux, and maximal chains in the weak Bruhat order of $S_n$ (research announcement), in these Proceedings.

26. C. Ehresmann, Sur la topologie de certains espaces homogènes, Ann. Math. 35 (1934), 396-443.

27. A.M. Garsia, Combinatorial methods in the theory of Cohen-Macaulay rings, Advances in Math. 38 (1980), 229-266.

28. G.Th. Guilbaud and P. Rosenstiehl, Analyse algébrique d'un scrutin, in "Ordres totaux finis", Gauthiers-Villars et Mouton, Paris, 1971, pp. 71-100.

29. L.H. Harper, Stabilization and the edgesum problem, Ars Combinatoria 4 (1977), 225-270.

30. G. Higman, Ordering by divisibility in abstract algebras, Proc. London Math. Soc. 2 (1952), 326-336.

31. H. Hiller, Geometry of Coxeter groups, Pitman, Boston and London, 1982.

32. C. Huneke and V. Lakshmibai, A characterization of Kempf varieties by means of standard monomials and its geometric consequences, preprint, 1982.

33. J.C. Jantzen, Moduln mit einem höchsten Gewicht, Lecture Notes in Math., Vol. 750, Springer, Berlin, 1979.

34. D. Kazhdan and G. Lusztig, Representations of Coxeter groups and Hecke algebras, Invent. Math. 53 (1979), 165-184.

35. D. Kazhdan and G. Lusztig, Schubert varieties and Poincaré duality, Proc. Sympos. Pure Math., Vol. 36, Amer. Math. Soc., Providence, R.I., 1980, pp. 185-203.

36. B. Kind and P. Kleinschmidt, Schälbare Cohen-Macaulay-Komplexe und ihre Parametrisierung, Math. Z. 167 (1979), 173-179.

37. V. Lakshmibai, C. Musili and C.S. Seshadri, Cohomology of line bundles on G/B, Ann. Scient. Éc. Norm. Sup. 7 (1974), 89-138.

38. V. Lakshmibai, C. Musili and C.S. Seshadri, Geometry of G/P, Bull. Amer. Math. Soc. 1 (1979), 432-435.

39. A. Lascoux and M.P. Schützenberger, Polynomes de Kazhdan et Lusztig pour les Grassmanniennes, Astérisque 87-88 (1981), 249-266.

40. A. Lascoux and M.P. Schützenberger, Symmetry and flag manifolds, Lecture Notes in Math., Vol. 996, Springer, Berlin, 1983, pp. 118-144.

41. E.L. Lehmann, Some concepts of dependence, Ann. Math. Statist. 37 (1966), 1137-1153.

42. R.A. Proctor, Interactions between combinatorics, Lie theory and algebraic geometry via the Bruhat orders, Ph. D. thesis, MIT, 1981.

43. R.A. Proctor, Classical Bruhat orders and lexicographic shellability, J. Algebra 77 (1982), 104-126.

44. I.G. Rosenberg, A semilattice on the set of permutations on an infinite set, Math. Nachrichten 60 (1974), 191-199.

45. G.-C. Rota, On the foundations of Combinatorial Theory I: Theory of Möbius functions, Z. Wahrsch. Verw. Gebiete 2 (1964), 340-368.

46. I.R. Savage, Contributions to the theory of rank order statistics — the "trend" case, Ann. Math. Statist. 28 (1957), 968-977.

47. I.R. Savage, Contributions to the theory of rank order statistics: Applications of lattice theory, Rev. Internat. Statist. Inst. 32 (1964), 52-64.

48. L. Solomon, A Mackey formula in the group ring of a Coxeter group, J. Algebra 41 (1976), 255-268.

49. R.P. Stanley, Binomial posets, Möbius inversion and permutation enumeration, J. Combinatorial Theory Ser. A 20 (1976), 336-356.

50. R.P. Stanley, Weyl groups, the hard Lefschetz theorem, and the Sperner property, SIAM J. Alg. Discrete Methods 1 (1980), 168-184.

51. R.P. Stanley, Combinatorics and Commutative Algebra, Birkhäuser, Boston, 1983.

52. R.P. Stanley, On the number of reduced decompositions of elements of Coxeter groups, Europ. J. Combinatorics, (to appear).

53. R. Steinberg, Lectures on Chevalley groups, Notes, Yale University, 1967.

54. J. Tits, Groupes et géométries de Coxeter, preprint, I.H.E.S., Paris, 1961.

55. J. Tits, Buildings of spherical type and finite BN-pairs, Lecture Notes in Math. No. 386, Springer, Berlin, 1974.

56. D.-N. Verma, Structure of certain induced representations of complex semisimple Lie algebras, Bull. Amer. Math. Soc. 74 (1968), 160-166.

57. D.-N. Verma, Möbius inversion for the Bruhat ordering on a Weyl group, Ann. Scient. Éc. Norm. Sup. 4 (1971), 393-398.

58. T. Yanagimoto and M. Okamoto, Partial orderings of permutations and monotonicity of a rank correlation statistic, Ann. Inst. Statist. Math. 21 (1969), 489-506.

59. J. Dixmier, Algèbres enveloppantes, Gauthier-Villars, Paris, 1974.

60. T.A. Springer, Quelques applications de la cohomologie d'intersection, Astérisque 92-93 (1982), 249-273.

DEPARTMENT OF MATHEMATICS
UNIVERSITY OF STOCKHOLM
Box 6701, 113 85 STOCKHOLM
SWEDEN

CURRENT ADDRESS:

DEPARTMENT OF MATHEMATICS
MASSACHUSETTS INSTITUTE OF TECHNOLOGY
CAMBRIDGE, MA 02139

Contemporary Mathematics
Volume 34, 1984

# MODULARLY COMPLEMENTED LATTICES AND SHELLABILITY

Dennis Stanton[1]
Michelle Wachs[2]

ABSTRACT. We study the class of geometric lattices in which every element has a modular complement. This class includes such fundamental lattices as the subset lattice, the subspace lattice, the partition lattice, and the Dowling lattice. Although all geometric lattices are shellable, this is not necessarily true for all order ideals of these lattices. We consider those lower order ideals which are generated by the complements of modular elements. The technique of recursive atom ordering is employed to establish the shellability of these ideals.

1. INTRODUCTION. A modularly complemented lattice is an upper semimodular lattice in which every element has a complement that is modular. This notion was discussed by Stonesifer [14] who presents some interesting properties of modularly complemented lattices. We will investigate another property, that of shellability. A poset is shellable if its maximal chains can be ordered in a certain favorable way. This property has strong consequences of a combinatorial, topological and algebraic nature (see e.g., [1,2,3,12]).

Since it is well known that all upper semimodular lattices are shellable, our goal is clearly not to establish shellability for the modularly complemented lattices. We restrict ourselves to certain order ideals of these lattices. Our main result is that the lower order ideal generated by the complements of a modular element is a shellable poset. For the lattice of subspaces of a vector space, these order ideals appear in Vogtman's work on homology stability [15]. She proves that these ideals are Cohen-Macaulay posets. This is a special case of our result since the vector space lattice is clearly modularly complemented and the Cohen-Macaulay property is a consequence

1980 Mathematics Subject Classification 05A99, 06A10, 06C10, 06C15.
[1]Partially supported by NSF grant MCS 83-00872.
[2]Partially supported by NSF grant MCS 81-03474.


0271-4132/84 $1.00 + $.25 per page

of shellability. These order ideals for the vector space lattice are also isomorphic to semilattices of bilinear forms which are used by Delsarte [6] and Stanton [13] in obtaining association schemes. In [4] Björner and Wachs show that the semilattice of bilinear forms is a shellable poset. Hence our result also generalizes this result.

In §3 we we discuss some examples and non-examples of modularly complemented lattices. In §4 we show that requiring that every element have a modular complement is equivalent to requiring just that the atoms have modular complements. The last section contains the result on shellability of order ideals.

2. PRELIMINARIES. The notation and terminology not otherwise defined can be found in [5 or 4].

Let L be a finite upper semimodular lattice with rank function r. An element $m \in L$ is <u>modular</u> in L if $(m \vee x) \wedge y = (m \wedge y) \vee x$ whenever $x \leq y$. This condition (in an upper semimodular lattice) is known to be equivalent to requiring that $(y \vee x) \wedge m = (m \wedge y) \vee x$ whenever $x \leq m$ and $y \in L$. Modular elements can also be characterized as those elements $m$ which satisfy $r(x) + r(m) = r(x \vee m) + r(x \wedge m)$ for all $x \in L$. Recall that a modular lattice consists entirely of modular elements. The lattice of subsets of a set and the lattice of subspaces of a vector space are the best known examples of modular lattices. The lattice of partitions of a set ordered by refinement is the standard example of a geometric lattice which is not modular. Recall that a <u>geometric lattice</u> is an upper semimodular lattice in which every element is a join of atoms or equivalently an upper semimodular lattice which is relatively complemented.

3. EXAMPLES. The geometric modular lattices are trivially examples of modularly complemented lattices. The partition lattice is another example. Indeed, the modular elements of the partition lattice are the partitions with at most one non-singleton block. To obtain a modular complement of an arbitrary partition $\pi$, form a block by taking exactly one element from each block of $\pi$ and then let the remaining elements form singleton blocks.

Stonesifer [14] shows that modularly complemented lattices are supersolvable. Therefore it is natural to look among the known examples of supersolvable lattices for additional examples of modularly complemented lattices. Stanley [11] gives two examples of supersolvable lattices which generalize the partition lattice. One example, the Dowling lattice, is always modularly complemented. The other example, the lattice of contractions of a supersolvable graph, turns out to be modularly complemented only when it is the partition lattice (or a direct product of partition lattices).

We should point out here that Kahn and Kung [17] have obtained a complete classification of the modularly complemented lattices since the completion of this work. They show that the modularly complemented lattices are direct products of Dowling lattices and projective geometries with the possible removal of certain atoms from the projective geometries. Theorems 3.2 and 4.5, presented below, now follow from this classification. However, the classification is not useful in proving our main result Theorem 5.3.

Let $V$ be a vector space of dimension $n$ over $GF(q)$. The Dowling lattice $D_n(q)$ is the lattice of subspaces of $V$ that are generated by vectors with exactly one or two non-zero entries. These lattices and a more general class of lattices were introduced by Dowling [7, 8]. Although the more general lattices also turn out to be modularly complemented, we will restrict our attention to $D_n(q)$. When $q = 2$, $D_n(q)$ becomes the partition lattice $\Pi_{n+1}$. Dowling shows that $D_n(q)$ is upper semimodular and that every subspace generated by vectors with exactly one non-zero entry is a modular element. It is easy to see that every element in $D_n(q)$ has such a subspace as a complement.

Let G be a graph (without loops) and let $\pi$ be a partition of the vertices of G such that the subgraphs induced by the blocks of $\pi$ are connected. A contraction of G is obtained from $\pi$ by identifying the vertices of each block. If we order the contractions of G by $H_1 > H_2$ if $H_1$ is a contraction of $H_2$, we get a geometric lattice, $L(G)$. Since a contraction of G can be thought of as a partition of the vertex set of G, it is clear that when G is the complete graph, $L(G)$ is the partition lattice.

PROPOSITION 3.1 (Stanley [10]) Let G be a doubly connected finite graph. Then $\pi \in L(G)$ is a modular element of $L(G)$ if and only if the following conditions hold.

(1) There is at most one non-singleton block B of $\pi$.

(2) Let H be the subgraph induced by the block B of (1). Let K be any connected component of the subgraph induced by G - B and let $H_1$ be the graph induced by the set of vertices in H which are adjacent to some vertex in K. Then $H_1$ is a complete subgraph of G. □

We now show that the lattice of contractions provides no additional examples of modularly complemented lattices.

THEOREM 3.2 The lattice of contractions of a doubly connected finite graph $G = (V,E)$ is modularly complemented if and only if G is complete.

Proof. If G is complete then $L(G)$ is the partition lattice which is modularly complemented.

Conversely, suppose $L(G)$ is modularly complemented. First we establish the following implication for all $a,b,c \in V$ :

$$(a,b),\ (a,c) \in E \Rightarrow (b,c) \in E\ . \tag{3.1}$$

Suppose $(b,c) \notin E$ . Let p be the shortest path from b to c which does not contain a. Since the length of p is greater than one, there is a vertex $d \neq c$ on p which is adjacent to b. Let $\pi_1$ be the atom of L(G) whose only non-singleton block is $\{a,b\}$. By Proposition 3.1, a modular complement of $\pi_1$ is either the partition with blocks $\{a\}$, $V - \{a\}$ or the partition with blocks $\{b\}$, $V - \{b\}$. Also, by Proposition 3.1, if the former holds then $(b,c) \in E$ and if the latter holds then $(a,d) \in E$. But our initial assumption was that $(b,c) \notin E$. Hence, $(a,d) \in E$.

Now let $\pi_2$ be the atom of L(G) whose only non-singleton block is $\{a,d\}$. Again by Proposition 3.1, the only modular complement of $\pi_2$ is the partition with blocks $\{d\}$, $V - \{d\}$. Let $e \neq b$ be adjacent to d on p. It follows from Proposition 3.1 that $(b,e) \in E$. This contradicts the minimality of the length of p. Hence (3.1) holds. A consequence of (3.1) is that G is complete. Indeed if $x, y \in V$ and p is any simple path from x to y of length greater than one then (3.1) allows p to be shortened until it has length one. □

If G is not doubly connected then L(G) is a direct product of the lattices of contractions of the maximal doubly connected subgraphs of G. If L(G) is modularly complemented then the factors of the direct product are also modularly complemented. This is because the factors are intervals of L(G) and intervals inherit the modularly complemented property (c.f. §4). We now have the following corollary.

<u>COROLLARY 3.3</u> The lattice of contractions of a finite graph is modularly complemented if and only if it is a product of partition lattices. □

4. ATOM MODULARLY COMPLEMENTED LATTICES. We now attempt to generalize the notion of modularly complemented lattices by requiring only that all atoms have modular complements. We call upper semimodular lattices with this property <u>atom-modularly complemented</u>. It turns out that this new class of lattices is really no more general than the original class. Before proving this we present some preliminary lemmas which appear in [10] and whose proofs are straightforward. Let L be an upper semimodular lattice.

<u>LEMMA 4.1.</u> If x is modular in L and $y \in L$ then $[y, y \vee x] \simeq [y \wedge x, x]$ . □

<u>LEMMA 4.2.</u> If x is modular in L and $y \in L$ then $x \wedge y$ is modular in $[\hat{0},y]$ and $x \vee y$ is modular in $[y,\hat{1}]$. □

<u>LEMMA 4.3.</u> If x is modular in L and y is modular in $[\hat{0},x]$ then y is modular in L. □

The following theorem was proved by Stonesifer [14] for modularly complemented lattices.

<u>THEOREM 4.4.</u> If L is atom-modularly complemented then every interval of L is

also atom-modularly complemented.

Proof. First we show that all intervals of the form $[\hat{0},b]$, $b \in L$ are atom-modularly complemented. Let $x$ be any atom of $[\hat{0},b]$. Then $x$ has a modular complement $y$ in $L$. By Lemma 4.2, $y \wedge b$ is modular in $[\hat{0},b]$. Since $x \wedge y \wedge b = \hat{0}$ and $x \vee (y \wedge b) = (x \vee y) \wedge b = b$, $y \wedge b$ is a modular complement of $x$ in $[\hat{0},b]$.

Now let $a$ be an atom of $L$ with modular complement $c$. Then $[a,\hat{1}] \simeq [\hat{0}, c]$ by Lemma 4.1. Hence $[a,\hat{1}]$ is atom-modularly complemented because of the previous paragraph. Combining this with the result of the previous paragraph gives that all intervals of the form $[a,b]$, where $a$ is an atom of $L$, are atom-modularly complemented. It follows by induction that all intervals are atom-modularly complemented. □

THEOREM 4.5. A lattice is atom-modularly complemented if and only if it is modularly complemented.

Proof. One direction of the theorem is trivial. Suppose $L$ is atom-modularly complemented. We prove that every element $b \in L$ has a modular complement in $L$, by induction on $r(b)$. If $r(b) = 0$ or $1$ then $b$ trivially has a modular complement.

Suppose $r(b) > 1$ and all elements of smaller rank than $b$ have modular complements. Let $b$ cover some $a \in L$. Then $a$ has a modular complement $m$. Since $b$ is an atom of $[a,\hat{1}]$, by Theorem 4.4 $b$ has a modular complement $s$ in $[a,\hat{1}]$. We will show that $c = s \wedge m$ is a modular complement of $b$ in $L$.

We begin by showing that $c$ is modular in $[\hat{0},m]$. Let $x, y \in [\hat{0},m]$ with $x < y$. We must establish,

$$(c \wedge y) \vee x = (c \vee x) \wedge y \quad . \tag{4.1}$$

We have,

$$\begin{aligned}
(c \wedge y) \vee x &= \big(c \wedge (y \vee (m \wedge a))\big) \vee x && (\text{since } m \wedge a = \hat{0}) \\
&= \big(c \wedge (m \wedge (y \vee a))\big) \vee x && (\text{since } m \text{ is modular}) \\
&= \big((c \wedge m) \wedge (y \vee a)\big) \vee x \\
&= \big((s \wedge m) \wedge (y \vee a)\big) \vee x \\
&= \big((s \wedge (y \vee a)) \wedge m\big) \vee x \\
&= \big((s \wedge (y \vee a)) \vee x\big) \wedge m && (\text{since } m \text{ is modular}) \quad .
\end{aligned}$$

Since $s \geq a$ and $y \vee a \geq a$, $s \wedge (y \vee a) \geq a$. Hence, $\big(s \wedge (y \vee a)\big) \vee x = \big(s \wedge (y \vee a)\big) \vee (x \vee a)$. Substituting this into the previous string of equations yields,

$$(c \wedge y) \vee x = \big((s \wedge (y \vee a)) \vee (x \vee a)\big) \wedge m$$

$$= ((s \vee (x \vee a)) \wedge (y \vee a)) \wedge m \quad \text{(since } s \text{ is modular in } [a,\hat{1}])$$

$$= ((s \vee x) \wedge m) \wedge ((y \vee a) \wedge m)$$

$$= ((s \wedge m) \vee x) \wedge (y \vee (a \wedge m)) \quad \text{(since } m \text{ is modular)}$$

$$= (c \vee x) \wedge y .$$

Hence (4.1) holds and c is modular in $[\hat{0},m]$. It follows from Lemma 4.3 that c is modular in L. We could also have used the rank condition to verify that c is modular in $[\hat{0},m]$.

We now show that c and b are complements. We have, $c \wedge b = m \wedge s \wedge b = m \wedge a = \hat{0}$. We also have $c \vee a = (m \wedge s) \vee a = (m \vee a) \wedge s = s$. Hence $s \leq c \vee b$. It follows that $c \vee b = c \vee b \vee s = \hat{1}$. □

COROLLARY 4.6. (Atom-)modularly complemented lattices are geometric. □

## 5. SHELLABILITY

In [1] a simple method was introduced by Björner for establishing the shellability of posets. This method, which is called edge-lexicographical shellability (EL-shellability) consists of labeling the edges of the Hasse diagram of a poset in a certain way. A slightly more general version of this method was formulated in [3] leading to the concept of chain-lexicographical shellability (CL-shellability). Many interesting classes of posets have been shown to be lexicographically shellable (see [1,2,3,4,9,16]). In proving our main result we use a recursive formulation of CL-shellability, which was introduced in [4].

DEFINITION 5.1 A graded poset P is said to admit a recursive atom ordering if the length of P is 1 or if the length of P is greater than 1 and there is an ordering $a_1, a_2, \ldots, a_t$ of the atoms of P which satisfies:

(i) For all $j = 1,2, \ldots, t$, $[a_j, \hat{1}]$ admits a recursive atom ordering in which the atoms of $[a_j, \hat{1}]$ that come first in the ordering are those that cover some $a_i$ where $i < j$.

(ii) For all $i < j$, if $a_i, a_j < y$ then there is a $k < j$ and an element z which covers $a_k$ and $a_j$ and is less than or equal to y.

If $a_1, a_2, \ldots, a_t$ is an ordering of the atoms of P that satisfies (i) and (ii) then $a_1, a_2, \ldots, a_t$ is said to be a recursive atom ordering.

PROPOSITION 5.2 [4] A poset P admits a recursive atom ordering if and only if P is CL-shellable. □

Let $\hat{P}$ denote the poset obtained from the poset P by adjoining a top element $\hat{1}$.

THEOREM 5.3 Let L be a modularly complemented lattice and let w be a modular element of $L$. If $P = P(L,w)$ is the order ideal $\{x \mid x \wedge w = \hat{0}\}$ then $\hat{P}$ is CL-shellable.

Proof. Since $w$ is modular, all its complements have the same rank. Since the complements are the maximal elements of $P$, $\hat{P}$ is graded.

We will use the notation $[a,\hat{1}]_L$ and $[a,\hat{1}]_{\hat{P}}$ to denote upper intervals in $L$ and $\hat{P}$ respectively. Let $a$ be an atom of $P$. Clearly $[a,\hat{1}]_{\hat{P}} = [a,\hat{1}]_L \cap \hat{P}$. We will first show that

$$[a,\hat{1}]_L \cap P = P([a,\hat{1}]_L, a \vee w). \tag{5.1}$$

This implies that all upper intervals of $\hat{P}$ are of the same form as $\hat{P}$ since by Lemma 4.2 $a \vee w$ is modular in $[a,\hat{1}]_L$. Hence we will be able to use induction on the length of $\hat{P}$.

Suppose $x \in [a,\hat{1}]_L \cap P$. Then since $w$ is modular, $x \wedge (a \vee w) = a \vee (x \wedge w) = a$. Hence $x \in P([a,\hat{1}]_L, a \vee w)$. Conversely, if $x \in P([a,\hat{1}]_L, a \vee w)$ then $a = x \wedge (a \vee w) = a \vee (x \wedge w)$. This means that $x \wedge w = a$ or $x \wedge w = \hat{0}$. Then the former implies that $a \leq w$ which contradicts $a \in P$. Hence $x \in [a,\hat{1}]_L \cap P$ and (5.1) holds.

Let $u$ be any modular complement of $w$ in $L$. We will prove the following assertion.

Assertion: Any atom ordering of $\hat{P}$ in which the atoms below $u$ come first is a recursive atom ordering.

Proof of assertion: We use induction on the length of $\hat{P}$. Let $\Omega$ be an atom ordering of $\hat{P}$ in which the atoms below $u$ come first. For each atom $a$ of $P$, we let $F(a)$ be the set of atoms of $[a,\hat{1}]_{\hat{P}}$ which cover some atom of $P$ which preceeds $a$ in $\Omega$. We now define another set of atoms of $[a,\hat{1}]_{\hat{P}}$ called $G(a)$.

Case 1. Suppose $a < u$. Let $G(a)$ be the set of atoms of $[a,\hat{1}]_{\hat{P}}$ which are less than $u$. The fact that $u$ is a complement of $a \vee w$ in $[a,\hat{1}]_L$ follows from

$$(a \vee w) \vee u = w \vee u = \hat{1} \ ,$$

and

$$\begin{aligned} u \wedge (a \vee w) &= (u \wedge w) \vee a \\ &= \hat{0} \vee a \\ &= a \ . \end{aligned}$$

Since $u$ is a modular complement of $a \vee w$ in $[a,\hat{1}]_{\hat{P}}$, it follows by induction that if the atoms of $G(a)$ come first in an atom ordering of $[a,\hat{1}]_{\hat{P}} = \hat{P}([a,\hat{1}]_L, a \vee w)$ then the atom ordering is recursive.

We must show that $F(a) \subseteq G(a)$ so that we can have the atoms of $F(a)$ come first followed by the atoms of $G(a) - F(a)$ and finally the remaining atoms. In this way we will have the atoms of $F(a)$ precede those atoms not in $F(a)$ and the atoms of G(a) precede those atoms not in G(a). Consequently Definition 5.1 (i) will be satisfied. Let $y \in F(a)$. Then there is an atom $b < y$ which precedes a in $\Omega$. Since $a < u$ and b precedes a, it follows that $b < u$. Thus $y = a \vee b < u$. Consequently $y \in G(a)$.

Case 2. Suppose $a \nless u$. Let $b = u \wedge (a \vee w)$. Then b is an atom since

$$\begin{aligned} r(b) &= r(u) + r(a \vee w) - r(u \vee a \vee w) \\ &= r(u) + r(w) + 1 - r(\hat{1}) \\ &= r(u \wedge w) + r(u \vee w) + 1 - r(\hat{1}) \\ &= r(\hat{0}) + r(\hat{1}) + 1 - r(\hat{1}) \\ &= 1 . \end{aligned}$$

By Theorem 4.4, b has a modular complement v in $[\hat{0},u]$. It turns out that $a \vee v$ is a modular complement of $a \vee w$ in $[a,\hat{1}]_L$. Indeed we have $u = v \vee b = v \vee (u \wedge (a \vee w)) = (v \vee (a \vee w)) \wedge u$. It follows that $u \leqslant v \vee (a \vee w)$. But we also have that $w < v \vee (a \vee w)$. Since $u \vee w = \hat{1}_L$, $v \vee (a \vee w) = \hat{1}_L$ which implies that $(a \vee v) \vee (a \vee w) = \hat{1}_L$. The equation $(a \vee v) \wedge (a \vee w) = a$ follows from,

$$\begin{aligned} r((a \vee v) \wedge (a \vee w)) &\leqslant r(a \vee v) + r(a \vee w) - r(a \vee v \vee a \vee w) \\ &= r(v) + 1 + r(w) + 1 - r(\hat{1}_L) \\ &= r(u) + r(w) + 1 - r(\hat{1}_L) \\ &= 1 . \end{aligned}$$

By Lemma 4.3, v is modular in L and by Lemma 4.2, $a \vee v$ is modular in $[a,\hat{1}]_L$. Hence $a \vee v$ is a modular complement of $a \vee w$ in $[a,\hat{1}]_L$.

We now let G(a) be the atoms of $[a,\hat{1}]_{\hat{P}}$ that are less than $a \vee v$. If the atoms of G(a) come first in an atom ordering of $[a,\hat{1}]_{\hat{P}}$ then again by induction the atom ordering is recursive. This time we will show that $G(a) \subseteq F(a)$. Let $y \in G(a)$. Then $y < a \vee v$ and hence $y \vee v = a \vee v$. We show that $y \wedge v$ is an atom as follows:

$$\begin{aligned} r(y \wedge v) &= r(y) + r(v) - r(y \vee v) \\ &= r(y) + r(v) - r(a \vee v) \\ &= r(y) + r(v) - (r(v) + 1) \\ &= r(y) - 1 \\ &= 1 . \end{aligned}$$

Hence $y \wedge v$ is an atom less than $y$ and less than $u$ which implies that $y \wedge v$ precedes $a$ in $\Omega$, which in turn implies that $y \in F(a)$.

We must now verify (ii) of Definition 5.1. Let $a,b$ be atoms of $P$ with $b$ preceding $a$ in $\Omega$.

Case 1. Suppose $a < u$. It follows that $b < u$. If $y > a,b$ then $y \geq a \vee b$. Since $a \vee b \leq u$, it follows that $a \vee b \in P$. Since $a \vee b$ clearly covers $a$ and $b$, we are done.

Case 2. Suppose $a \nless u$. Let $y > a,b$, and $y \in \hat{P}$. If $a \vee b \in P$, just as in Case 1 we have verified (ii). So we can assume $a \vee b \notin P$ and that $y = \hat{1}_{\hat{P}}$. Because the atoms below $u$ are first in $\Omega$, we can assume that $b < u$ to verify (ii). Since $a \vee b \notin P$, we have $w \wedge (a \vee b) \neq \hat{0}$. Clearly $a \vee b \nleq w$, so $w \wedge (a \vee b)$ is an atom. This implies that $w \vee (a \vee b)$ covers $w$. But since $a \wedge w = \hat{0}$, $w \vee a$ also covers $w$, so $w \vee a = w \vee (a \vee b)$. This clearly implies $b \leq w \vee a$. However $u \wedge (w \vee a)$ is also an atom and $b \leq u$, $b \leq w \vee a$ implies $b = u \wedge (w \vee a)$. Therefore $b$ is uniquely determined by $u$, $w$, and $a$. So any atom $b' < u$ with $b' \neq b$ satisfies $b' \vee a \in P$. Thus $b'$ satisfies (ii) because $b' \vee a \leq \hat{1}_{\hat{P}} = y$. □

In closing we would like to raise the question of whether Theorem 5.3 can be extended to a larger class of lattices, perhaps the supersolvable geometric lattices. We also do not know whether Theorem 5.3 holds for upper order ideals $\{x | x \vee w = \hat{1}\}$.

## BIBLIOGRAPHY


1. A. Björner, "Shellable and Cohen-Macaulay partially ordered sets", Trans. Amer. Math. Soc., 260 (1980), 159-183.

2. A. Björner, A.M. Garsia, and R.P. Stanley, "An introduction to Cohen-Macaulay partially ordered sets", Ordered Sets (I. Rival, ed.), Reidel, Dordrecht, 1982, pp. 583-615.

3. A. Björner and M. Wachs, "Bruhat order on Coxeter groups and shellability", Adv. in Math., 43 (1982), 87-100.

4. A. Björner and M. Wachs, "On lexicographically shellable posets", Trans. Amer. Math. Soc., 277 (1983), 323-341.

5. G. Birkhoff, Lattice Theory, 3rd ed., Amer. Math. Soc. Colloq. Publ., vol. 25, Amer. Math. Soc., Providence, RI, 1967.

6. P. Delsarte, "Association schemes and t-designs in regular semilattices", J. Combin Theory Ser. A, 20 (1976), 230-243.

7. T.A. Dowling, "A q-analog of the partition lattice", A Survey of Combinatorial Theory (J.N. Srivastava, ed.), North-Holland, Amsterdam, 1973, pp. 101-115.

8. T.A. Dowling, "A class of geometric lattices based on finite groups", J. Combin. Theory Ser. B, 14 (1973), 61-86.

9. C. Huneke and V. Lakshmibai, "A characterization of Kempf varieties by means of standard monomials and its geometric consequences," to appear.

10. R.P. Stanley, "Modular elements of geometric lattices", Algebra Universalis, 1 (1971), 214-217.

11. R.P. Stanley, "Supersolvable lattices", Algebra Universalis, 2 (1972), 197-217.

12. R.P. Stanley, "Cohen-Macaulay complexes", Higher Combinatorics (M. Aigner, ed.), Reidel, Dordrecht/Boston, 1977, pp 51-62.

13. D. Stanton, "A partially ordered set and q-Krawtchouk polynomials", J. Combin. Theory Ser. A, 30 (1981), 276-284.

14. J.R. Stonesifer, "Modularly complemented geometric lattices", Discrete Math., 32 (1980), 85-88.

15. K. Vogtmann, "Spherical posets and homology stability for $O_{n,n}$", Topology, 20 (1981), 119-132.

16. M. Wachs, "Quotients of Coxeter complexes and buildings with linear diagram", Europ. J. of Combinatorics, to appear.

17. J. Kahn and J.P.S. Kung, "A classification of modularly complemented geometric lattices", preprint.

SCHOOL OF MATHEMATICS
UNIVERSITY OF MINNESOTA
MINNEAPOLIS, MN 55455

DEPARTMENT OF MATHEMATICS
UNIVERSITY OF MIAMI
CORAL GABLES, FL 33124

Current address of second author:
Department of Electrical Engineering and Computer Sciences
University of California, San Diego
La Jolla, CA 92093

Contemporary Mathematics
Volume 34, 1984

# COUNTING FACES AND CHAINS IN POLYTOPES AND POSETS

Margaret M. Bayer[1] and Louis J. Billera[1]

ABSTRACT. The purpose of this paper is to provide an overview of a class of problems concerning the enumeration of faces in convex polytopes or general triangulated spheres, and of chains in certain related posets. The degree to which we refine the objects being counted may vary; for example, we may count the number of chains consisting of $k$ different faces of a given polytope, or we may instead ask for the number of chains of faces having dimensions given by a prespecified k-set of integers. Throughout, the unifying problem will be to determine all the affine linear relations satisfied by the numbers in question. The best known of all such relations is the Euler equation, $f_0-f_1+f_2 = 2$, which relates the number of vertices, edges and 2-faces of any 3-polytope.

While the paper is mostly expository, sections 4, 6 and 7 discuss new results. A listing of the section headings follows.



## 1. f-VECTORS OF CONVEX POLYTOPES.

By a <u>convex polytope</u> $P$ we mean the convex hull of a finite point set in a real Euclidean space. Equivalently, $P$ can be defined as the bounded intersection of finitely many closed half-spaces. By a face $F$ of $P$ we mean the intersection of $P$ with a hyperplane having the property that $P$ is contained in one of its closed half-spaces. Thus, the empty set is always a face of $P$, and we call $P$ a face of $P$ (whether or not it arises in the above manner). All other faces will be called <u>proper</u> faces, and they are finite in number. Each face of a polytope $P$ is again a polytope.

[1]Partially supported by NSF grant MCS81-02353 at Cornell University.

We define the dimension of a polytope $P$, $\dim P$, to be the dimension of $\mathrm{aff}(P)$, its affine hull, and say that $P$ is a d-polytope if $\dim P = d$. In this case each face of $P$, except $P$ itself, has dimension less than $d$. For each $i = -1,0,1,\dots,d-1$, let $f_i(P)$ denote the number of i-dimensional faces of $P$. In particular, $f_{-1}(P) = 1$ counts the empty face, $f_0(P)$ is the number of vertices, $f_1(P)$ is the number of edges and $f_{d-1}(P)$ is the number of facets of $P$. We denote by $f(P)$ the vector $(f_{-1}(P),f_0(P),\dots,f_{d-1}(P))$, called the f-vector of $P$. For a comprehensive treatment of the theory of convex polytopes and, in particular, of f-vectors see [15], [23] or [34]. For a survey of the latter topic which includes a discussion of the more recent results, see [29].

Let $f(P^d)$ denote the set of all f-vectors of d-polytopes. There is considerable interest in describing the set $f(P^d)$ exactly, but this remains unsettled in general. However, we can describe $\mathrm{aff}(f(P^d))$ and certain inequalities satisfied by each $f \in f(P^d)$. First, each $f \in f(P^d)$ satisfies the Euler Equation

$$f_0 - f_1 + f_2 - \dots \pm f_{d-1} = 1 - (-1)^d,$$

and, further, this equation specifies the affine hull, namely

$$\mathrm{aff}(f(P^d)) = \{(f_{-1},f_0,\dots,f_{d-1}) : f_{-1} = 1,\ f_0 - f_1 + \dots = 1 - (-1)^d\}.$$

(We will give a proof of the latter assertion in Section 3.)

The inequalities are somewhat harder to describe. To this end, consider the moment curve in $R^d$ given by $x(t) = (t,t^2,t^3,\dots,t^d)$ and choose real numbers $t_1 < t_2 < \dots < t_n$, with $n > d$. Define $C(n,d)$ to be the convex hull of $V = \{x(t_1),x(t_2),\dots,x(t_n)\}$. While the actual polytope obtained by this procedure depends on the choices of the $t_i$'s, it is known that its combinatorial structure, in particular, its f-vector, is independent of the $t_i$'s. We use the symbol $C(n,d)$ to refer to this combinatorial type. (In general, when we refer to a polytope, we will be concerned only with its combinatorial type, that is, its face lattice.) It is easily seen that $C(n,d)$ is a simplicial d-polytope, that is, each facet (and thus, each proper face) is a simplex (an $(r-1)$-polytope having just $r$ vertices).

One of the most remarkable properties of the polytope $C(n,d)$ is that it is neighborly, that is, each pair of vertices forms an edge of $C(n,d)$. In fact, for $k = 1,\dots,[d/2]$, the convex hull of <u>any</u> k-subset of $V$ is a face of $C(n,d)$. (Here $[x]$ denotes the largest integer less than or equal to $x$.) Thus among all d-polytopes with $n$ vertices, $C(n,d)$ clearly has the maximum number of i-faces for $i = 0,1,\dots,[d/2]-1$. That $C(n,d)$ has the maximum number of i-faces, among all d-polytopes with $n$ vertices, for <u>all</u> $i$ is the content of the <u>Upper Bound Theorem</u>, first formulated by Motzkin [36] and proved by McMullen [31]. Thus we have that for all d-polytopes $P$ with $n$ vertices

$$f(P) \leq f(C(n,d)).$$

Since the number of i-faces of $C(n,d)$ is known for each $i$ as a function of $n$ and $d$, this gives upper bounds for each $f_i(P)$ in terms of $n = f_0(P)$ (and $d$), and thus inequalities which must be satisfied for each $f \in f(P^d)$. By the above discussion we have

$$f_i(C(n,d)) = \binom{n}{i+1}$$

for $i = 0,1,\dots,[d/2]-1$. See [23] or [34] for a general expression for the remaining coordinates of $f(C(n,d))$ and its derivation.

A complete description of $f(P^d)$ has remained elusive for general $d$. There do not seem to be even reasonable conjectures as to a final set of conditions. However, if one restricts to the case of simplicial polytopes, then the situation is considerably better understood. In fact, the set $f(P_s^d)$ of all f-vectors of simplicial d-polytopes is completely known, being specified entirely by a list of linear equations, linear inequalities and nonlinear inequalities. We will describe these in turn. First note that by the usual polyhedral polarity [23], to each d-polytope $P$ there corresponds another d-polytope $P^*$ which has the property that $f_i(P) = f_{d-1-i}(P^*)$. In particular, when $P$ is simplicial, then $P^*$ is <u>simple</u>, that is, each vertex is on precisely $d$ facets. Thus, describing $f(P_s^d)$ is equivalent to describing the f-vectors of all simple d-polytopes. Further, one proves the Upper Bound

Theorem by first showing that the maximum number of faces must occur in simplicial polytopes, and then proving the Upper Bound Theorem for simplicial polytopes.

First, we note that each $f \in f(P_s^d)$ satisfies the Dehn-Sommerville Equations

$$E_k^d\colon \sum_{j=k}^{d-1} (-1)^j \binom{j+1}{k+1} f_j = (-1)^{d-1} f_k$$

for $k = -1,0,1,\ldots,d-1$. The equation $E_{-1}^d$ is just the Euler equation. It is known that $[(d+1)/2]$ of these equations are independent, and they completely determine $\mathrm{aff}(f(P_s^d))$. Thus the dimension of $\mathrm{aff}(f(P_s^d))$ is $[d/2]$. We treat these questions in detail in sections 2 and 3.

To be able to describe the remaining conditions, we must apply a change of variables to the space of f-vectors, first used by Sommerville [40], which recasts the Dehn-Sommerville equations in a particularly simple form. If $f = f(P)$ for a d-polytope $P$, then we define the h-vector of $P$ to be the vector $h(P) = (h_0,h_1,\ldots,h_d)$, where for each $i$

$$h_i = \sum_{j=0}^{i} (-1)^{i-j} \binom{d-j}{d-i} f_{j-1}.$$

Note that $h_0(P) = 1$ since $f_{-1}(P) = 1$. (Also, note the dependence of $h$ on $d = \dim P$.) These relations can be inverted to give

$$f_j = \sum_{i=0}^{j+1} \binom{d-i}{d-j-i} h_i.$$

Thus $f_j$ is a non-negative linear combination of $h_0,\ldots,h_{j+1}$, and so an inequality of the form $h(P) \leq h(P')$ implies the corresponding inequality $f(P) \leq f(P')$. In fact, the proof of the Upper Bound Theorem proceeds by showing $h(P) \leq h(C(n,d))$ for the simplicial d-polytopes with $n$ vertices. In terms of the h-vector of $P$, the Dehn-Sommerville equations become $h_i = h_{d-i}$, for $i = 0,1,\ldots,[d/2]$. (See [34] or [35] where our $h_{k+1}$ corresponds to their $g_k^d(P)$.)

Let $h(P_s^d)$ denote the set of h-vectors of simplicial d-polytopes. By the above discussion, knowing $h(P_s^d)$ is equivalent to knowing $f(P_s^d)$. We can now describe the set of linear inequalities satisfied by all $f \in f(P_s^d)$. They were first proposed by McMullen and Walkup in the form of a Generalized Lower Bound Conjecture, which stated that $h_{i+1} \geq h_i$, for $i = 0,1,\ldots,[d/2]-1$. Thus, in light of the Dehn-Sommerville equations, these inequalities imply that the h-vector must be unimodal. In terms of the $f_i$'s, $h_{i+1} \geq h_i$ implies a lower bound on $f_i$ as a linear function of the $f_j$'s for $j < i < [d/2]$; these lower bounds imply the lower bounds given in the so-called Lower Bound Theorem proved by Barnette [6] (see [35]).

To complete the description, we must establish a last bit of notation. For positive integers $h$ and $i$, we note that $h$ can always be written uniquely in the form

$$h = \binom{n_i}{i} + \binom{n_{i-1}}{i-1} + \ldots + \binom{n_j}{j}.$$

where $n_i > n_{i-1} > \ldots > n_j \geq j \geq 1$. (Choose $n_i$ to be the largest integer with $h \geq \binom{n_i}{i}$, etc.) Define the ith pseudopower of $h$ to be

$$h^{\langle i \rangle} = \binom{n_i+1}{i+1} + \binom{n_{i-1}+1}{i} + \ldots + \binom{n_j+1}{j+1}.$$

Put $0^{\langle i \rangle} = 0$ for all $i$.

We state the nonlinear inequalities on the components of the h-vector (and thus the f-vector) together with the earlier conditions in the form of a characterization of $h(P_s^d)$.

THEOREM (MCMULLEN'S CONDITIONS). An integer vector $h = (h_0,h_1,\ldots,h_d)$ is the h-vector of a simplicial convex d-polytope if and only if the following three conditions hold:

(i) $h_i = h_{d-i}$, $i = 0,1,\ldots,[d/2]$,

(ii) $h_{i+1} \geq h_i$, $i = 0,1,\ldots,[d/2]-1$, and

(iii) $h_0 = 1$ and $h_{i+1}-h_i \leq (h_i-h_{i-1})^{\langle i \rangle}$, $i = 1,\ldots,[d/2]-1$.

This characterization was conjectured in 1971 by McMullen [33], [34], and proved by him for $d \leq 5$ and for the case of d-polytopes having $n$ vertices where $d < n < d+3$. The sufficiency of these conditions was proved by Billera and Lee [11], [12]; the proof of necessity was given by Stanley [47]. The proof of sufficiency depends heavily on insights provided by earlier work of Stanley [43] in which the Upper Bound Theorem was extended to general triangulations of spheres by means of techniques of commutative algebra. The proof of necessity extends this earlier work, introducing powerful new techniques from algebraic geometry. See [10] and [50] for overviews of these developments.

Throughout the paper we will use the following notational conventions. $N$ will denote the set of natural numbers. A point $z \in N^{r+1}$ will have coordinates $(z_0, z_1, \ldots, z_r)$. If $z, w \in N^{r+1}$, the inequality $z \leq w$ means $z_i \leq w_i$ for $0 \leq i \leq r$. If further for some $i$, $z_i < w_i$ we will write $z < w$. For $r \in N$ write $\langle r \rangle = \{0, 1, \ldots, r-1\}$. For $S \subseteq \langle r \rangle$, we sometimes denote $\langle r \rangle \setminus S$ by $\tilde{S}$. It is convenient to assign the following values to binomial coefficients: $\binom{n}{0} = 0$ if $n < 0$, $\binom{n}{-1} = 0$ if $n \neq -1$, $\binom{-1}{-1} = 1$ and $\binom{n}{m} = 0$ if $0 \leq n < m$.

2. DEHN-SOMMERVILLE EQUATIONS. In 1905 Dehn conjectured the existence of $[d/2]$ linear relations on the f-vectors of simplicial d-polytopes. These equations, which form part of the McMullen conditions, were discovered and proved by Sommerville in 1927 [40]. They were largely forgotten until Klee reproved them in 1963 in a more general context [26]. They are referred to as the Dehn-Sommerville equations. We give here Sommerville's original proof. First we need the definition of an interval in the face lattice of a polytope.

Let $F_1$ and $F_2$ be i- and k-faces, respectively, of a d-polytope $P$, and suppose $F_1 \subseteq F_2$. (Here we allow $F_1 = \emptyset$ or $F_2 = P$.) Define the <u>interval</u> $[F_1, F_2]$ to be the set of faces $G$ of $P$ such that $F_1 \subseteq G \subseteq F_2$. $[F_1, F_2]$ is ordered by inclusion, and is isomorphic to the face lattice of a (k-i-1)-polytope. We will assume the Euler relation holds for any polytope. (See [23] for a proof.)

THEOREM 2.1. Dehn-Sommerville Equations. If $f(P) = (f_{-1}, f_0, \ldots, f_{d-1})$ is the f-vector of a simplicial d-polytope $P$, then for $-1 \leq k \leq d-2$,

$$f_k = \sum_{j=k}^{d-1} (-1)^{d-1-j} \binom{j+1}{k+1} f_j.$$

PROOF. Let $F$ be a k-face of $P$, and write $P_F$ for the (d-k-1)-polytope with face lattice $[F,P]$. Applying Euler's formula to $P_F$ we get

$$1 = \sum_{i=-1}^{d-k-2} (-1)^{d-k-i} f_i(P_F).$$

If we sum this equation over all k-faces of $P$ we get

$$f_k(P) = \sum_{i=-1}^{d-k-2} (-1)^{d-k-i} \sum_{\dim F=k} f_i(P_F).$$

The i-faces of $P_F$ correspond to (k+i+1)-faces of $P$ containing $F$. So

$$\sum_{\dim F=k} f_i(P_F) = \text{the number of pairs } F^k \subseteq F^{k+i+1}$$

$$= \sum_{\dim G=k+i+1} f_k(G),$$

where $F^k$ is a k-face, and $F^{k+i+1}$ is a (k+i+1)-face of P. Now, since $P$ is simplicial, $G$ is a (k+i+1)-simplex and $f_k(G) = \binom{k+i+2}{k+1}$. So

$$f_k(P) = \sum_{i=-1}^{d-k-2} (-1)^{d-k-i} \sum_{\dim G=k+i+1} \binom{k+i+2}{k+1}$$

$$= \sum_{i=-1}^{d-k-2} (-1)^{d-k-i} \binom{k+i+2}{k+1} f_{k+i+1}(P)$$

$$= \sum_{j=k}^{d-1} (-1)^{d-1-j} \binom{j+1}{k+1} f_j(P). \quad \square$$

Note that Euler's formula is the Dehn-Sommerville equation with $k = -1$. The equation for $k = d-2$ is $2f_{d-2} = df_{d-1}$. This simply says that each (d-2)-face is on exactly two facets, and each facet has exactly $d$ (d-2)-faces. Sommerville noted that in terms of the h-vector of a polytope the Dehn-Sommerville equations have a very nice form.

COROLLARY 2.2. If $h(P) = (h_0,h_1,\ldots,h_d)$ is the h-vector of a simplicial d-polytope $P$, then for $0 \le r \le d$, $h_r = h_{d-r}$.

PROOF. Let $E_k^d$ be the Dehn-Sommerville equation:

$$f_k = \sum_{j=k}^{d-1} (-1)^{d-1-j}\binom{j+1}{k+1}f_j.$$

For $0 \le r \le d$ we take the following linear combination of the equations: $\sum_{i=0}^{r} (-1)^i\binom{d-i}{d-r}E_{i-1}^d$. On the left-hand side we get

$$\sum_{i=0}^{r} (-1)^i\binom{d-i}{d-r}f_{i-1} = (-1)^r h_r.$$

On the right-hand side we get

$$\sum_{i=0}^{r} (-1)^i\binom{d-i}{d-r}\sum_{j=i-1}^{d-1} (-1)^{d-1-j}\binom{j+1}{i}f_j = \sum_{i=0}^{r} (-1)^i\binom{d-i}{d-r}\sum_{j=i}^{d} (-1)^{d-j}\binom{j}{i}f_{j-1}$$

$$= \sum_{j=0}^{d} (-1)^{d-j}f_{j-1}\sum_{i=0}^{j} (-1)^i\binom{d-i}{d-r}\binom{j}{i}.$$

We use the identity $\sum_{i=0}^{n} (-1)^i\binom{n}{i}\binom{i+m}{t} = (-1)^n\binom{m}{t-n}$ to simplify the right-hand sum:

$$\sum_{i=0}^{j} (-1)^i\binom{d-i}{d-r}\binom{j}{i} = \sum_{s=0}^{j} (-1)^{s-i}\binom{j}{s}\binom{d-j+s}{r} = \binom{d-j}{d-r-j}.$$

So the right-hand side of $\sum_{i=0}^{r} (-1)^i\binom{d-i}{d-r}E_{i-1}^d$ is

$$\sum_{j=0}^{d} (-1)^{d-j}f_{j-1}\binom{d-j}{r} = \sum_{j=0}^{d-r} (-1)^{d-j}\binom{d-j}{r}f_{j-1} = (-1)^r h_{d-r}.$$

Thus the combination of $E_k^d$ gives $h_r = h_{d-r}$. □

Since $h_0 = 1$, we see immediately that at most $[d/2]$ of the $h_i$, and thus of the $f_i$, can be independent. It is easy to show that the equations $h_r = h_{d-r}$ are actually equivalent to the equations $E_k^d$. In the next section we will do this by showing that the affine span of the f-vectors of simplicial d-polytopes is determined by the equations $h_r = h_{d-r}$. Although Sommerville observed this form of the equations, he did not realize the significance of the h-vector, which was extremely important in the discovery and proof of the McMullen conditions (see [34], [45], [47] and [12]). Since the links of simplices in homology spheres are again homology spheres, Sommerville's proof of the Dehn-Sommerville equations extends to this more general case as well.

3. SPANNING THE EULER HYPERPLANE AND THE DEHN-SOMMERVILLE SPACE. In this section we describe some operations on convex polytopes and their effect on the f- and h-vectors. We use these to give simple direct proofs that the Euler equation and the Dehn-Sommerville equations are the only affine linear equations satisfied by the f-vectors of convex polytopes and simplicial convex polytopes, respectively. Note first that for any d-polytope $Q$, the Euler equation is equivalent to the relation $h_d(Q) = 1$.

If $Q$ is a d-polytope, the pyramid on $Q$, $P(Q)$, is the convex $(d+1)$-polytope formed by taking the convex hull of $Q$ with a point not in the affine span of $Q$. In terms of f-vectors we have

$$f_i(P(Q)) = f_i(Q) + f_{i-1}(Q) \quad \text{for} \quad i \leq d-1$$

and

$$f_d(P(Q)) = 1 + f_{d-1}(Q),$$

with the convention that $f_j(Q) = 0$ if $j < -1$. (See [23, §4.2].) The following is due to Sommerville [40].

PROPOSITION 3.1. For any d-polytope $Q$,

$$h(P(Q)) = (h(Q),1).$$

PROOF. Since the Euler equation for $P(Q)$ gives $h_{d+1}(P(Q)) = 1$, we need show $h_i(P(Q)) = h_i(Q)$ for $i \leq d$. Since $P(Q)$ is a $(d+1)$-polytope, if $i \leq d$,

$$\begin{aligned} h_i(P(Q)) &= \sum_{j=0}^{i} (-1)^{i-j}\binom{d+1-j}{d+1-i} f_{j-1}(P(Q)) \\ &= \sum_{j=0}^{i} (-1)^{i-j}\binom{d+1-j}{d+1-i}[f_{j-1}(Q) + f_{j-2}(Q)] \\ &= \sum_{j=0}^{i} (-1)^{i-j}\binom{d+1-j}{d+1-i} f_{j-1}(Q) + \sum_{j=0}^{i-1} (-1)^{i-j-1}\binom{d-j}{d+1-i} f_{j-1}(Q) \\ &= \sum_{j=0}^{i} (-1)^{i-j}\binom{d-j}{d-i} f_{j-1}(Q) = h_i(Q). \quad \square \end{aligned}$$

Note that $P(Q)$ is not simplicial unless $Q$ is itself a simplex since $Q$ is a facet of $P(Q)$. In this case $P(Q)$ is again a simplex, and we get by induction

COROLLARY 3.2. If $Q$ is a simplex then

$$h(Q) = (1,1,\ldots,1). \quad \square$$

For a d-polytope $Q$, the <u>bipyramid</u> over $Q$, $B(Q)$, is defined to be the convex $(d+1)$-polytope formed by taking the convex hull of $Q$ with a line segment which meets $Q$ in a relative interior point of each. For example, the bipyramid over an interval is a square and the bipyramid over a square is an octahedron. For the f-vectors, we have [23, §4.3]

$$f_i(B(Q)) = f_i(Q) + 2f_{i-1}(Q) \quad \text{for} \quad i \leq d-1$$

and

$$f_d(B(Q)) = 2f_{d-1}(Q).$$

In terms of h-vectors, we have the following

PROPOSITION 3.3. For any d-polytope $Q$,

$$h(B(Q)) = (h(Q),0) + (0,h(Q)).$$

PROOF. Again, $h_{d+1}(B(Q)) = 1 = h_d(Q)$, so we must show $h_i(B(Q)) = h_i(Q) + h_{i-1}(Q)$ for $i \leq d$. By the proof of Proposition 3.1,

$$\begin{aligned} h_i(B(Q)) &= \sum_{j=0}^{i} (-1)^{i-j}\binom{d+1-j}{d+1-i}[f_{j-1}(Q) + 2f_{j-2}(Q)] \\ &= h_i(Q) + \sum_{j=0}^{i} (-1)^{i-j}\binom{d-(j-1)}{d-(i-1)}f_{j-2}(Q) \\ &= h_i(Q) + h_{i-1}(Q). \quad \square \end{aligned}$$

Note that if $Q$ is simplicial, then so is $B(Q)$. For example, since the h-vector of the interval is $(1,1)$, that of the square is $(1,2,1) = (1,1,0) + (0,1,1)$, and that of the octahedron is $(1,3,3,1) = (1,2,1,0) + (0,1,2,1)$. While we do not in this section make use of the bipyramid operation, it will prove useful later on.

To define the final operation, let $Q$ be a simplicial d-polytope and $F$ a proper face of $Q$. If $H$ is a ((d-1)-dimensional) hyperplane containing $Q$ in one of its closed half spaces, then a point $x \notin H$ is said to be <u>beneath</u> $H$ if it is on the same side of $H$ as $Q$, and <u>beyond</u> $H$ otherwise. Now let $F_1,\ldots,F_k$ be all the facets ((d-1)-dimensional faces) of $Q$ which contain $F$. Let $x$ be a point which is beyond the hyperplanes generated by these $F_i$'s and beneath the hyperplanes generated by any other facets. (A point $x \notin Q$ sufficiently close to the centroid of $F$ will do.) Define the stellar subdivision of the face $F$ in $Q$, $st(F,Q)$, to be the (simplicial) d-polytope which is the convex hull of $Q \cup \{x\}$. (See [18] where this operation is described for nonsimplicial $Q$ as well.)

To describe $st(F,Q)$ combinatorially, let $\Delta$ be the boundary complex of $Q$, and let $\sigma$ be the set of vertices of $F$. Then the boundary complex of $st(F,Q)$ is the complex $st(\sigma,\Delta)$, the stellar subdivision of simplex $\sigma$ in $\Delta$, where

$$st(\sigma,\Delta) = (\Delta\backslash\sigma) \cup \overline{x}\cdot\partial\sigma\cdot\ell k_\Delta\sigma.$$

Here $\Delta\backslash\sigma = \{\tau \in \Delta \mid \tau \not\supseteq \sigma\}$, $\ell k_\Delta\sigma$ is the link of $\sigma$ in $\Delta$, defined by

$$\ell k_\Delta\sigma = \{\tau \in \Delta \mid \tau \cap \sigma = \emptyset, \quad \tau \cup \sigma \in \Delta\},$$

$\partial\sigma$ is the complex of proper subsets of $\sigma$, $\overline{x}$ denotes the complex consisting of $\{x\}$ and $\emptyset$ and $\cdot$ denotes the join of simplicial complexes. (See [18].)

The complex $\ell k_\Delta\sigma$ is also the boundary complex of a polytope, the one whose face lattice is isomorphic to the interval $[F,Q]$ in the face lattice of $Q$. This polytope will be of dimension $k-1$ if $\dim F = d-k$; let $h([F,Q])$ denote its h-vector. For vectors $a = (a_0,a_1,\ldots,a_k)$, $b = (b_0,b_1,\ldots,b_\ell)$, let $a*b = (c_0,c_1,\ldots,c_{k+\ell})$ denote their convolution, where $c_i = \sum_{j=0}^{i} a_j b_{i-j}$. The following is proved in [28; Proposition 2.10.1].

LEMMA 3.4. If $Q$ is a simplicial d-polytope and $F$ is a (d-k)-face of $Q$, then

$$h(st(F,Q)) = h(Q) + \underbrace{(0,1,1,\ldots,1,0)}_{d-k+2}*h([F,Q]).$$

We wish to apply this when $Q$ is a simplex. In this case, $[F,Q]$ is the face lattice of a (k-1)-simplex, and so

$$h([F,Q]) = \underbrace{(1,1,\ldots,1)}_{k}.$$

So when $Q$ is a simplex and $F$ is a $(d-k)$-face of $Q$, we have

$$h(st(F,Q)) = h(Q) + \underbrace{(0,1,\ldots,1,0)}_{d-k+2}*\underbrace{(1,\ldots,1)}_{k}.$$

We denote the polytopes $st(F,Q)$ in this case by $T_k^d$; they are all the simplicial d-polytopes with $d+2$ vertices (see [23, §6.1]). For convenience, let $T_0^d$ denote the d-simplex.

PROPOSITION 3.5. For $0 \leq k \leq [d/2]$,

$$h(T_k^d) = (1,2,3,\ldots,k,k+1,k+1,\ldots,k+1,k,\ldots,3,2,1)$$

PROOF. It is enough to note in the expression above for $h(st(F,Q))$ that

$$h(Q) = (1,1,\ldots,1)$$

and $(0,1,\ldots,1,0)*(1,\ldots,1)$ is just the vector of (reverse) diagonal sums of the matrix

$$k\left.\left(\begin{array}{cccccc} 0 & 1 & \cdots & 1 & 0 \\ 0 & 1 & \cdots & 1 & 0 \\ \cdot & \cdot & & \cdot & \cdot \\ \cdot & \cdot & & \cdot & \cdot \\ \cdot & \cdot & & \cdot & \cdot \\ 0 & 1 & \cdots & 1 & 0 \end{array}\right)\right. \quad \overbrace{\qquad}^{d-k+2}$$

starting in the upper left corner, i.e. 0, 0+1, 0+1+1, etc. □

We can now prove

THEOREM 3.6. The f-vectors of the simplicial d-polytopes $T_k^d$, $0 \leq k \leq [d/2]$ span the Dehn-Sommerville subspace, that is, the set

$$\{f(T_k^d) \mid 0 \leq k \leq [d/2]\}$$

is affinely independent.

PROOF. Since $f_{-1}(T_k^d) = 1$ for each $k$, it is enough to show the matrix of these f-vectors has full row rank. But since the transformation from $f$ to $h$ is invertible, it is enough to consider the matrix of h-vectors. By Proposition 3.5, this matrix is

$$\begin{pmatrix} 1 & 1 & 1 & \ldots\ldots & 1 & 1 & 1 \\ 1 & 2 & 2 & \ldots\ldots & 2 & 2 & 1 \\ 1 & 2 & 3 & \ldots\ldots & 3 & 2 & 1 \\ \cdot & \cdot & \cdot & & \cdot & \cdot & \cdot \\ \cdot & \cdot & \cdot & & \cdot & \cdot & \cdot \\ \cdot & \cdot & \cdot & & \cdot & \cdot & \cdot \\ 1 & 2 & 3 & \ldots [d/2]+1 \ldots & 3 & 2 & 1 \end{pmatrix},$$

and it is easily seen to have independent rows. □

To demonstrate a basis for the Euler hyperplane, we introduce another class of polytopes. Define $T_k^{d,r}$ to be the r-fold pyramid over the (d-r)-polytope $T_k^{d-r}$, where $0 \leq r \leq d-2$ and $1 \leq k \leq [\frac{1}{2}(d-r)]$ (that is, the result of performing the pyramid operation r times, beginnning with $T_k^{d-r}$). These polytopes constitute all the d-polytopes with d+2 vertices (see [23; §6.1]). It is straightforward, using Propositions 3.1 and 3.5 to write down the vectors $h(T_k^{d,r})$. However, for our purposes, we need consider only the case $k = 1$.

THEOREM 3.7. The f-vectors of the d-polytopes $T_1^{d,r}$, $0 \leq r \leq d-2$, together with that of the d-simplex $T_0^d$, span the Euler hyperplane.

PROOF. Again, it is enough to show that the matrix of h-vectors has full row rank. By Propositions 3.1 and 3.5, this matrix (with $h(T_0^d)$ first) is easily seen to be

$$\begin{pmatrix} 1 & 1 & \ldots\ldots & & & 1 & 1 \\ 1 & 2 & 1 & 1 & \ldots & 1 & 1 \\ 1 & 2 & 2 & 1 & \ldots & 1 & 1 \\ \cdot & \cdot & \cdot & & & \cdot & \cdot \\ \cdot & \cdot & \cdot & & & \cdot & \cdot \\ \cdot & \cdot & \cdot & & & \cdot & \cdot \\ 1 & 2 & 2 & \ldots\ldots & & 2 & 1 \end{pmatrix}. \quad \square$$

We conclude that the Euler relation is the only linear relation holding for f-vectors of all polytopes. For similar results for general Euler-like relations, see [39].

## 4. LABELED SIMPLICIAL COMPLEXES.

DEFINITIONS. We start with some definitions pertaining to simplicial complexes.

Let $\Delta$ be a simplicial complex with $n$ vertices, $v_1, v_2, \ldots, v_n$. Associated with $\Delta$ is its Stanley-Reisner ring $A_\Delta$. Take the polynomial ring in $n$ indeterminates corresponding to the vertices: $K[x_1, x_2, \ldots, x_n]$. (We will assume throughout that $K$ is the field of rational numbers, although much of what we do works for more general fields.) Define the support of a monomial $m = \Pi_{i=1}^{n} x_i^{r_i}$ to be $\text{supp}(m) = \{v_i : r_i > 0\}$. Let $I_\Delta$ be the ideal generated by those monomials $m$ whose supports are not faces of $\Delta$. Then the Stanley-Reisner ring of $\Delta$ is $A_\Delta = K[x_1, \ldots, x_n]/I_\Delta$. Note that as a K-vector space, $A_\Delta$ is generated by those monomials $m$ whose supports are faces of $\Delta$.

Let $A$ be a K-algebra. An N-grading of $A$ is a decomposition of $A$ as a direct sum of K-vector spaces $A = \sum_{i \in N} A_i$, such that $\forall\, i,j \in N$, $A_i A_j \subseteq A_{i+j}$. (Throughout this paper, $\sum$ will denote direct sum.) More generally, if $M$ is a commutative monoid with operation $+$, $A$ is M-graded if $A = \sum_{m \in M} A_m$ with $A_m A_n \subseteq A_{m+n}$ $\forall\, m,n \in M$. We will assume our graded rings $A$ are finitely generated K-algebras, so that each graded piece $A_m$ is a finite-dimensional K-vector space, and that $A_0 = K$. In this case, $A_0 A_m = KA_m = A_m$ for all $m \in M$. So $A$ is an M-graded K-vector space. We can then define the Hilbert function of $A$ to be $H(A,m) = \dim_K A_m$ for $m \in M$. In the case where $M = N^{r+1}$ we define a generating function for the Hilbert function as follows. Let $t_0, t_1, t_2, \ldots, t_r$ be indeterminates and for $z = (z_0, z_1, z_2, \ldots, z_r) \in N^{r+1}$ write $t^z = t_0^{z_0} t_1^{z_1} t_2^{z_2} \cdots t_r^{z_r}$. Then the Hilbert (or Poincaré) series for $A$ is $P(A,t) = \sum_{z \in N^{r+1}} H(A,z) t^z$. The Hilbert series of $A_\Delta$ contains important combinatorial information about $\Delta$.

A labeling of a simplicial complex $\Delta$ is a partition of the vertex set of $\Delta$ into subsets: $V(\Delta) = V_0 \cup \ldots \cup V_r$. The vertices in $V_i$ are said to be labeled $i$. If the labeling is such that each maximal face of $\Delta$ contains exactly one vertex from such $V_i$, then $\Delta$ is said to be completely balanced [46]. In this case $\Delta$ is a pure simplicial r-complex, that is, a simplicial complex in which every maximal face has $r+1$ vertices. Maximal faces are then called facets.

A combinatorially interesting class of simplicial complexes comes from partially ordered sets (posets). We assume all posets have least and greatest elements, $\hat{0}$ and $\hat{1}$. For $P$ a finite poset, we form a simplicial complex $\Delta(P)$: $\Delta(P)$ has as vertices the elements of $P$ (except $\hat{0}$ and $\hat{1}$), and a set of elements $\{x_1,x_2,\ldots,x_k\}$ is a face of $\Delta(P)$ if and only if $\hat{0} < x_{\sigma(1)} < x_{\sigma(2)} < \ldots < x_{\sigma(k)} < \hat{1}$ for some permutation $\sigma$ of $\{1,2,\ldots,k\}$. We call simplicial complexes arising in this way poset complexes. If we wish to stress the poset giving rise to the complex, we call $\Delta(P)$ the order complex of $P$.

We give here a characterization of poset complexes due to Stanley [46]. A cycle of length $k$ of a complex $\Delta$ is a sequence of vertices $v_1,v_2,\ldots,v_k,v_{k+1} = v_1$, allowing repetitions, where $\{v_i,v_{i+1}\}$ is an edge of $\Delta$ and no pair of vertices occurs twice in the same order. Such a cycle has a triangular chord if for some $i$, $\{v_i,v_{i+2}\}$ (subscripts modulo $k$) is an edge of $\Delta$. A simplicial complex $\Delta$ is the order complex of a finite poset if and only if (i) any minimal set of vertices not forming a face of $\Delta$ has two elements; and (ii) every odd cycle of $\Delta$ has a triangular chord (that is, the 1-skeleton of $\Delta$ is a comparability graph [20]).

A poset $P$ is called ranked if for every $x \in P$ all maximal chains up to $x$, $\hat{0} < x_0 < \ldots < x_k = x$, have the same length $k+1$. We then call $k$ the rank of $x$, written $r(x)$. We make the convention $r(\hat{0}) = -1$ and define $r(P)$ to be $r(\hat{1})$. (Note that this rank function corresponds to the usual rank function shifted down by 1; it corresponds to the usual rank in $P\backslash\{\hat{0}\}$.) A labeling of the vertices of $\Delta(P)$ with the ranks of the corresponding elements in $P$ makes $\Delta(P)$ completely balanced, since every maximal chain contains exactly one element of each rank. Conversely, if $P$ is any finite poset whose order complex is completely balanced then every maximal chain in $P$ has the same length, so $P$ is ranked.

If $\Delta$ is a simplicial complex labeled by $0,1,2,\ldots,r$, define, for each $z = (z_0,\ldots,z_r) \in N^{r+1}$, $f_z(\Delta)$ to be the cardinality of the set $\Delta_z = \{\sigma \in \Delta: |\sigma \cap V_i| = z_i$ for each $i$, $0 \leq i \leq r\}$. Then the total number of j-faces of

$\Delta$ is $f_j(\Delta) = \sum_{\substack{z \\ \sum z_i = j+1}} f_z(\Delta)$. In the case where the labeling makes $\Delta$ completely balanced, we adopt a notation that will prove useful later. In this case, $f_z(\Delta) = 0$ unless $z_i \leq 1$ for all $i$. Write $Z = \operatorname{supp} z = \{i: z_i = 1\}$ and define $\Delta_Z = \Delta_z$ and $f_Z(\Delta) = f_z(\Delta)$; then $f_j(\Delta) = \sum_{\substack{T \subseteq \langle r+1 \rangle \\ |T| = j+1}} f_T(\Delta)$. If $\Delta$ is completely balanced, we define $h_S(\Delta)$ in analogy to $h_i$:

$$h_S(\Delta) = \sum_{T \subseteq S} (-1)^{|S|-|T|} f_T(\Delta).$$

Letting $d = r+1$, it is easy to check that

$$\sum_{\substack{S \subseteq \langle d \rangle \\ |S| = i}} h_S(\Delta) = \sum_{j=0}^{i} (-1)^{i-j} \binom{d-j}{d-i} f_{j-1}(\Delta) = h_i(\Delta).$$

Because of the relationships between the numbers $f_S$ and $f_i$, $h_S$ and $h_i$, we will refer to the vectors $(f_S)_{S \subseteq \langle d \rangle}$ and $(h_S)_{S \subseteq \langle d \rangle}$ as the <u>extended</u> f-vector and h-vector.

If $P$ is a poset of rank $d$ (i.e., $r(\hat{1}) = d$), then $\Delta(P)$ is completely balanced with labels $\{0,1,\ldots,d-1\}$. For $T = \{i_1,\ldots,i_k\} \subseteq \{0,1,\ldots,d-1\}$, $f_T(\Delta(P))$ is the number of chains of $P$ of the form $\hat{0} < x_{i_1} < \ldots < x_{i_k} < \hat{1}$, where $r(x_j) = j$. In this case we will often write $f_T(P) = f_T(\Delta(P))$, and $h_S(P) = h_S(\Delta(P))$.

An important special case of ranked poset complexes occurs when $P$ is the lattice of faces of a convex d-polytope $Q$. (Here our choice of rank function leads to $r(P) = \dim Q$.) In this case $\Delta(P)$ is the complete barycentric subdivision of the polytope $Q$ and is itself a convex polytope [18]. Each vertex of $\Delta(P)$ corresponds to a face of $Q$, its label being the dimension of that face. Here, $f_T(P)$ is the number of chains of faces of $Q$ having precisely the dimensions in $T$. In particular $f_i(Q) = f_{\{i\}}(P)$, so information on the extended f-vector of $P$ will yield information on $f(Q)$. In the remainder of this paper, especially in section 7, we will extend the methods and results of sections 2 and 3 to the study of extended f-vectors.

Another labeled complex associated with $Q$ is the minimal subdivision $\sigma(Q)$ discussed in [7; section 2.5]. It is the result of performing stellar subdivisions on all the non-simplex faces of $Q$ in order of decreasing dimension and so by [18] is a simplicial polytope. Vertices of $\sigma(Q)$ are labeled 0 if they are vertices of $Q$; otherwise they are labeled with the dimension of the face that they subdivide. This is not a balanced labeling (e.g. each original edge has two vertices with 0 labels and there are no vertices with label 1), but the results of this section will apply here.

Another way to describe the complex $\sigma(Q)$ is as follows. If $V_Q$ is the set of vertices of $Q$, then the vertex set $V$ of $\sigma(Q)$ is

$$V_Q \cup \{v_F\colon F \text{ a nonsimplex face of } Q\}$$

where the $v_F$'s are new symbols. Each $v \in V$ is labeled with the appropriate dimension. Let

$$\phi = F_0 \subset F_1 \subset F_2 \subset \dots \subset F_k$$

be a chain of faces of $Q$; a simplex in $\sigma(Q)$ can be defined as follows. Let $F_i$, $0 \le i \le k$ be the largest face in the chain which is a simplex. Then

$$F_i \cup \{v_{F_{i+1}},\dots,v_{F_k}\}$$

is a simplex in $\sigma(Q)$; by considering all chains we obtain all simplices in $\sigma(Q)$. For example, if $Q$ is the pyramid over a square base, $\sigma(Q)$ is an octahedron with one vertex labeled 2, the rest labeled 0.

We now describe an $N^{r+1}$-grading on the ring $A_\Delta$ associated with a complex $\Delta$ labeled by $0,1,\dots,r$. If the generator $x$ in $A_\Delta$ corresponds to a vertex $v$ of $\Delta$ with label $i$, let the degree of $x$ (deg $x$) be $e_i$, the ith unit vector in $N^{r+1}$. Multiplication of monomials in $A_\Delta$ results in addition of degrees in $N^{r+1}$. So if we write $(A_\Delta)_z$ for the subspace of $A_\Delta$ spanned by monomials of degree $z \in N^{r+1}$, then $(A_\Delta)_z(A_\Delta)_w \subseteq (A_\Delta)_{z+w}$. Thus $A_\Delta = \sum_{z\in N^{r+1}} (A_\Delta)_z$ defines a grading on $A_\Delta$. Stanley [46] derives the Hilbert function of $A_\Delta$ with respect to the $N^{r+1}$-grading. (Although he states the proposition for balanced complexes the proof uses only the fact that the complex is labeled.)

PROPOSITION 4.1. (Stanley). If $\Delta$ is a simplicial complex labeled by $0,1,\ldots,r$, then for all $w \in N^{r+1}$,

$$H(A_\Delta, w) = \sum_{z \in N^{r+1}} f_z(\Delta) \prod_{i=0}^{r} \binom{w_i - 1}{z_i - 1}. \quad \square$$

Recall that we use the conventions $\binom{n}{0} = 0$ if $n < 0$, $\binom{n}{-1} = 0$ if $n \neq -1$, and $\binom{-1}{-1} = 1$.

Let us apply Proposition 4.1 in the case where $\Delta$ is completely balanced. In this case $f_z(\Delta) = 0$ if $z_i > 1$ for some $i$. On the other hand the product $\Pi_{i=0}^{r} \binom{w_i - 1}{z_i - 1}$ is nonzero if and only if $z \leq w$ and, for all $i$, $z_i = 0$ if and only if $w_i = 0$. Together these mean that the term $f_z(\Delta) \Pi_{i=0}^{r} \binom{w_i - 1}{z_i - 1}$ is nonzero if and only if $z$ is given by

$$z_i = \begin{cases} 1 & \text{if } w_i \geq 1 \\ 0 & \text{if } w_i = 0. \end{cases}$$

For this $z$, $\Pi_{i=0}^{r} \binom{w_i - 1}{z_i - 1} = 1$, so $H(A_\Delta, w) = f_z(\Delta) = f_{\text{supp } w}(\Delta)$.

AN EXACT SEQUENCE FOR SPHERES. This subsection proves that a certain sequence of graded vector spaces associated with a homology sphere is exact. The exact sequence is given without proof by Danilov [17], who referred to a related result of Kouchnirenko [27]. The proof here is based on that of Kouchnirenko. The exact sequence enables us to prove generalizations of the Dehn-Sommerville equations for labeled simplicial spheres.

Any set $L$ of faces of a simplicial complex determines a complex $\overline{L}$ consisting of all faces of elements of $L$. Recall that for $\sigma$ a face of a complex $\Delta$ the <u>link</u> of $\sigma$ in $\Delta$ is $\ell k_\Delta \sigma = \{\tau \in \Delta : \tau \cap \sigma = \emptyset, \tau \cup \sigma \in \Delta\}$; define the <u>star</u> of $\sigma$ in $K$ to be $\text{star}_\Delta \sigma = \{\tau \in \Delta : \sigma \subseteq \tau\}$. Then $\overline{\text{star}_\Delta \sigma}$ is the join $\ell k_\Delta \sigma \cdot \overline{\{\sigma\}} = \{\tau \cup \rho : \tau \in \ell k_\Delta \sigma, \rho \subseteq \sigma\}$. Also define the <u>boundary</u> of $\text{star}_\Delta E$ to be $(\text{star}_\Delta \sigma)^\bullet = \overline{\text{star}_\Delta \sigma} \backslash \text{star}_\Delta \sigma = \{\tau \in \overline{\text{star}_\Delta \sigma} : \sigma \not\subseteq \tau\}$. A $(d-1)$-dimensional complex $\Delta$ is said to be a <u>homology</u> $(d-1)$-<u>sphere</u> if for each $k$-face $\sigma \in \Delta$, $\ell k_\Delta \sigma$ is a $(d-k-2)$-dimensional complex having the rational homology of a $(d-k-2)$-sphere, $-1 \leq k \leq d-1$.

Let $\Delta$ be a homology (d-1)-sphere, and let $A_\Delta$ be its Stanley-Reisner ring. As a vector space basis for $A_\Delta$ we choose the set of monomials whose supports are faces of $\Delta$. Let $M$ be the semigroup of all monomials in $\{x_1,x_2,\ldots,x_n\}$ $(n = f_0(\Delta))$, including the "empty" monomial 1, with standard monomial multiplication. Then $A_\Delta$ is an M-graded algebra, and if supp m is a face of $\Delta$, we write $A_\Delta(m)$ for the subspace of $A_\Delta$ generated by m.

Let $\Delta_j$ be the set of j-faces of $\Delta$ $(-1 \le j \le d-1)$, and for $\sigma \in \Delta_j$, $\sigma = \{v_{i_0},v_{i_1},\ldots,v_{i_j}\}$, let $A_\sigma = K[x_{i_0},x_{i_1},\ldots,x_{i_j}]$. Define $C_j = \sum_{\sigma\in\Delta_j} A_\sigma$. $C_j$ with componentwise multiplication is then a K-algebra. A typical element of $C_j$ is written $(a_\sigma)_{\sigma\in\Delta_j}$. For $m \in M$ the <u>homogeneous elements of degree</u> <u>m</u> in $C_j$ are the elements $(q_\sigma m)_{\sigma\in\Delta_j}$, where $\forall_\sigma$ $q_\sigma \in K$ and $q_\sigma = 0$ unless supp $m \subseteq \sigma$. If we write $C_j(m)$ for the set of homogeneous elements of degree m in $C_j$, then $C_j = \sum_{m\in M} C_j(m)$, and for $m,n \in M$, $C_j(m)\cdot C_j(n) \subseteq C_j(mn)$, so this defines an M-grading of $C_j$. Note that $C_j(m) = 0$ for $j < \dim(\text{supp } m)$.

If A and B are M-graded vector spaces and $g: A \to B$ is a linear transformation, we say g is <u>homogeneous with respect to M</u> (or M-homogeneous or a homomorphism of M-graded vector spaces) if $\forall_{m\in M}\ g(A_m) \subseteq B_m$. In the context defined above, a linear transformations $g: C_j \to C_{j-1}$ is M-homogeneous if $g(C_j(m)) \subseteq C_{j-1}(m)$.

THEOREM 4.2. For $\Delta$ a homology (d-1)-sphere, $A_\Delta$, $C_j$ and M defined as above, there exist M-homogeneous linear transformations $\partial_i$ such that the sequence

$$0 \longrightarrow A_P \xrightarrow{\partial_d} C_{d-1} \xrightarrow{\partial_{d-1}} C_{d-2} \xrightarrow{\partial_{d-2}} \cdots$$
$$\xrightarrow{\partial_1} C_0 \xrightarrow{\partial_0} C_{-1} \longrightarrow 0$$

is exact.

PROOF. First we define $\partial_j$, $0 \le j < d-1$. To simplify notation write $m_\sigma = (0,\ldots,0,m,0,\ldots,0)$, the element of $C_j$ with monomial m in the $\sigma$ component and zeros elsewhere. As a K-vector space, $C_j$ has basis $\{m_\sigma: \sigma \in \Delta_j, \text{ supp } m \subseteq \sigma\}$. We define the $\sigma'$ component $(\sigma' \in \Delta_{j-1})$ of $\partial_j(m_\sigma)$ by analogy to the differential on the ordered chain complex of $\Delta$.

Assume the vertices of $\Delta$ are ordered once and for all $v_1, v_2, \ldots, v_n$. Write $\sigma = \{v_{i_0}, v_{i_1}, \ldots, v_{i_j}\}$ with $i_0 < i_1 < \ldots < i_j$. If $\sigma' \subseteq \sigma$ and $\dim \sigma' = j-1$ then $\sigma' = \{v_{i_0}, v_{i_1}, \ldots, v_{i_j}\} \backslash \{v_{i_q}\}$. In this case, define $s(\sigma,\sigma') = (-1)^q$. If $\sigma' \nsubseteq \sigma$ let $s(\sigma,\sigma') = 0$. Now define the $\sigma'$ component of $\partial_j(m_\sigma)$ to be

$$(\partial_j(m_\sigma))_{\sigma'} = \begin{cases} s(\sigma,\sigma')m, & \text{if } \operatorname{supp} m \subseteq \sigma' \\ 0, & \text{otherwise.} \end{cases}$$

This map is clearly M-homogeneous.

Next we show the exactness of the sequence $0 \to A_\Delta \to C_{d-1} \to \ldots \to C_0 \to C_{-1} \to 0$ (by showing that the sequence "restricted to each $m \in M$" is exact. Let $m \in M$, $m \neq 1$; let $\sigma = \operatorname{supp} m$, $k = \dim \sigma$. At $m$ we get the sequence $C_{d-1}(m) \to C_{d-2}(m) \to \ldots \to C_{k+1}(m) \to C_k(m) \to 0$ or

$$(4.3) \qquad \sum_{\substack{\tau\in\Delta_{d-1}\\ \sigma\subseteq\tau}} Km \to \sum_{\substack{\tau\in\Delta_{d-2}\\ \sigma\subseteq\tau}} Km \to \ldots \to \sum_{\substack{\tau\in\Delta_{k+1}\\ \sigma\subseteq\tau}} Km \to Km \to 0.$$

The proof that this part of the sequence is exact will also produce the map $\partial_d\colon A_\Delta(m) = Km \to C_{d-1}(m)$.

We claim that the restricted sequence (4.3) is a chain complex isomorphic to the relative chain complex $C(\overline{\operatorname{star}_\Delta\sigma})/C((\operatorname{star}_\Delta\sigma)^\cdot)$ over K. $C_j(m)$ is generated as a K-vector space by elements $m_\tau$, where $\tau \in \Delta_j$, $\sigma \subseteq \tau$. Such $\tau$ are precisely the j-dimensional elements of $\overline{\operatorname{star}_\Delta\sigma} \backslash (\operatorname{star}_\Delta\sigma)^\cdot$; thus they form a basis for $C_j(\overline{\operatorname{star}_\Delta\sigma})/C_j((\operatorname{star}_\Delta\sigma)^\cdot)$.

We use this correspondence to define the linear map $g_j\colon C_j(m) \to C_j(\overline{\operatorname{star}_\Delta\sigma})/C_j((\operatorname{star}_\Delta\sigma)^\cdot)$ such that $g_j(m_\tau) = \tau$; this map is clearly invertible. We want the following square to commute:

$$\begin{array}{ccc} C_j(m) & \xrightarrow{\quad \partial_j \quad} & C_{j-1}(m) \\ \Big\downarrow g_j & & \Big\downarrow g_{j-1} \\ C_j(\overline{\operatorname{star}_\Delta\sigma})/C_j((\operatorname{star}_\Delta\sigma)^\cdot) & \xrightarrow{\ d_j\ } & C_{j-1}(\overline{\operatorname{star}_\Delta\sigma})/C_{j-1}((\operatorname{star}_\Delta\sigma)^\cdot). \end{array}$$

Here $\partial_j$ means $\partial_j | C_j(m)$, and $d_j$ is the differential of the relative chain complex. But the $\partial_j$ were defined precisely to make this diagram commute. We have

$$g_{j-1}(\partial_j(m_\tau)) = g_{j-1}\Big( \sum_{\substack{\tau' \in \Delta_{j-1} \\ \text{supp } m \subseteq \tau' \subseteq \tau}} s(\tau,\tau') m_{\tau'} \Big)$$

$$= \sum_{\substack{\tau' \in \Delta_{j-1} \\ \text{supp } m \subseteq \tau' \subseteq \tau}} s(\tau,\tau')\tau'.$$

On the other hand,

$$d_j(g_j(m_\tau)) = d_j(\tau) = \sum s(\tau,\tau')\tau',$$

the summation being over $\tau' \in \overline{\text{star}_\Delta\sigma} \backslash (\text{star}_\Delta\sigma)^\bullet$ with $\dim \tau' = j-1$. By the preceding paragraph this shows $g_{j-1} \circ \partial_j = d_j \circ g_j$.

So $g_j$ is an isomorphism between chain complexes. By [24; Corollary 2.10.12],

$$H_j(\overline{\text{star}_\Delta\sigma}, (\text{star}_\Delta\sigma)^\bullet) = \widetilde{H}_{j-k-1}(\ell k_\Delta\sigma) = \begin{cases} K & \text{if } j = d-1 \\ 0 & \text{else.} \end{cases}$$

Thus for $j \leq d-2$, $\text{Im}(\partial_{j+1} | C_{j+1}(m)) = \text{Ker}(\partial_j | C_j(m))$. Now we are ready to define $\partial_d$. Let $z$ be a generator of $H_{d-1}(\overline{\text{star}_\Delta\sigma}, (\text{star}_\Delta\sigma)^\bullet)$, so $g_{d-1}^{-1}(z)$ generates $\text{Ker}(\partial_{d-1} | C_{d-1}(m))$. Define $\partial_d(qm) = qg_{d-1}^{-1}(z)$, $q \in K$. Then $\partial_d | A_\Delta(m)$ is an injection and $\text{Im}(\partial_d | A_\Delta(m)) = \text{Ker}(\partial_{d-1} | C_{d-1}(m))$. So for $m \in M$, $m \neq 1$, and $k = \dim \text{supp } m$, the sequence

$$0 \to A_\Delta(m) \to C_{d-1}(m) \to \cdots \to C_{k+1}(m) \to C_k(m) \to 0$$

is exact.

Now we deal with the case $m = 1$, for which $\text{supp } m = \emptyset \subseteq \tau$ for all faces $\tau$ of $\Delta$. We want exactness of the sequence

$$0 \to A_\Delta(1) \to C_{d-1}(1) \to C_{d-2}(1) \to \cdots \to C_0(1) \to C_{-1}(1) \to 0$$

or

$$0 \to K \to \sum_{\tau \in \Delta_{d-1}} K \to \sum_{\tau \in \Delta_{d-2}} K \to \cdots \to \sum_{\tau \in \Delta_0} K \to K \to 0.$$

Just as before we get an isomorhism between $C(1)$ and a simplicial chain complex. Let $C(\Delta)$ be the simplicial chain complex associated with $\Delta$, and augment $C(\Delta)$ with $C_0(\Delta) \overset{\varepsilon}{\to} C_{-1}(\Delta) = K \to 0$ (here $\varepsilon(v) = 1$ for all vertices $v$). $C_j(1)$ is generated by $\{1_\tau : \tau \in \Delta_j\}$; a basis for $C_j(\Delta)$ is just $\Delta_j$. So if we define $g_j(1_\tau) = \tau$ for $0 \leq j \leq d-1$, then the same calculation as in the previous case shows that the square

$$\begin{array}{ccc} C_j(1) & \xrightarrow{\partial_j} & C_{j-1}(1) \\ \downarrow g_j & & \downarrow g_{j-1} \\ C_j(\Delta) & \xrightarrow{d_j} & C_{j-1}(\Delta) \end{array}$$

commutes. Now, defining $g_{-1}(q) = q \in C_{-1}(\Delta)$ we check that the square

$$\begin{array}{ccc} C_0(1) & \xrightarrow{\partial_0} & C_{-1}(1) \\ \downarrow g_0 & & \downarrow g_{-1} \\ C_0(\Delta) & \xrightarrow{\varepsilon} & C_{-1}(\Delta) \end{array}$$

commutes. We have $\partial_0(1_\tau) = 1$ for all $\tau \in \Delta_0$, so

$$g_{-1}(\partial_0(1_\tau)) = g_{-1}(1) = 1 = \varepsilon(\tau) = \varepsilon(g_0(1_\tau)).$$

Now $\tilde{H}_i(\Delta) = 0$ for $i \neq d-1$ so the sequence

$$C_{d-1}(1) \to C_{d-2}(1) \to \dots \to C_0(1) \to C_{-1}(1) \to 0$$

is exact. Finally we define $\partial_d(1)$. We know $\mathrm{Ker}(\partial_{d-1}|C_{d-1}(1)) \cong \tilde{H}_{d-1}(\Delta) \cong K$. Let $z$ be a generator of $\tilde{H}_{d-1}(\Delta)$, and for $q \in K$ define $\partial_d(q) = qg_{d-1}^{-1}(z)$. As in the case $m \neq 1$ we get that the sequence

$$0 \to A_\Delta(1) \to C_{d-1}(1) \to \dots \to C_0(1) \to C_{-1}(1) \to 0$$

is exact.

Since the maps $\partial_i$ are homogeneous,

$$\mathrm{Im}\,\partial_j = \textstyle\sum_{m\in M} \mathrm{Im}(\partial_j|C_j(m)) \text{ and } \mathrm{Ker}\,\partial_j = \textstyle\sum_{m\in M} \mathrm{Ker}(\partial_j|C_j(m)),$$

so the exactness of the restricted sequences implies the exactness of the sequence

$$0 \longrightarrow A_\Delta \xrightarrow{\partial_d} C_{d-1} \xrightarrow{\partial_{d-1}} \cdots \xrightarrow{\partial_1} C_0 \xrightarrow{\partial_0} C_{-1} \longrightarrow 0. \square$$

REMARK. Consider the exact sequence restricted to $m$, where $\sigma = \operatorname{supp} m$ is a k-face of $\Delta$. Then by the additivity of vector space dimension $\dim_K A_\Delta(m) = \sum_{j=k}^{d-1} (-1)^{d-1-j}\dim_K C_j(m)$. Now $\dim_K A_\Delta(m) = 1$ and $\dim_K C_j(m) = |\{\tau \in \Delta_j : \sigma \subseteq \tau\}| = f_{j-k-1}(\ell k_\Delta \sigma)$. Recall that $\ell k_\Delta \sigma$ has the homology of a (d-k-2)-sphere. The equation on dimensions just says

$$1 = \sum_{j=k}^{d-1} (-1)^{d-1-j} f_{j-k-1}(\ell k_\Delta \sigma) = \sum_{i=-1}^{d-k-2} (-1)^{d-k-i-2} f_i(\ell k_\Delta \sigma) \quad \text{or}$$

$$1 - (-1)^{d-k-1} = \sum_{i=0}^{d-k-2} (-1)^{d-k-i} f_i(\ell k_\Delta \sigma).$$

This is equivalent to Euler's formula for $\ell k_\Delta \sigma$: multiplying it by $(-1)^{d-k}$ we get

$$1 - (-1)^{d-k-1} = \sum_{i=0}^{d-k-2} (-1)^i f_i(\ell k_\Delta \sigma).$$

GRADINGS ON RINGS. We show that by assigning certain gradings to the rings $A_\Delta$ and $C_j$ the exact sequence of Theorem 4.2 will give interesting numerical criteria for spheres. We continue to denote by $M$ the semigroup of monomials in $x_1, x_2, \ldots, x_n$. Let $M'$ be a commutative monoid, and let $\phi: M \to M'$ be a monoid homomorphism. Assume further that $\phi^{-1}(x)$ is finite for every $x \in M'$. Then we can decompose the rings associated with a simplicial complex $\Delta$ as follows:

$$A_\Delta = \sum_{x \in M'} A_\Delta(x) = \sum_{x \in M'} \Big( \sum_{m \in \phi^{-1}(x)} Km \Big),$$

$$C_j = \sum_{x \in M'} C_j(x) = \sum_{x \in M'} \Big( \sum_{m \in \phi^{-1}(x)} C_j(m) \Big).$$

The finiteness condition above allows us to define the Hilbert functions with respect to $M'$; for $x \in M'$

$$H(A_\Delta, x) = \dim_K A_\Delta(x), \quad H(C_j, x) = \dim_K C_j(x).$$

The maps $\partial_i$ of Theorem 4.2 are homogeneous with respect to this grading by

$M'$, so the exact sequence gives

$$H(A_\Delta,x) = \sum_{j=-1}^{d-1} (-1)^{d-1-j} H(C_j,x).$$

We apply this in the case where $\Delta$ is a labeled homology $(d-1)$-sphere, $M' = N^{r+1}$, and $\phi$ is the degree map defined prior to Proposition 4.1 ($\phi$ is induced by the labeling of the vertices of the sphere). In this case it is convenient to write the relation in terms of Hilbert series (in $r+1$ variables $(t_0,\ldots,t_r) = t$):

$$P(A_\Delta,t) = \sum_{j=-1}^{d-1} (-1)^{d-1-j} P(C_j,t),$$

where, by Proposition 4.1, the left hand side can be written

$$P(A_\Delta,t) = \sum_{w \in N^{r+1}} \sum_{z \leq w} f_z(P) \prod_{i=0}^{r} \binom{w_i-1}{z_i-1} t^w.$$

Now we wish to compute $P(C_j,t)$. Recall that $C_j = \sum A_\sigma$, the summation being over all $j$-faces $\sigma$ of $\Delta$. $A_\sigma$ inherits the $N^{r+1}$ grading in the obvious way, and $P(C_j,t) = \sum P(A_\sigma,t)$. Now $A_\sigma$ is just the polynomial ring in variables corresponding to the vertices of $\sigma$. For $0 \leq i \leq r$, let $z_i = |\sigma \cap V_i|$; since $\dim \sigma = j$, $\sigma$ has $j+1$ vertices, so $\sum_{i=0}^{r} z_i = j+1$. The coefficient of $t^w$ in the series $1/\Pi_{i=0}^{r} (1-t_i)^{z_i}$ is the number of monomials of degree $w$ in $A_\sigma$. So

$$P(C_j,t) = \sum_{\substack{\sigma \\ \dim \sigma = j}} P(A_\Delta,t) = \sum_{\substack{z \\ \sum z_i = j+1}} \sum_{\sigma \in \Delta_z} P(A_\sigma,t)$$

$$= \sum_{\substack{z \\ \sum z_i = j+1}} f_z(\Delta) / \prod_{i=0}^{r} (1-t_i)^{z_i}$$

$$= \sum_{\substack{z \\ \sum z_i = j+1}} f_z(\Delta) \prod_{i=0}^{r} \left( \sum_{k=0}^{\infty} \binom{k+z_i-1}{z_i-1} t_i^k \right).$$

So the relation on Hilbert series becomes

$$\sum_{w\in N^{r+1}} \sum_{z\leq w} f_z(\Delta) \prod_{i=0}^{r} \binom{w_i-1}{z_i-1} t^w$$

$$= \sum_{j=0}^{d} (-1)^{d-j} \sum_{\substack{z\in N^{r+1} \\ \sum z_i=j}} f_z(\Delta) \prod_{i=0}^{r} \left(\sum_{k=0}^{\infty} \binom{k+z_i-1}{z_i-1} t_i^k\right).$$

By equating coefficients we get

THEOREM 4.4. Let $\Delta$ be a homology (d-1)-sphere with vertices labeled by $\{0,1,\ldots,r\}$. Then for all $w \in N^{r+1}$

$$\sum_{z\leq w} f_z(\Delta) \prod_{i=0}^{r} \binom{w_i-1}{z_i-1} = \sum_{z\in N^{r+1}} (-1)^{d-\sum z_i} f_z(\Delta) \prod_{i=0}^{r} \binom{w_i+z_i-1}{z_i-1}. \quad \square$$

An alternate proof of Theorem 4.4 has been suggested (privately) by Stanley. It is based on a result of his [48; Theorem 7.1] and avoids the use of the exact sequence of Theorem 4.2. We feel that this sequence is of independent interest, apart from the current application.

COROLLARY 4.5. The f-vector of a homology (d-1)-sphere satisfies the Dehn-Sommerville equations:

$$f_k = \sum_{j=k}^{d-1} (-1)^{d-j-1} \binom{j+1}{k+1} f_j, \quad -1 \leq k \leq d-2.$$

PROOF. We apply Theorem 4.4 to the case $r = 0$, but we must be careful with notation. Conventionally the f-vector of a sphere $\Delta$ is $f(\Delta) = (f_0, f_1, \ldots, f_{d-1})$, where $f_i$ is the number of faces of dimension $i$. In the vector-subscripted notation $f_{(i)}(\Delta)$ is the number of faces of cardinality $i$, so $f_{(i)}(\Delta) = f_{i-1}$. With this translation, Theorem 4.4 with $w = (q+1)$ and $q \geq 0$ gives

$$\sum_{k=0}^{q} \binom{q}{k} f_k = \sum_{j=0}^{d-1} (-1)^{d-j-1} \binom{q+1+j}{j} f_j. \tag{4.6}$$

For $q = -1$ the theorem gives the Euler equation (which is the Dehn-Sommerville equation for $k = -1$): $1 = \sum_{j=-1}^{d-1} (-1)^{d-j-1} f_j$. We now use (4.6) and induction to prove the Corollary. Assume the kth Dehn-Sommerville equation holds for $-1 \leq k \leq q-1$ $(q \geq 0)$. By (4.6)

$$f_q = \sum_{j=0}^{d-1} (-1)^{d-j-1} \binom{q+1+j}{j} f_j - \sum_{k=0}^{q-1} \binom{q}{k} f_k$$

$$= \sum_{j=0}^{d-1} (-1)^{d-j-1} \binom{q+1+j}{j} f_j - \sum_{k=0}^{q-1} \binom{q}{k} \sum_{j=k}^{d-1} (-1)^{d-j-1} \binom{j+1}{k+1} f_j$$

(by the induction hypothesis)

$$= \sum_{j=0}^{d-1} (-1)^{d-j-1} \sum_{k=0}^{q} \binom{j+1}{k+1}\binom{q}{k} f_j - \sum_{j=0}^{d-1} (-1)^{d-j-1} \sum_{k=0}^{q-1} \binom{j+1}{k+1}\binom{q}{k} f_j$$

$$= \sum_{j=0}^{d-1} (-1)^{d-j-1} \binom{j+1}{q+1} f_j = \sum_{j=q}^{d-1} (-1)^{d-j-1} \binom{j+1}{q+1} f_j.$$

The next-to-last equality follows from the combinatorial identity $\binom{m+1+j}{j} = \sum_{k=0}^{m} \binom{j+1}{k+1}\binom{m}{k}$ which, in turn, follows easily by considering an (m+1+j)-set partitioned into a (j+1)-set and an m-set. □

REMARK. Conversely, the equations (4.6) are implied by the Dehn-Sommerville equations as follows:

$$\sum_{k=0}^{q} \binom{q}{k} f_k = \sum_{k=0}^{q} \binom{q}{k} \sum_{j=k}^{d-1} (-1)^{d-j-1} \binom{j+1}{k+1} f_j$$

$$= \sum_{j=0}^{d-1} (-1)^{d-j-1} \sum_{k=0}^{q} \binom{q}{k}\binom{j+1}{k+1} f_j$$

$$= \sum_{j=0}^{d-1} (-1)^{d-j-1} \binom{q+1+j}{j} f_j.$$

The following corollary has been proved for certain classes of poset complexes in [14] and [49]. Further results for these complexes are derived in Section 7.

COROLLARY 4.7. If $\Delta$ is a completely balanced homology (d-1)-sphere, then for all $S \subseteq \{0,1,\dots,d-1\}$

$$f_S(\Delta) = \sum_{\substack{T \\ S \subseteq T \subseteq \langle d \rangle}} (-1)^{d-|T|} f_T(\Delta)$$

or, equivalently, for all $S$,

$$h_S(\Delta) = h_{\tilde{S}}(\Delta).$$

PROOF. Apply Theorem 4.4 to $\Delta$ with $w \in N^d$ such that $w_i \leq 1$ for all $i$.

(4.8) $$\sum_{z\leq w} f_z(\Delta) \prod_{i=0}^{d-1} \binom{w_i-1}{z_i-1} = \sum_{z\in N^d} (-1)^{d-\Sigma z_i} f_z(\Delta) \prod_{i=0}^{d-1} \binom{w_i+z_i-1}{z_i-1}.$$

Now $\binom{w_i-1}{z_i-1} = \binom{w_i+z_i-1}{z_i-1} = 0$ if $w_i = 1$ and $z_i = 0$. On the left-hand side, then, the only nonzero term is $f_w$. For the right-hand side note that $f_z(\Delta) = 0$ if $z_i > 1$ for some $i$. So the vectors $z$ making nonzero contribution on the right hand side are those for which $w \leq z \leq (1,1,\ldots,1)$. For such $z$, $\Pi_{i=0}^{d-1}\binom{w_i+z_i-1}{z_i-1}$ is the product of terms $\binom{w_i}{0} = 1$ (when $z_i = 1$) and terms $\binom{-1}{-1} = 1$ (when $z_i = w_i = 0$). So (4.8) says for all $w \leq (1,1,\ldots,1)$,

$$f_w(\Delta) = \sum_{w\leq z\leq(1,1,\ldots,1)} (-1)^{d-\Sigma z_i} f_z(\Delta).$$

In the notation introduced earlier for completely balanced complexes this says that for all $S \subseteq \{0,1,\ldots,d-1\}$

$$f_S(\Delta) = \sum_{\substack{T\\ S\subseteq T\subseteq \langle d\rangle}} (-1)^{d-|T|} f_T(\Delta).$$

Now substituting the resulting expression for $f_S(\Delta)$ into the definition of $h_S(\Delta)$ we get:

$$\begin{aligned} h_S(\Delta) &= \sum_{T\subseteq S} (-1)^{|S|-|T|} f_T(\Delta) \\ &= \sum_{T\subseteq S} (-1)^{|S|-|T|} \sum_{U\supseteq T} (-1)^{d-|U|} f_U(\Delta) \\ &= \sum_{U\subseteq \langle d\rangle} (-1)^{d-|S|-|U|} f_U(\Delta) \sum_{T\subseteq U\cap S} (-1)^{|T|}. \end{aligned}$$

If $U \cap S \neq \emptyset$, $\sum_{T\subseteq U\cap S} (-1)^{|T|} = (1-1)^{|U\cap S|} = 0$, so

$$h_S(\Delta) = \sum_{U\subseteq \tilde{S}} (-1)^{d-|S|-|U|} f_U(\Delta) = h_{\tilde{S}}(\Delta).$$

Similarly, the equations $h_S(\Delta) = h_{\tilde{S}}(\Delta)$ imply the equations $f_S(\Delta) = \sum_{T\supseteq S} (-1)^{d-|T|} f_T(\Delta)$. Assume $h_S(\Delta) = h_{\tilde{S}}(\Delta)$. Then

$$\sum_{T\supseteq S} (-1)^{d-|T|} f_T(\Delta) = \sum_{T\supseteq S} (-1)^{d-|T|} \sum_{U\subseteq T} h_U(\Delta)$$

$$= \sum_{U\subseteq \langle d\rangle} h_U(\Delta) \Big( \sum_{T\supseteq S\cup U} (-1)^{d-|T|} \Big).$$

If $S \cup U \neq \{0,1,\ldots,d-1\}$ then $\sum_{T\supseteq S\cup U} (-1)^{d-|T|} = \sum_{V\subseteq \widetilde{(S\cup U)}} (-1)^{|V|} = 0.$ So

$$\sum_{T\supseteq S} (-1)^{d-|T|} f_T(\Delta) = \sum_{U\supseteq \tilde{S}} h_U(\Delta) = \sum_{U\supseteq \tilde{S}} h_{\tilde{U}}(\Delta) = \sum_{V\subseteq S} h_V(\Delta) = f_S(\Delta). \quad \square$$

We note here that the Dehn-Sommerville equations in the form $h_i = h_{d-i}$ follow immediately from the equations $h_S = h_{\tilde{S}}$ and the fact that $h_i = \sum_{|S|=i} h_S$.

A <u>Gorenstein complex</u> is the join of a homology sphere with a simplex [14], [44]. If $\Delta$ is a completely balanced Gorenstein complex, then the homology sphere is itself a completely balanced complex (as is the simplex -- on a disjoint set of labels). If $S$ is the label set for the homology sphere and $T \subseteq \{0,\ldots,d-1\}$ then it follows from Corollary 4.7 that

$$h_T(\Delta) = \begin{cases} 0 & \text{if } T \not\subseteq S \\ h_{S\setminus T}(\Delta) & \text{if } T \subseteq S. \end{cases}$$

Corollary 4.7 can also be obtained by ring theoretic techniques as in [51]. Alternatively, in the case of certain poset complexes, Stanley has given an elementary proof using the Möbius function [49]. We describe these proofs in the next section.

Finally, we state for completeness the result of applying Theorem 4.4 when $\Delta = \sigma(Q)$, the minimal subdivision of the d-polytope $Q$. The proof is fairly direct and can be found in [7]. Recall that the label set is $\{0,1,\ldots,d-1\}$, although the label 1 never appears.

COROLLARY 4.9. Let $Q$ be a d-polytope and $\Delta = \sigma(Q)$. Let $w = (w_0,\ldots,w_{d-1}) \in N^d$ be such that $w_1 = 0$ and $w_i \leq 1$ for $i \geq 2$. Then

$$f_w(\Delta) = \sum_{z\geq w} (-1)^{d-\Sigma z_i} \binom{z_1}{w_1} f_z(\Delta). \quad \square$$

## 5. SOME OTHER PROOFS OF $h_S = h_{\tilde{S}}$

In this section we describe other approaches to the proof of Corollary 4.7 in general or for special cases. We attempt here to give the flavor of the arguments and refer to the literature for important details.

Suppose first that $\Delta$ is a completely balanced homology (d-1)-sphere. We sketch a ring theoretic proof of Corollary 4.7 based on a proof of $h_i = h_{d-i}$ for Gorenstein complexes given by Stanley in [51]. We assume here the terminology of [46].

By [46; Corollary 4.2], $A_\Delta$ has a system of parameters $\theta_0,\ldots,\theta_{d-1}$ which is homogeneous in the $N^d$ grading on $A_\Delta$ $(d = r+1)$; in fact, $\deg \theta_i = e_i$, the $i^{th}$ unit vector in $N^d$. Thus the ring $B = A_\Delta/(\theta_0,\ldots,\theta_{d-1})$ inherits an $N^d$ grading; by [46; Proposition 3.2 and proof of Theorem 4.4] we have that $B = \sum_{S\subseteq\langle d\rangle} B_S$ (direct sum) where $\dim_K B_S = h_S$, $B_S\cdot B_T \subseteq B_{S\cup T}$ if $S\cap T = \emptyset$ and $B_S\cdot B_T = 0$ otherwise. Let $B_+ = \sum_{\substack{S\subseteq\langle d\rangle\\ S\neq\emptyset}} B_S$; then since $A_\Delta$ is Gorenstein (e.g. [44], [51]), it follows that the subspace

$$C = \{x \in B\colon xB_+ = 0\}$$

has K-dimension 1 [51].

SECOND PROOF OF COROLLARY 4.7. First we recall that $h_{\langle d\rangle}(\Delta) = h_d(\Delta) = 1$ by the Euler relation for $\Delta$. Thus $B_{\langle d\rangle} \neq 0$. But $B_{\langle d\rangle}\cdot B_S = 0$ for all $S \neq \emptyset$, so $B_{\langle d\rangle} \subseteq C$; since $\dim_K C = 1$, $B_{\langle d\rangle} = C$.

We next claim that if $T \subseteq \langle d\rangle$ and $x_T \in B_T$, $x_T \neq 0$, then $x_T\cdot B_{\langle d\rangle\setminus T} \neq 0$. Since $B_\phi = K$, if the claim fails for some $T$, we must have $T \neq \langle d\rangle$. Suppose $T$ is a maximal set for which $x_T\cdot B_{\langle d\rangle\setminus T} = 0$ for some $x_T \in B_T$, $x_T \neq 0$. For $R \subseteq \langle d\rangle\setminus T$, $R \neq \emptyset$, and $x_R \in B_R$, $x_T x_R \in B_{T\cup R}$ so if $x_T x_R \neq 0$, then by the maximality of $T$ we have $x_T x_R B_{\langle d\rangle\setminus(T\cup R)} \neq 0$. But $x_R B_{\langle d\rangle\setminus(T\cup R)} \subseteq B_{\langle d\rangle\setminus T}$ contradicting $x_T\cdot B_{\langle d\rangle\setminus T} = 0$. Thus we conclude $x_T B_R = 0$ for each $R \subseteq \langle d\rangle\setminus T$. On the other hand, if $R\cap T \neq \emptyset$, then $x_T B_R = 0$ as well. So $x_T B_+ = 0$, i.e., $x_T \in C = B_{\langle d\rangle}$. But $x_T \in B_T$, $T \neq \langle d\rangle$, and so $x_T \in B_T \cap B_{\langle d\rangle} = 0$; this contradiction proves the claim.

Now, for each $S \subseteq \langle d\rangle$ we have a pairing of K-vector spaces

$$B_S \times B_{\langle d\rangle\setminus S} \to B_{\langle d\rangle} = K$$

defined by ring multiplication, which gives us a linear map

$$B_S \to \mathrm{Hom}_K(B_{\langle d\rangle\setminus S}, K).$$

The claim allows us to conclude that this map is an injection, and so for all $S$,

$$\dim_K B_S \leq \dim_K B_{\langle d\rangle\setminus S},$$

completing the proof. □

Suppose now that $\Delta$ is a <u>shellable</u> completely balanced homology (d-1)-sphere, i.e. there is an ordering $\sigma_1, \sigma_2, \ldots, \sigma_k$ of all the (d-1)-simplices of $\Delta$ so that for $2 \leq i \leq k$, $(\sigma_1 \cup \sigma_2 \cup \ldots \cup \sigma_{i-1}) \cap \sigma_i$ is a nonempty union of (d-2)-faces of $\sigma_i$. The following interpretation of the $h_S(\Delta)$ for shellable completely balanced $\Delta$ is a special case of [46; Proposition 3.3], which extends the interpretation of $h_i$ in [31].

PROPOSITION 5.1. Let $\Delta$ be a completely balanced (d-1)-complex, and suppose $\sigma_1, \sigma_2, \ldots, \sigma_k$ is a shelling of $\Delta$. For each $i$, $1 \leq i \leq k$, define $\tau_i$ to be the unique minimal face of $\sigma_i$ not contained in $\sigma_1 \cup \sigma_2 \cup \ldots \cup \sigma_{i-1}$. Let $S_i$ be the subset of $\{0,1,\ldots,d-1\}$ which labels $\tau_i$. Then for any $S \subseteq \{0,1,\ldots,d-1\}$, $h_S(\Delta)$ is the number of $i$ for which $S_i = S$. □

To prove Corollary 4.7 in the special case of shellable completely balanced homology spheres (these are always spheres; for example, barycentric subdivisions of polytopes [18], [16]) we need the fact that if $\sigma_1, \ldots, \sigma_k$ is a shelling of a homology sphere $\Delta$, then so is $\sigma_k, \ldots, \sigma_1$ [28; Proposition 3.3.11]. Then if $\tau_i$ is the unique minimal face of $\sigma_i$ not contained in $\sigma_1 \cup \ldots \cup \sigma_{i-1}$, then $\sigma_i \setminus \tau_i$ has the same property for $\sigma_k \cup \ldots \cup \sigma_{i+1}$. If $\tau_i$ has label set $S_i$ then $\sigma_i \setminus \tau_i$ has label set $\langle d\rangle \setminus S_i$ and so

$$\begin{aligned} h_S &= \left|\{i: \tilde{S}_i = S\}\right| \\ &= \left|\{i: S_i = \tilde{S}\}\right| = h_{\tilde{S}}, \end{aligned}$$

the first equality coming from the shelling $\sigma_k,\dots,\sigma_1$, the last from the shelling $\sigma_1,\dots,\sigma_k$.

Finally, we wish to prove the equations $h_S = h_{\tilde{S}}$ for an interesting class of poset complexes. Recall that the Möbius function of a poset $P$ is an integer-valued function $\mu$ on $P\times P$ defined by $\mu(x,y) = 0$ if $x \not\leq y$, $\mu(x,x) = 1$ for all $x$ and $\sum_{x\leq y\leq z} \mu(x,y) = 0$ if $x < z$ [21]. A ranked poset $P$ is said to be Eulerian if its Möbius function satisfies $\mu(x,y) = (-1)^{r(y)-r(x)}$ for all $x \leq y$ in $P$. (Recall $\hat{0},\hat{1} \in P$.) We will write $\mu_P$ when confusion may arise as to the appropriate poset.

For any ranked poset $P$, we have $\mu(P) \equiv \mu_P(\hat{0},\hat{1}) = \chi(\Delta) - 1$, where $\chi(\Delta)$ is the Euler characteristic of the complex $\Delta = \Delta(P)$, and it follows that

$$h_{\langle d\rangle}(\Delta) = \sum_{T\subseteq\langle d\rangle} (-1)^{d-|T|} f_T(\Delta) = (-1)^{d-1}\mu(P),$$

where $d = r(P) = r(\hat{1})$. (See, for example, [21] or [46].) By considering the "rank-selected" sub-poset

$$P_S = \{\hat{0},\hat{1}\} \cup \{x \in P\colon r(x) \in S\}$$

for $S \subseteq \langle d\rangle$, we obtain

$$h_S(\Delta) = (-1)^{|S|-1}\mu(P_S),$$

where $\mu(P_S) = \mu_{P_S}(\hat{0},\hat{1})$. By [4; Lemma 4.6] we can write

$$\mu(P_S) = \sum_{\substack{x_1<\dots<x_k \\ r(x_i)\notin S}} (-1)^k \mu(\hat{0},x_1)\mu(x_1,x_2)\dots\mu(x_k,\hat{1})$$

where the sum ranges over chains $x_0 = \hat{0} < x_1 < \dots < x_k < 1$ in $P_{\tilde{S}}$ $(k \geq 0)$ and $\mu = \mu_P$ on the right hand side.

Now suppose $P$ is Eulerian; we show $h_S(\Delta(P)) = h_{\tilde{S}}(\Delta(P))$. The following proof is due to Stanley [49; Proposition 2.2]. We have, with $\Delta = \Delta(P)$

$$\begin{aligned} h_S(\Delta) &= (-1)^{|S|-1}\mu(P_S) \\ &= (-1)^{|S|-1} \sum_{\substack{x_1<\dots<x_k \\ r(x_i)\notin S}} (-1)^k \mu(\hat{0},x_1)\dots\mu(x_k,\hat{1}) \\ &= (-1)^{|S|-1} \sum_{\substack{x_1<\dots<x_k \\ r(x_i)\notin S}} (-1)^{k+r(\hat{1})-r(\hat{0})} \\ &= (-1)^{|S|+d} \sum_{T\subseteq\tilde{S}} (-1)^{|T|} f_T(\Delta) \\ &= h_{\tilde{S}}(\Delta) \end{aligned}$$

(recall the convention that $r(\hat{0}) = 1$).

We will treat the case of Eulerian posets in detail in section 7 where we will find all the linear relations holding among the numbers $f_S(\Delta)$. The face lattice of a convex polytope is an Eulerian poset [30], [38]. We will, in fact, get all the linear relations holding for the chain numbers of an arbitrary convex polytope.

## 6. AFFINE SPAN FOR COMPLETELY BALANCED SPHERES

We consider in this section the question of whether the equations given by Corollary 4.7 are the only linear relations on the extended f-vector which hold for all competely balanced homology spheres. We give an affirmative answer by exhibiting a set of completely balanced simplicial <u>polytopes</u> whose extended h-vectors span the space determined by the relations $h_S = h_{\tilde{S}}$ along with the trivial relation $h_\phi = 1$.

We will define an operation that subdivides a face of a completely balanced complex, resulting in another such complex. This notion of a completely balanced stellar subdivision is then used to create the desired basis in the same way that we produced a basis for the Dehn-Sommerville space in Section 3 by performing stellar subdivisions on a simplex. To this end, we define a class of polytopes to play the role of simplices.

Suppose in $R^d$ we have d mutually orthogonal segments, $[v_i,w_i]$ $(0 \leq i \leq d-1)$, that intersect at a single point interior to each of the segments. The convex hull of these segments is a simplicial polytope, called the d-crosspolytope $Q^d$ or just Q (see [23; §4.3]). Alternatively, Q is the polar to the d-cube, and its boundary complex can be viewed as the order complex of the rank d poset having 2 elements of each rank, any two elements of different rank being comparable.

For $0 \leq k \leq d-1$, the k-faces of Q are determined by sets F of k+1 of the points $\{v_i\} \cup \{w_i\}$ where no pair $\{v_i,w_i\}$ is in F. In particular, the facets consist of exactly one element from each pair $\{v_i,w_i\}$, $0 \leq i \leq d-1$. Thus, the sets $V_i = \{v_i,w_i\}$ $(0 \leq i \leq d-1)$ partition the vertices of Q, making Q a completely balanced simplicial complex. For $S \subseteq \{0,1,\ldots,d-1\}$, the S-labeled faces of Q are the sets $\{y_i : i \in S\}$ where $y_i \in V_i = \{v_i,w_i\}$. There are clearly $2^{|S|}$ such sets, i.e., $f_S(Q) = 2^{|S|}$. Also,

$$h_S(Q) = \sum_{T \subseteq S} (-1)^{|S|-|T|} f_T(Q)$$

$$= \sum_{T \subseteq S} (-1)^{|S|-|T|} 2^{|T|} = (2-1)^{|S|} = 1$$

for all $S \subseteq \{0,1,\ldots,d-1\}$. This makes the crosspolytope the completely balanced analog of the simplex, which, as we have seen has $h_i = 1$ for all i, $0 \leq i \leq d$.

Suppose F is a j-face of Q labeled by some subset $X \subseteq \{0,1,\ldots,d-1\}$ $(|X| = j+1)$. Then $\ell k_Q F$ is a complex whose facets consist of exactly one element from each $V_i$, for $i \in \{0,1,\ldots,d-1\} \setminus X$. So $\ell k_Q F$ is itself a (d-j-1) -crosspolytope with vertices labeled by $\{0,1,\ldots,d-1\} \setminus X$.

For a completely balanced complex $\Delta$, we define the completely balanced subdivision of a face $\sigma \in \Delta$ as follows. Suppose $\Delta$ is labeled with $0,1,\ldots,r$ and suppose for ease of exposition that $\sigma$ has label set $\{0,1,\ldots,k\}$, $k \leq r$. Define $\Delta_0 = st(\sigma,\Delta)$, the usual stellar subdivision of complexes discussed in Section 3. Assign the new vertex introduced in $\Delta_0$ the label 0. If $k > 0$, $\Delta_0$ is not completely balanced; there will be a unique edge having two vertices with label 0. Denote this edge by $\sigma_0$ and define $\Delta_1 = st(\sigma_0, \Delta_0)$. Label the new vertex in $\Delta$ with a 1; if $k > 1$, then there is a unique edge with two vertices labeled 1. Continuing this process, the final complex in the sequence, $\Delta_k$, will be completely balanced, and we define this complex to be the desired subdivision of $\sigma$ in $\Delta$. A proof that this procedure is well-defined is included in the proof of Theorem 3.1 in [9]. It is clear from this definition that if $\Delta$ is the boundary complex of a simplicial polytope, then so is $\Delta_k$.

To view this subdivision another way, consider first the case $k = r$. Let $\hat{\Delta}$ be the (boundary complex of the) (r+1)-crosspolytope and choose a maximal simplex $\hat{\sigma} \in \hat{\Delta}$; $\hat{\sigma}$ is completely labeled with $0,1,\ldots,r$. Then $\Delta_r$ is the complex $(\Delta\backslash\sigma) \cup (\hat{\Delta}\backslash\hat{\sigma})$, where we identify the corresponding faces of $\sigma$ and $\hat{\sigma}$. In the case $k < r$, take $\hat{\Delta}$ to be a (k+1)-crosspolytope, $\hat{\sigma} \in \hat{\Delta}$ a k-simplex (with labels $0,1,\ldots,k$). The balanced subdivision $\Delta_k$ is the union of $\Delta\backslash\sigma$ and $(\hat{\Delta}\backslash\hat{\sigma})\cdot \ell k_{\Delta}\sigma = (\hat{\Delta}\cdot \ell k_{\Delta}\sigma)\backslash\hat{\sigma}$, again identifying the corresponding faces of $\sigma$ and $\hat{\sigma}$. See [9] for details.

Now let $Q$ be the d-crosspolytope. (For the remainder of this section, when we refer to a (simplicial) polytope $Q$ we will mean its boundary complex; faces of $Q$ will be considered as sets of vertices.) For each subset $X \subseteq \langle d\rangle$, $X \neq \emptyset$, let $P^X$ denote the completely balanced simplicial polytope which results from a completely balanced subdivision of a face of $Q$ having label set $X$. Note that if $|X| = 1$ then $P^X = Q$.

THEOREM 6.1. For each $S \subseteq \langle d\rangle$,

$$h_S(P^X) = \begin{cases} 1 & \text{if } S \cap X = \emptyset \text{ or } X \subseteq S \\ 2 & \text{otherwise.} \end{cases}$$

PROOF. Suppose $|X| = k+1$ and let $F$ be the face of $Q$ which is subdivided. Let $\hat{Q}$ be the (k+1)-crosspolytope; $F'$ a facet of $\hat{Q}$ labeled by $X$; and $Q' = \hat{Q}\cdot \ell k_Q F$. By the above discussion (with $\hat{\Delta} = \hat{Q}$ and $\hat{\sigma} = F'$) the faces of $P^X$ are either faces of $Q$ not containing $F$ or faces of $Q'$ not containing $F'$; faces of the latter type get vertices with labels in $X$

from $\hat{Q}$, the others from $\ell k_Q F$.

Thus for any $T \subseteq \{0,1,\ldots,d-1\}$,

$$f_T(P^X) = f_T(Q\backslash F) + f_{T\backslash X}(\ell k_Q F)(f_{T\cap X}(\hat{Q}) - f_{T\cap X}(F')).$$

(Here $f_{T\cap X}(F') = 1$ denotes the number of faces of the simplex $F'$ having label set $T \cap X$.) If $X \not\subseteq T$ then $f_T(Q\backslash F) = f_T(Q)$; if $X \subseteq T$ then

$$\begin{aligned} f_T(Q\backslash F) &= f_T(Q) - f_{T\backslash X}(\ell k_Q F) f_X(F) \\ &= f_T(Q) - f_{T\backslash X}(\ell k_Q F). \end{aligned}$$

So if we write $\chi(X \subseteq T) = \begin{cases} 1 & \text{if } X \subseteq T \\ 0 & \text{if } X \not\subseteq T \end{cases}$, then we get

$$\begin{aligned} f_T(P^X) &= f_T(Q) + f_{T\backslash X}(\ell k_Q F)(f_{T\cap X}(\hat{Q}) - f_{T\cap X}(F') - \chi(X \subseteq T)) \\ &= 2^{|T|} + 2^{|T\backslash X|}(2^{|T\cap X|} - 1 - \chi(X \subseteq T)). \end{aligned}$$

So

$$\begin{aligned} h_S(P^X) &= \sum_{T\subseteq S} (-1)^{|S|-|T|} f_T(P^X) \\ &= \sum_{T\subseteq S} (-1)^{|S|-|T|}(2^{|T|} + 2^{|T|} - 2^{|T\backslash X|} - 2^{|T\backslash X|}\chi(X \subseteq T)). \end{aligned}$$

Now $\sum_{T\subseteq S} (-1)^{|S|-|T|} 2^{|T|} = 1$;

$$\begin{aligned} \sum_{T\subseteq S} (-1)^{|S|-|T|} 2^{|T\backslash X|} &= \sum_{U\subseteq S\cap X} \sum_{V\subseteq S\backslash X} (-1)^{|S|-|U|-|V|} 2^{|V|} \\ &= \sum_{U\subseteq S\cap X} (-1)^{|S\cap X|-|U|} = (1-1)^{|S\cap X|} = \chi(S \cap X = \emptyset); \end{aligned}$$

and

$$\begin{aligned} &\sum_{T\subseteq S} (-1)^{|S|-|T|} 2^{|T\backslash X|} \chi(X \subseteq T) \\ &\qquad = \chi(X \subseteq S) \sum_{V\subseteq S\backslash X} (-1)^{|S|-|X|-|V|} 2^{|V|} = \chi(X \subseteq S). \end{aligned}$$

So

$$\begin{aligned} h_S(P^X) &= 2 - \chi(S \cap X = \emptyset) - \chi(X \subseteq S) \\ &= \begin{cases} 1 & \text{if } S \cap X = \emptyset \text{ or } X \subseteq S \\ 2 & \text{else.} \end{cases} \end{aligned}$$

(Note that when $|X| > 0$ we cannot have $S \cap X = \emptyset$ and $X \subseteq S$ simultaneously.) □

Let $C^d$ be the set of all completely balanced homology (d-1)-spheres and $(h_S(C^d))$ the set of the vectors $(h_S(\Delta))_{S\subseteq\langle d\rangle} \in N^{2^d}$ for $\Delta \in C^d$.

THEOREM 6.2. $\dim \operatorname{aff}(h_S(C^d)) = 2^{d-1}-1$.

PROOF. For all $\Delta \in C^d$ we have $h_\emptyset = 1$ and (by Corollary 2.4) $h_S(\Delta) = h_{\tilde{S}}(\Delta)$ for $S \subseteq \{0,1,\ldots,d-1\}$. These are clearly independent linear equations, and there are $2^{d-1}+1$ of them, so $\dim \operatorname{aff} h_S(C^d) \leq 2^d - (2^{d-1}+1) = 2^{d-1}-1$. The rest of the proof will consist in showing the other inequality by demonstrating $2^{d-1}$ affinely independent vectors in $h_S(C^d)$.

We first define a lexicographic order on the subsets of $\{0,1,\ldots,d-1\}$ as follows. If $S = \{s_1,\ldots,s_k\}$, $s_1 < s_2 < \ldots < s_k$, and $T = \{t_1,\ldots,t_{k'}\}$, $t_1 < t_2 < \ldots < t_{k'}$, then $S < T$ if $k < k'$ or if $k = k'$ and for some $j \leq k$, $s_j < t_j$, while for $i < j$, $s_i = t_i$. In this ordering $S < T$ if and only if $\tilde{S} > \tilde{T}$, so the complement of the nth subset is the $(2^d-n+1)$st subset.

Now define a $(2^{d-1}\times 2^{d-1})$ matrix $A$ with columns indexed by the first $2^{d-1}$ subsets of $\{0,1,\ldots,d-1\}$ in increasing order, and the rows indexed by the last $2^{d-1}$ subsets of $\{0,1,\ldots,d-1\}$ arranged in decreasing order. If $S$ is one of the first $2^{d-1}$ subsets and $X$ is one of the last $2^{d-1}$ subsets, then the $(X,S)$ entry in $A$ is

$$a_{X,S} = 2 - h_S(P^X) = \begin{cases} 1 & \text{if } S \cap X = \emptyset \text{ or } X \subseteq S \\ 0 & \text{else,} \end{cases}$$

where $P^X$ is given by Theorem 6.1. Note that for $X$ and $S$ within the range defined, $S < X$, so $X \not\subseteq S$.

The matrix $A$ is lower triangular with ones along the diagonal. To see this, let $1 \leq q < n \leq 2^{d-1}$, and let $X$ be the $(2^d-q+1)$st set (the set indexing row $q$) and $S$ the nth set. Then $\tilde{X}$ is the qth set, so $\tilde{X} < S$. This implies $S \not\subseteq \tilde{X}$, i.e., $S \cap X \neq \emptyset$. So $a_{X,S} = 0$. The diagonal elements of $A$ are $a_{\tilde{S},S} = 1$. Thus, rank $A = 2^{d-1}$. This says that the polytopes $P^X$, as $X$ ranges over the last $2^{d-1}$ subsets of $\{0,1,\ldots,d-1\}$, have affinely independent vectors $(h_S) \in N^{2^{d-1}}$ (S ranges over the first $2^{d-1}$ subsets of $\{0,1,\ldots,d-1\}$). But then their complete h-vectors $(h_S(P^X))_{S\subseteq\langle d\rangle}$ must be affinely independent. Thus $\dim \operatorname{aff} h(C^d) \geq 2^{d-1}-1$; combined with the other inequality, this gives the desired result. □

Since the basis constructed for the proof of Theorem 6.2 actually consists of polytopes, we have the following.

COROLLARY 6.3. The dimension of the affine span of the extended h-vectors of completely balanced simplicial d-polytopes is $2^{d-1}-1$.

We note, finally, that various nonlinear conditions are known to hold for the extended h-vectors of the more general balanced Cohen-Macaulay complexes [46]. In particular, it is known that if $\Delta$ is such a complex, then $h_S(\Delta) = f_S(\Lambda)$, where $\Lambda$ is another labeled complex having the property that each simplex has <u>at most</u> one vertex of each label. We call a complex having this property <u>colored</u>, and note that it need not be a pure complex, i.e., maximal simplices may be of different sizes.

Conversely, if $\Lambda$ is any colored complex, then $f_S(\Lambda) = h_S(\Delta)$ for some completely balanced simplicial Cohen-Macaulay complex $\Delta$, in fact for a shellable completely balanced complex. We sketch a proof of this originally suggested by Björner. Order the faces of $\Lambda$ by cardinality, starting with $\phi$ and continuing with the vertices (in any order), and so on. Suppose $\dim \Lambda = r$. Define $r+1$ <u>new</u> vertices, labeled $0,1,\ldots,r$, and define a complex $\Delta$ on these new vertices plus those of $\Lambda$. The maximal faces of $\Delta$ will be $(r+1)$-sets consisting of a face $\lambda$ of $\Lambda$ augmented by those new vertices whose labels do not appear on the vertices of $\lambda$. Call this facet $\lambda^*$. The order on $\Lambda$ defines an order on the maximal faces of $\Delta$ which is easily seen to be a shelling order. In fact, the minimal face of $\lambda^*$ not in the union of those facets preceding $\lambda^*$ is $\lambda$, and so we get $h_S(\Delta) = f_S(\Lambda)$ by a direct application of Proposition 5.1.

It seems to be a difficult problem to characterize numerically the extended f-vectors of colored complexes. It seems necessary to have such a characterization before one could hope to extend McMullen's conditions to a characterization of the extended h-vectors of completely balanced polyhedral complexes.

7. EULERIAN POSET COMPLEXES. In this section we take a further look at a special class of completely balanced complexes: the order complexes of Eulerian posets. Recall that a poset $P$ is called Eulerian if for every $x < y$ in $P$, $\mu(x,y) = (-1)^{r(y)-r(x)}$. We shall see that this means that Euler's formula holds for intervals in the poset, and so, in particular, the order complexes of these posets are Eulerian manifolds [26].

THEOREM 7.1. Let $P$ be an Eulerian poset of rank $d$, and $S \subseteq \{0,1,\ldots,d-1\}$. If $\{i,k\} \subseteq S \cup \{-1,d\}$ and $S$ contains no $j$ such that $i < j < k$, then

$$\sum_{j=i+1}^{k-1} (-1)^{j-i-1} f_{S\cup j}(P) = f_S(P)(1 - (-1)^{k-i-1})$$

PROOF. Let $C$ be a chain in $P$ with rank set $S$. Let $x$ be the element of $C$ with rank $i$ ($\hat{0}$ if $i = -1$), and $y$ the element with rank $k$ ($\hat{1}$ if $i = d$). For $i \leq j \leq k$ write $f_j(x,y)$ for the number of rank $j$ elements of $P$ between $x$ and $y$. Since $P$ is Eulerian, $\mu(x,y) = (-1)^{k-i}$, and we use the fact that $\sum_{x \leq z < y} \mu(x,z) = 0$ to get

$$\begin{aligned}(-1)^{k-i} = \mu(x,y) &= -\sum_{x \leq z < y} \mu(x,z) \\ &= -\sum_{x \leq z < y} (-1)^{r(z)-i} \\ &= -\sum_{j=i}^{k-1} \sum_{\substack{x \leq z < y \\ r(z)=j}} (-1)^{j-i} \\ &= -\sum_{j=i}^{k-1} (-1)^{j-i} f_j(x,y).\end{aligned}$$

Then, since $f_i(x,y) = 1$ we get a form of Euler's equation:

$$1 - (-1)^{k-i-1} = \sum_{j=i+1}^{k-1} (-1)^{j-i-1} f_j(x,y).$$

Summing over all S-chains $C$ we get (here $x$ and $y$ depend on $C$)

$$f_S(P)(1-(-1)^{k-i-1}) = \sum_{C \text{ an } S\text{-chain}} \sum_{j=i+1}^{k-1} (-1)^{j-i-1} f_j(x,y)$$

$$= \sum_{j=i+1}^{k-1} (-1)^{j-i-1} \sum_{C \text{ an } S\text{-chain}} f_j(x,y)$$

$$= \sum_{j=i+1}^{k-1} (-1)^{j-i-1} f_{S\cup j}(P). \quad \square$$

The motivating example of an Eulerian poset is the face lattice of a polytope [30], [38]. For arbitrary polytopes, the equations of this theorem are the analogs of the Dehn-Sommerville equations, which hold for the f-vectors of simplicial polytopes. Note that taking $S = \emptyset$, $i = -1$, $k = d$, the equation given by Theorem 7.1 is Euler's formula.

We now analyze the dependencies among the variables $f_S$ given by the equations of Theorem 7.1.

PROPOSITION 7.2. For $d \geq 1$, let $\Psi^d$ be the set of subsets $S \subseteq \{0,1,\ldots,d-2\}$ such that $S$ contains no two consecutive integers. Then for all $T \subseteq \{0,1,\ldots,d-1\}$ such that $T \notin \Psi^d$, there is a nontrivial linear relation expressing $f_T(P)$ in terms of $f_S(P)$, $S \in \Psi^d$, which holds for all Eulerian posets $P$ of rank $d$. The cardinality of $\Psi^d$ is $c_d$, the dth Fibonacci number ($c_d = c_{d-1} + c_{d-2}$, $c_1 = 1$, $c_2 = 2$).

PROOF. Order the subsets of $\{0,1,\ldots,d-1\}$ by increasing cardinality, and within cardinality lexicographically. Thus, if $S = \{s_1,s_2,\ldots,s_k\}$ and $V = \{v_1,v_2,\ldots,v_{k'}\}$, then $S < V$ if and only if either $k < k'$ or $k = k'$ and for some $j$, $1 \leq j \leq k$, $s_j < v_j$ while for $1 \leq i < j$, $s_i = v_i$. If $T \notin \Psi^d$ then for some $k$, $1 \leq k \leq d$, $\{k-1,k\} \subseteq T \cup \{d\}$. Let $S = T\setminus\{k-1\}$, and $i = \max\{j \in T \cup \{-1\}: j < k-1\}$. Then Theorem 7.1 for these $S$, $i$ and $k$ says

$$f_T(P) = \sum_{j=i+1}^{k-2} (-1)^{k-j} f_{S\cup j}(P) + f_S(P)(1-(-1)^{k-i-1}).$$

All the subscripts appearing on the right-hand side of this equation are less than $T$ in the lexicographic order. Repeating the process for any subscript not in $\Psi^d$ we eventually get the desired linear relation.

To compute $|\Psi^d|$, note that any element $S$ of $\Psi^d$ is one of two types: either $d-2 \notin S$ or $d-2 \in S$. In the first case $S \in \Psi^{d-1}$; in the second case $d-3 \notin \Psi$, so $S\backslash\{d-2\} \in \Psi^{d-2}$. Thus $|\Psi^d| = |\Psi^{d-1}| + |\Psi^{d-2}|$; it is easy to see $|\Psi^1| = 1$, $|\Psi^2| = 2$, so the proposition is proved. □

Adding the relation $f_\phi = 1$ we get that the dimension of the affine span of the extended f-vectors of Eulerian posets is at most $c_d-1$. In fact, this upper bound is the actual dimension, and its value gives us a hint as to the proof. We need to exhibit $c_d$ affinely independent extended f-vectors; it turns out we can do this within the class of d-polytopes. Since $c_d = c_{d-1}+c_{d-2}$ we try to use bases for aff $f_S(P^{d-1})$ and aff $f_S(P^{d-2})$ to create a basis for aff $f_S(P^d)$. To do this we use the operations of taking the pyramid and bipyramid over a polytope. We will use the convention that the symbol $P$ alone stands for the "0-dimensional" polytope, i.e., a single point. An ordered string or word made up of the symbols $B$ and $P$, and ending in $P$, stands for the polytope obtained by taking successive pyramids and bipyramids over $P$ in the order indicated by the word. (Alternatively, one could consider the first $P$ to denote taking the pyramid over the empty polytope.) Thus $P^2 = PP$ is an interval, $P^3$ is a triangle, $BP^2$ is a square and $B^2P^2$ is an octahedron. We choose to write words ending in $P^2$ to avoid redundancy, since $BP = P^2$. Clearly the dimension of the polytope is one less than the length of the word.

PROPOSITION 7.3. For $d \geq 1$, let $\Omega^d$ be the set of d-polytopes named by words of length $d+1$ in $B$ and $P$ that end in $P^2$ and contain no two adjacent $B$'s. The extended f-vectors of elements of $\Omega^d$ are affinely independent, and $\Omega^d$ contains $c_d$ elements.

SKETCH OF PROOF. First we count the elements of $\Omega^d$. Consider the two types of words in $\Omega^d$: those beginning with $P$ and those beginning with $B$. Words of the first type are of the form $PQ'$, where $Q'$ is any word in $\Omega^{d-1}$.

Words of the second type must start with BP (since $B^2$ is not allowed) and thus are of the form BPQ", where Q" is any word in $\Omega^{d-2}$. So $|\Omega^d| = |\Omega^{d-1}| + |\Omega^{d-2}|$; i.e., the cardinality of $\Omega^d$ satisfies the Fibonacci recursion. Since $|\Omega^1| = |\{P^2\}| = 1$, $|\Omega^2| = |\{P^3, BP^2\}| = 2$, we get $|\Omega^d| = c_d$.

The proof that the extended f-vectors of elements in $\Omega^d$ are independent is difficult because the effect on the extended f-vector of taking a pyramid or bipyramid is not easily described. It is relatively easy, however, to describe the faces of a pyramid or bipyramid in terms of the faces of the original polytope. In particular, all the faces of a polytope in $\Omega^d$ are in $(\cup_{i=1}^{d-1} \Omega^i) \cup \{\phi, P\}$, which we will call $M^{d+1}$. (Recall P is a single point.) The idea of the proof is to work with the $c_d \times c_d$ matrix $A^d$, whose typical entry is $a_{QM}$ where for $Q \in \Omega^d$, $M \in M^d$, $a_{QM}$ is the number of faces of Q of combinatorial type M. One can find an invertible transformation that takes $A^d$ to the matrix whose rows are the vectors $(f_S(Q))_{S \in \Psi^d}$ for $Q \in \Omega^d$. Given this it suffices to show $A^d$ is non-singular. The proof of this latter fact makes crucial use of the recursive construction of $\Omega^d$. It consists in exhibiting row and column operations that have the following effect.

$$\begin{pmatrix} A^{d-1} & * \\ 0 & I \end{pmatrix} \to \begin{pmatrix} A^{d-1} & * \\ A^{d-2} & \text{shifted} \end{pmatrix} \to A^d.$$

Then rank $A^d$ = rank $A^{d-1}$ + rank $A^{d-2}$ $= |\Omega^{d-1}| + |\Omega^{d-2}| = |\Omega^d| = c_d$. So the extended f-vectors of the elements of $\Omega^d$ are affinely independent. For complete details, see [9]. □

Together Propositions 7.2 and 7.3 give us the following result.

THEOREM 7.4. For $d \geq 1$

$$\dim \text{aff}\{(f_S(P))_{S \subseteq \langle d \rangle} : P \text{ is an Eulerian poset of rank } d\}$$
$$= \dim \text{aff}\{(f_S(P))_{S \subseteq \langle d \rangle} : P \text{ is a d-polytope}\} = c_d - 1$$

where $c_d$ is the dth Fibonacci number. □

In particular, the extended f-vectors of Eulerian posets (or polytopes) are contained in a proper subspace of the affine span of the extended f-vectors of completely balanced homology spheres. In other words the equations $h_S = h_{\tilde{S}}$ for Eulerian posets are dependent on the equations given by Theorem 7.1.

Already at dimension 4, the f-vectors of polytopes have not been characterized. The results of this section show that the extended f-vectors of 4-polytopes are determined linearly by the values of $f_0, f_1, f_2$ and $f_{\{0,2\}}$ (here we have dropped the set brackets on $f_{\{i\}}$, because it coincides with $f_i$ in the original f-vector).

It is interesting to note here that for simplicial polytopes (and hence for simple polytopes) the extended f-vector is linearly determined by the usual f-vector, and so the affine hull of these remains $[d/2]$-dimensional. To see this, suppose that $S = \{i_1,\ldots,i_s\}$, $s \geq 2$, $i_1 < i_2 < \ldots < i_s < d$. Then for a simplicial d-polytope $P$,

$$f_S(P) = f_{i_s}(P) \cdot f_{S \setminus \{i_s\}}(T^{i_s}), \tag{7.5}$$

where $T^{i_s}$ denotes the $i_s$-dimensional simplex. Thus the difference between $c_d - 1$ and $[d/2]$ can be thought of as a crude indication of how special simplicial (or simple) polytopes are.

Further, by an extension of the usual argument of "pulling vertices" [23; p. 80], one can show that, for a fixed number of vertices, the number of chains (for any dimension set) will be maximized by a simplicial polytope. Thus by (7.5) and the Upper Bound Theorem, we get that, for d-polytopes with $n$ vertices, the numbers $f_S(P)$ are simultaneously maximized (for all $S$) by the cyclic polytope $C(n,d)$.

8. CONCLUDING REMARKS. The results on f-vectors and extended f-vectors surveyed in this paper serve to make combinatorial distinctions among different classes of complexes. The f-vectors of simplicial d-polytopes span a $[d/2]$-dimensional affine subspace of the Euler hyperplane (the (d-1)-space spanned by the f-vectors of all d-polytopes). The extended f-vectors of the (complete

barycentric subdivisions of) simplicial d-polytopes still have dimension $[d/2]$. They are properly contained in the $(c_d-1)$-dimensional space spanned by the extended f-vectors of rank $d$ Eulerian posets (alternatively, of complete barycentric subdivisions of arbitrary d-polytopes). This in turn is a subspace of the $(2^d-1)$-dimensional space determined by the extended f-vectors of completely balanced homology $(d-1)$-spheres.

Determining affine dimensions is the first step towards the goal of characterizing the f-vectors of these classes of complexes. The characterization of the f-vectors of simplicial polytopes was motivated by the particular form $h_i = h_{d-i}$ of the Dehn-Sommerville equations. It is natural,therefore, to look at the generalizations of the Dehn-Sommerville equations surveyed in this paper for clues towards nonlinear conditions on the numbers of chains. Of primary interest is characterizing the extended f-vectors of arbitrary d-polytopes. Perhaps what is needed is a change of variables from the $f_S$, which will put the equations of Theorem 7.1 into a simple form (the transformation to the extended h-vector does not do this).

Part of the motivation for considering extended f-vectors is to derive information on the original f-vectors themselves. The linear equations obtained do not help with this problem. We mention here a conjecture which attempts to generalize the Dehn-Sommerville equations in another way.

CONJECTURE. If $f(P) = (f_0,f_1,\dots,f_{d-1})$ is the f-vector of a d-polytope $P$, then for $0 \leq k \leq d-2$

$$f_k \geq \sum_{j=k}^{d-1} (-1)^{d-1-j}\binom{j+1}{k+1} f_j.$$

The inequality holds for $k = d-2$ and, equivalently, for $k = d-3$. The conjecture is true for polytopes of dimension $\leq 4$; for simple polytopes (and, of course, simplicial polytopes, for which the relations are equalities); and for prisms on simplicial polytopes. If the inequalities hold for some polytope, then they hold for the pyramid and bipyramid over that polytope. (For details see [8].) Note that Theorem 3.7 says that the f-vectors of all d-polytopes are spanned by f-vectors of (r-fold) pyramids over simplicial (d-r)-polytopes and simplicial d-polytopes. All such polytopes satisfy the inequalities of the conjecture. Thus, the conjecture is related to the following question: can the f-vector of any polytope be written as a linear combination of the f-vectors of simplicial polytopes and the f-vectors of pyramids over simplicial polytopes with nonnegative coefficients on the latter?

A resolution of the conjecture would be relevant to the question of characterizing the f-vectors of d-polytopes. We note that the f-vectors of 3-polytopes are characterized by the inequality of the conjecture and its "polar" (obtained by interchanging $f_i$ with $f_{d-1-i}$), the inequalities $f_i \geq \binom{d+1}{i+1}$, and the Euler equation [23; p. 190]. That is

$$f(P^3) = \{(f_0,f_1,f_2)\colon f_0-f_1+f_2 = 2,\ \ f_0 \geq 4,\ f_2 \geq 4,\ \ 3f_2 \leq 2f_1,\ \ 3f_0 \leq 2f_1\}.$$

## BIBLIOGRAPHY


[1] M. Aigner, Combinatorial Theory, Springer-Verlag, New York, 1979.

[2] M.F. Atiyah and I.G. Macdonald, Introduction to Commutative Algebra, Addison-Wesley, Reading, Massachusetts, 1969.

[3] K. Baclawski, "Cohen-Macaulay ordered sets," J. Alg. 63 (1980), 226-258.

[4] K. Baclawski, "Cohen-Macaulay connectivity and geometric lattices," Europ. J. Combin. 3 (1982), 293-305.

[5] K. Baclawski and A. Garsia, "Combinatorial decompositions of a class of rings," Adv. in Math. 39 (1981), 155-184.

[6] D.W. Barnette, "A proof of the lower bound conjecture for convex polytopes," Pacific J. Math. 46 (1973), 349-354.

[7] M.M. Bayer, Facial Enumeration in Polytopes, Spheres and Other Complexes, Ph.D. Thesis, Cornell University, 1983.

[8] M.M. Bayer, "A conjecture concerning the f-vectors of polytopes," Ars Combinatoria (to appear).

[9] M.M. Bayer and L.J. Billera, "Generalized Dehn-Sommerville relations for polytopes, spheres and Eulerian partially ordered sets," (to appear).

[10] L.J. Billera, "Polyhedral theory and commutative algebra," Mathematical Programming - Bonn 1982, The State of the Art, A. Bachem, M. Grötschel and B. Korte, eds., Springer-Verlag, 1983.

[11] L.J. Billera and C.W. Lee, "Sufficiency of McMullen's conditions for f-vectors of simplicial polytopes," Bull. Amer. Math. Soc. (New Series) 2 (1980), 181-185.

[12] L.J. Billera and C.W. Lee, "A proof of the sufficiency of McMullen's conditions for f-vectors of simplicial convex polytopes," J. Combinatorial Theory (A) 31 (1981), 237-255.

[13] A. Björner, "Shellable and Cohen-Macaulay partially ordered sets," Trans. Amer. Math. Soc. 260 (1980), 159-183.

[14] A. Björner, A. Garsia and R. Stanley, "An introduction to Cohen-Macaulay partially ordered sets," Ordered Sets, I. Rival, ed., D. Reidel, Boston, 1982, 583-615.

[15] A. Brønsted, An Introduction to Convex Polytopes, Graduate Texts in Mathematics, No. 90, Springer-Verlag, New York, 1983.

[16] H. Bruggesser and P. Mani, "Shellable decompositions of cells and spheres," Math. Scand. 29 (1971), 197-205.

[17] V.I. Danilov, "The geometry of toric varieties," Russian Math. Surveys 33:2 (1978), 97-154; translated from Uspekhi Mat. Nauk. 33:2 (1978), 85-134.

[18] G. Ewald and G.C. Shephard, "Stellar subdivisions of boundary complexes of convex polytopes," Math. Ann. 210 (1974), 7-16.

[19] A. Garsia, "Combinatorial methods in the theory of Cohen-Macaulay rings," Adv. in Math. 38 (1980), 229-266.

[20] P. Gilmore and A.J. Hoffman, "A characterization of comparability graphs and of interval graphs," Canad. J. Math. 16 (1964), 539-548.

[21] C. Greene, "The Möbius function of a partially ordered set," Ordered Sets, I. Rival, ed., D. Reidel, Boston, 1982, 555-581.

[22] C. Greene and D. Kleitman, "Proof techniques in the theory of finite sets," Studies in Combinatorics, G.-C. Rota, ed., Math. Assoc. of America, Washington, 1978, 22-79.

[23] B. Grünbaum, Convex Polytopes, Wiley Interscience, New York, 1967.

[24] P.J. Hilton and S. Wylie, Homology Theory, Cambridge Univ. Press, 1960.

[25] M. Hochster, "Rings of invariants of tori, Cohen-Macaulay rings generated by monomials, and polytopes," Annals of Math. 96 (1972), 318-337.

[26] V. Klee, "A combinatorial analogue of Poincaré's duality theorem," Can. Jour. Math. 16 (1964), 517-531.

[27] A.G. Kouchnirenko, "Polyèdres de Newton et nombres de Milnor," Invent. Math. 32 (1976), 1-31.

[28] C.W. Lee, Counting the Faces of Simplicial Convex Polytopes, Ph.D. Thesis, Cornell University, 1981.

[29] C.W. Lee, "Characterizing the numbers of faces of a simplicial convex polytope," in Convexity and Related Combinatorial Geometry, D.C. Kay and M. Breen, eds., Marcel Dekker, New York and Basel, 1982.

[30] B. Lindström, "On the realization of convex polytopes, Euler's formula and Mobius functions," Aequationes Math. 6 (1971), 235-240.

[31] P. McMullen, "The maximum numbers of faces of a convex polytope," Mathematika 17 (1970), 179-184.

[32] P. McMullen, "The minimum number of faces of a convex polytope," J. London Math. Soc. (2)3 (1971), 350-354.

[33] P. McMullen, "The numbers of faces of simplicial polytopes," Israel J. Math. 9 (1971), 559-570.

[34] P. McMullen and G.C. Shephard, Convex Polytopes and the Upper Bound Conjecture, London Math. Soc. Lecture Notes Series, Vol. 3, 1971.

[35] P. McMullen and D.W. Walkup, "A generalized lower-bound conjecture for simplicial polytopes," Mathematika 18 (1971), 264-273.

[36] T.S. Motzkin, "Comonotone curves and polyhedra," Abstract 111, Bull. Amer. Math. Soc. 63 (1957), p. 35.

[37] G. Reisner, "Cohen-Macaulay quotients of polynomial rings," Advances in Math. 21 (1976), 30-49.

[38] G.-C. Rota, "On the combinatorics of the Euler characteristic," in Studies in Pure Mathematics, L. Mirsky, ed., Academic Press, London, 1971.

[39] G.T. Sallee, "Polytopes, valuations and the Euler relation," Canad. J. Math. 20 (1968), 1412-1424.

[40] D.M.Y. Sommerville, "The relations connecting the angle-sums and volume of a polytope in space of n dimensions," Proc. Roy. Soc. London, ser. A 115 (1927), 103-119.

[41] R. Stanley, "Combinatorial reciprocity theorems," Advances in Math. 14 (1974), 194-253.

[42] R. Stanley, "Cohen-Macaulay rings and constructible polytopes," Bull. Am. Math. Soc. 81 (1975) 133-135.

[43] R. Stanley, "The Upper Bound Conjecture and Cohen-Macaulay rings," Studies in Applied Math. 54 (1975), 135-142.

[44] R. Stanley, "Cohen-Macaulay complexes," in Higher Combinatorics, M. Aigner, ed., D. Reidel, Boston, 1977, 51-62.

[45] R. Stanley, "Hilbert functions of graded algebras," Advances in Math. 28 (1978), 57-83.

[46] R. Stanley, "Balanced Cohen-Macaulay complexes," Trans. Amer. Math. Soc. 249 (1979), 139-157.

[47] R. Stanley, "The number of faces of a simplicial convex polytope," Advances in Math. 35 (1980), 236-238.

[48] R. Stanley, Interactions Between Commutative Algebra and Combinatorics, Progress in Mathematics Series, Birkhäuser, Boston, 1983 (to appear).

[49] R. Stanley, "Some aspects of groups acting on finite posets," J. Combinatorial Theory (A) 32 (1982), 132-161.

[50] R. Stanley, "The number of faces of simplicial polytopes and spheres," Discrete Geometry and Convexity, J. Goodman, et. al., eds., N.Y. Academy of Sciences, 1984 (to appear).

[51] R. Stanley, "An introduction to combinatorial commutative algebra," Proceedings of the Silver Jubilee Conference on Combinatorics, University of Waterloo (to appear).

[52] B. Teissier, "Variétés toriques et polytopes," Séminaire Bourbaki, 1980/81, no. 565, Lecture Notes in Mathematics, Springer-Verlag, Berlin, Heidelberg, 1981.

[53] D.W. Walkup, "The lower bound conjecture for 3- and 4-manifolds," Acta Math. 125 (1970), 75-107.

DEPARTMENT OF MATHEMATICS
NORTHEASTERN UNIVERSITY
BOSTON, MASSACHUSETTS 02115

and

SCHOOL OF OPERATIONS RESEARCH
CORNELL UNIVERSITY
ITHACA, NEW YORK 14853

Contemporary Mathematics
Volume 34, 1984

# THE COMBINATORICS OF $(k,\ell)$-HOOK SCHUR FUNCTIONS

Jeffrey B. Remmel[1]

ABSTRACT. The concept of a $(k,\ell)$-hook Schur function $HS_\lambda(x_1,\ldots,x_k;y_1,\ldots,y_\ell)$ was introduced by Berele and Regev and occurs naturally in the representation theory of Lie superalgebras. A number of symmetric function identities for the $HS_\lambda(\overline{x};\overline{y})$'s have been proven in the literature. One can give bijective proofs of such identities by using two distinct generalizations of the Robinson-Schensted correspondence for two sets of variables. We give a survey of how these two generalizations of the Robinson-Schensted correspondence have been used and prove two new identities for the $HS_\lambda(\overline{x};\overline{y})$'s.

## INTRODUCTION.

The concept of a $(k,\ell)$-hook Schur function of shape $\lambda$, $HS_\lambda(x_1,\ldots,x_k;y_1,\ldots,y_\ell)$, was introduced by Berele and Regev [1] in their study of the representations of Lie superalgebras. Briefly in terms of skew symmetric functions we may define

$$(0.1)\qquad HS_\lambda(x_1,\ldots,x_k;y_1,\ldots,y_\ell) = \sum_{\beta\subseteq\lambda} S_\beta(x_1,\ldots,x_k)S_{(\lambda/\beta)'}(y_1,\ldots,y_\ell)$$

and in terms of $\Lambda$-ring notation $HS_\lambda(x_1,\ldots,x_k;y_1,\ldots,y_\ell) = S_\lambda((x_1+\ldots+x_k)-(y_1+\ldots+y_\ell))$. In [1], Berele and Regev proved a number of symmetric function identities for the $HS_\lambda(\overline{x};\overline{y})$'s by partly algebraic and partly combinatorial means including an elegant factorization theorem for those $HS_\lambda(\overline{x};\overline{y})$'s where $\lambda$ contains the $k\times\ell$ rectangle; see equation 1.5 to follow. In [10], the author gave a completely bijective proof of this factorization theorem. Moreover in [2], Berele and the author generalized the classical Cauchy identities for Schur functions to the setting of $(k,\ell)$-hook Schur functions and provided bijective proofs of such identities. The main

---

1980 Mathematics Subject Classification. 05A15, 05A19, 17A70, 17B10, 20C30
[1] Partially supported by NSF Grant 82-02333.

combinatorial tool one needs to construct the bijections in [2] and [10] is a generalization of the Robinson-Schensted correspondence to two sets of variables. A rather surprising fact is that [2] and [10] employ two different generalizations of the Robinson-Schensted correspondence and that each correspondence has special properties which are well suited to prove a class of identities which is different from that of the other correspondence. Thus the combinatorics of $(k,\ell)$-hook Schur functions provide for rich applications for the techniques of tableaux theory and give rise to many interesting combinatorial identities which generalize classical Schur function identities.

The main purpose of this paper is to give an overview of the combinatorics of $(k,\ell)$-hook Schur functions that have been developed so far and to give combinatorial proofs of two more identities for $(k,\ell)$-hook Schur functions which have not appeared in the literature. In particular, we shall generalize two formulas of Littlewood [8] for the plethysms $S_2 \otimes S_n$ and $S_{1^2} \otimes S_n$ and prove combinatorially that the multiplication of $(k,\ell)$-hook Schur functions obeys the Littlewood-Richardson rule just as the multiplication of usual Schur functions.

The outline of this paper is as follows. In section 1, we shall establish notation and give a list of the various symmetric function identities for the $HS_\lambda(\overline{x};\overline{y})$ we shall discuss and their relationship to the classical Schur function identities. In section 2, we shall give the two generalizations of the Robinson-Schensted correspondence alluded to above and discuss their key properties. Then in section 3, we shall illustrate how these two correspondences have been used in the literature to construct bijections. Finally in sections 4 and 5, we shall prove the new results mentioned above.

## §1. Notations and identities.

Given a partition $\lambda = (0 < \lambda_1 \leq \ldots \leq \lambda_k)$, the Ferrers diagram of shape $\lambda$ is the set of left justified rows of cells or squares with $\lambda_1$ cells in the top row, $\lambda_2$ cells in the next row, etc. For example,

$F_{(1,3,3,4)} =$ 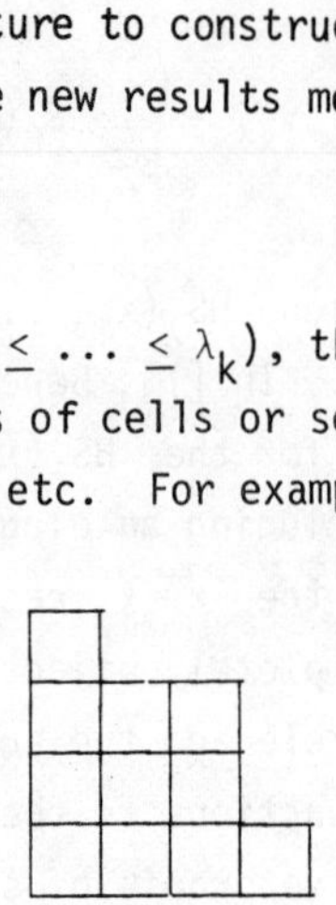

We shall call the cell in the $i^{th}$ row of $F_\lambda$ and the $j^{th}$ column of $F_\lambda$, the $(i,j)^{th}$ cell of $F_\lambda$ where the rows are labeled in order from bottom to top and the columns are labeled in order from left to right. Given two partitions $\lambda = (\lambda_1,\ldots,\lambda_k)$ and $\mu = (\mu_1,\ldots,\mu_\ell)$, we write $\mu \le \lambda$ if $\ell \le k$ and $\mu_i \le \lambda_{k-\ell+i}$ for $i = 1,\ldots,\ell$. If $\mu \le \lambda$, the skew diagram of shape $\lambda/\mu$ consists of the squares of $F_\lambda$ that remain after we remove the squares of $F_\mu$. For example,

$F_{(1,2,2,3)/(1,2)} =$ 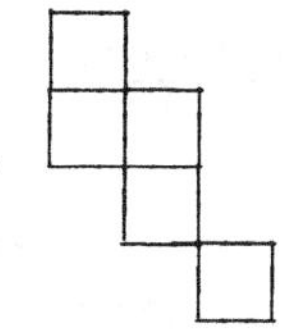

In what follows we shall consider fillings of skew diagrams $F_{\lambda/\mu}$ with two types of numbers, namely the numbers $1,2,\ldots$, which we shall call <u>regular</u> numbers and the numbers $1',2',\ldots$, which we shall call <u>primed</u> numbers. A <u>column strict tableau</u> $T_1$ of shape $\lambda/\mu$ is a filling of the cells of $F_{\lambda/\mu}$ with either all regular numbers or all primed numbers such that the numbers in each row weakly increase from left to right and the numbers in each column strictly increase from bottom to top. A <u>row strict tableau</u> $T$ of shape $\lambda/\mu$ is a filling of the cells of $F_{\lambda/\mu}$ with either all regular numbers or all primed numbers such that the numbers in each row strictly increase from left to right and the numbers in each column weakly increase from bottom to top. For example, the tableaux $T_1$ and $T_2$ in Figure 1 are column strict and row strict respectively.

Figure 1

$T_1 =$

| | | | |
|---|---|---|---|
| 5 | 5 | | |
| 3 | 4 | 5 | |
| 2 | 2 | 4 | |
| 1 | 1 | 2 | 3 |

$T_2 =$

| | | | |
|---|---|---|---|
| 2 | 3 | | |
| 2 | 3 | 4 | |
| 1 | 2 | 4 | |
| 1 | 2 | 3 | 5 |

A <u>$(k,\ell)$-semistandard tableau</u> $T$ of shape $\lambda/\mu$ is a filling of $F_{\lambda/\mu}$ with numbers from the set $\{1,\ldots,k, 1',\ldots,\ell'\}$ such that (i) the <u>regular part</u> of $T$, i.e., the cells of $F_{\lambda/\mu}$ which contain regular numbers form a column strict tableau $T_r$ of shape $\beta/\mu$ where $\beta$ is some partition such that $\mu \le \beta \le \lambda$ and (ii) the <u>primed part</u> of $T$, i.e., the cells of $F_{\lambda/\mu}$ which contain primed numbers form a row strict tableau $T_p$ of shape $\lambda/\beta$. For example, $T$ in Figure 2 is a (3,4)-semistandard tableau of shape (1,2,4,4).

Figure 2

T =

| | | | |
|---|---|---|---|
| 1' | | | |
| 1' | 2' | | |
| 2 | 2 | 3' | 4' |
| 1 | 1 | 3 | 4' |

A standard tableau S of shape $\lambda/\mu$ is a column strict tableau of shape $\lambda/\mu$ in which the numbers $1,\ldots,n$ occur where $n = |\lambda/\mu|$ = number of cells in $F_{\lambda/\mu}$. For emphasis, we shall refer to column strict (row strict, etc.) tableaux T of shape $\lambda/\mu$ where $\mu \neq \phi$ as skew tableaux and to column strict (row strict, etc.) tableaux of shape $\lambda$ as simply tableaux.

We let $\mathcal{T}_{(k,\ell)}(\lambda/\mu)$ denote the set of all $(k,\ell)$-semistandard (skew) tableaux of shape $\lambda/\mu$. Thus $\mathcal{T}_{(k,0)}(\lambda/\mu)$ is the set of all column strict skew tableaux of shape $\lambda/\mu$ filled with regular numbers $\{1,\ldots,k\}$. Note that $\mathcal{T}_{(k,0)}(\lambda/\mu) \subseteq \mathcal{T}_{(k,\ell)}(\lambda/\mu)$. Given $T \in \mathcal{T}_{(k,\ell)}(\lambda)$ or for that matter any filling T of $F_\lambda$ with the numbers from $\{1,\ldots,k,1',\ldots,\ell'\}$, we define the type of T, $\tau(T)$, as $\tau(T) = (\nu_1,\ldots,\nu_k;\mu_1,\ldots,\mu_\ell)$ where $\nu_i$ = number of times i occurs in T and $\mu_j$ = number of times j' occurs in T. We define the weight of such a T, $\omega(T)$, by $\omega(T) = x_1^{\nu_1}\cdots x_k^{\nu_k} y_1^{\mu_1}\cdots y_\ell^{\mu_\ell}$. For example, for T of Figure 2, $\tau(T) = (2,2,1;2,1,1,2)$ and $\omega(T) = x_1^2x_2^2x_3y_1^2y_2y_3y_4^2$. Similarly for $T_1$ of Figure 1, $\omega(T_1) = x_1^2x_2^3x_3^2x_4^2x_5^3$. We then can define the $(k,\ell)$-hook Schur function $HS_\lambda(x_1,\ldots,x_k;y_1,\ldots,y_\ell)$ by

$$(1.1) \qquad HS_\lambda(x_1,\ldots,x_k;y_1,\ldots,y_\ell) = \sum_{T\in\mathcal{T}_{(k,\ell)}(\lambda)} \omega(T)$$

and the $(k,\ell)$-hook skew Schur function $HS_{\lambda/\mu}(x_1,\ldots,x_k;y_1,\ldots,y_\ell)$ by

$$(1.2) \qquad HS_{\lambda/\mu}(x_1,\ldots,x_k;y_1,\ldots,y_\ell) = \sum_{T\in\mathcal{T}_{k,\ell}(\lambda/\mu)} \omega(T) .$$

The definition of $(k,\ell)$-semistandard tableaux and $(k,\ell)$-hook Schur functions via (0.1) was first given by Berele and Regev in [1] where they showed that $(k,\ell)$-hook Schur functions correspond naturally to irreducible representations of Lie superalgebras. The notions of $(k,\ell)$-semistandard skew tableaux and (k,ℓ)-hook skew Schur functions are the obvious extensions of their definitions to skew shapes $\lambda/\mu$. The reason for calling the $HS_\lambda$'s (k,ℓ)-hook Schur functions is due to the fact that $HS_\lambda(x_1,\dots,x_k;y_1,\dots,y_\ell) = 0$ unless the shape $\lambda$ fits inside the "(k,ℓ)-hook" consisting of all cells in the first k rows and $\ell$ columns (see Figure 3).

Figure 3

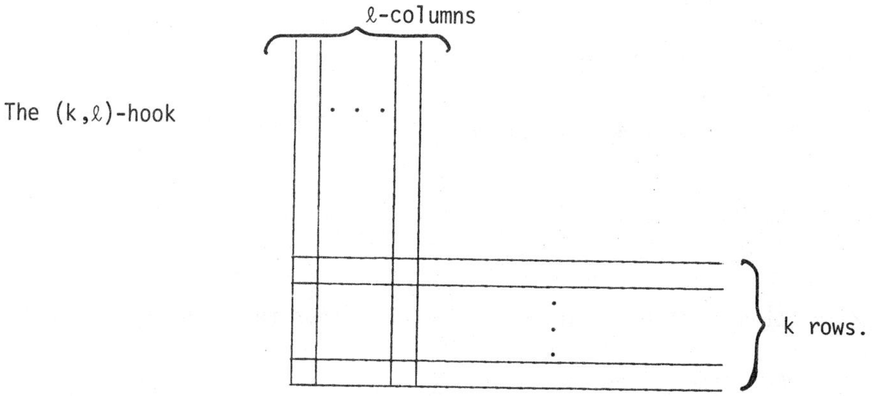


We shall also note that (k,0)-hook Schur functions and skew Schur functions are simply the usual Schur functions $S_\lambda(x_1,\dots,x_k)$ and skew Schur functions $S_{\lambda/\mu}(x_1,\dots,x_k)$. Moreover it is clear that $HS_\lambda(\overline{x},\overline{y})$ can be expressed in terms of skew Schur functions in the x's and the y's by

$$(1.3)\qquad HS_\lambda(x_1,\dots,x_k;y_1,\dots,y_\ell) = \sum_{\beta\leq\lambda} S_\beta(x_1,\dots,x_k)S_{\lambda'/\beta'}(y_1,\dots,y_\ell),$$

where for a partition $\lambda$, $\lambda'$ denotes the conjugate partition corresponding to $\lambda$, i.e., the partition induced by the columns of $F_\lambda$.

Next we shall state a number of symmetric function identities for the $HS_\lambda$'s which all have combinatorial significance.

$$(1.4)\qquad HS_{(\ell)^k}(x_1,\dots,x_k;y_1,\dots,y_\ell) = \prod_{i=1}^{k}\prod_{j=1}^{\ell}(x_i+y_j).$$

(Here $(\ell)^k$ is the partition with $k$ parts of size $\ell$, i.e., the partition corresponding to the $k\times\ell$ rectangle k □ .)
ℓ

(1.5) If $\lambda \geq (\ell)^k$ and is of the form $(\beta_1,\ldots,\beta_j,\ \ell+\alpha_1,\ldots,\ell+\alpha_k)$, i.e.

$\lambda =$ 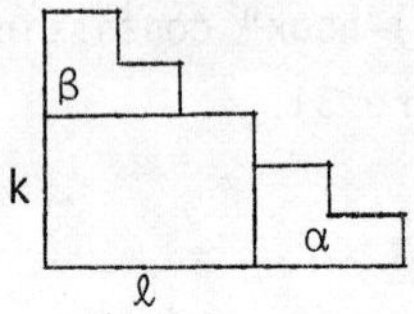

where $\beta = (\beta_1,\ldots,\beta_j)$ and $\alpha = (\alpha_1,\ldots,\alpha_k)$, then

$$HS_\lambda(x_1,\ldots,x_k;\ y_1,\ldots,y_\ell)$$

$$= HS_{(\ell)^k}(x_1,\ldots,x_k;\ y_1,\ldots,y_\ell)\ \ S_\alpha(x_1,\ldots,x_k)\ \ S_{\beta'}(y_1,\ldots,y_\ell)$$

(1.6) $HS_\lambda(y_1,\ldots,y_\ell;\ x_1,\ldots,x_k) = HS_{\lambda'}(x_1,\ldots,x_k;\ y_1,\ldots,y_\ell)$

(i.e., interchanging the roles of the x's and y's turns a $(\ell,k)$-hook Schur function of shape $\lambda$ into a $(k,\ell)$-hook Schur function of shape $\lambda'$).

(1.7) (Analogue of the Cauchy identity).

$$\sum_\lambda HS_\lambda(x_1,\ldots,x_{k_1};\ s_1,\ldots,s_{\ell_1})\ HS_\lambda(y_1,\ldots,y_{k_2};\ t_1,\ldots,t_{\ell_2})$$

$$= \prod_{i,j}^{k_1,k_2} \frac{1}{1-x_iy_j} \prod_{i,j}^{\ell_1,\ell_2} \frac{1}{1-s_it_j} \prod_{i,j}^{k_1,\ell_2} (1+x_it_j) \prod_{i,j}^{k_2,\ell_1} (1+y_is_j)$$

(1.8) $$\sum_{\lambda\ \text{even}} HS_\lambda(x_1,\ldots,x_k;\ y_1,\ldots,y_\ell)$$

$$= \prod_{i\leq j} \frac{1}{1-x_ix_j} \prod_{i<j} \frac{1}{1-y_iy_j} \prod_{i,j}^{k,\ell} (1+x_iy_j)$$

where we say a partition $\lambda = (\lambda_1 \leq \ldots \leq \lambda_k)$ is <u>even</u> if $\lambda_i$ is even for $i=1,\ldots,k$.

(1.9) (Analogue of Jacobi-Trudi identity).

$$HS_{\lambda/\mu}(\bar{x};\bar{y}) = \det \| HS_{\hat{\lambda}_i-\hat{\mu}_j}(\bar{x};\bar{y}) \|$$

where if $\lambda = (\lambda_1,\ldots,\lambda_k)$, $\hat{\lambda} = (\lambda_1,\ \lambda_2+1,\ldots,\lambda_i+i-1,\ldots,\lambda_k+k-1)$.

(1.10) If $S_\nu(\bar{x})\ S_\mu(\bar{x}) = \sum_\lambda g^\lambda_{\mu,\nu}\ S_\lambda(\bar{x})$, then

$$HS_\nu(\bar{x};\bar{y})\ HS_\mu(\bar{x};\bar{y}) = \sum_\lambda g^\lambda_{\mu,\nu}\ HS_\lambda(\bar{x};\bar{y}).$$

(i.e., the products of $(k,\ell)$-hook Schur functions obey the usual Littlewood-Richardson Rule).

Identity 1.4 is a restatement of an identity due to Littlewood in terms of $(k,\ell)$-hook Schur functions (see also MacDonald [9], pg. 37). The remarkable factorization theorem for $HS_\lambda(x_1,..,x_k;\ y_1,...,y_\ell)$ where $\lambda$ contains the $k \times \ell$ rectangle given in 1.5 is due to Berele and Regev [1] who also noted 1.6. The first completely bijective proofs of 1.4, 1.5, and 1.6 are due to Remmel [10]. The analogue of the Cauchy identity is due to Berele and Remmel in [2] where they gave a bijective proof by generalizing the Knuth correspondence [5] between nonnegative integer valued matrices and pairs of column strict tableaux. One nice fact about 1.7 is that it specializes to give both classical Cauchy identities. That is, if we set $\ell_1 = \ell_2 = 0$ in 1.7, we get

1.11 $$\sum_\lambda S_\lambda(\bar{x})\ S_\lambda(\bar{y}) = \prod_{i,j} \frac{1}{1-x_iy_j}\ .$$

Similarly if we set $\ell_1 = k_2 = 0$ and note that $HS_\lambda(0,..,0;\ t_1,..,t_{\ell_2})$ $= S_{\lambda'}(t_1,...,t_{\ell_2})$, then we get

1.12 $$\sum_\lambda S_\lambda(\bar{x})\ S_{\lambda'}(\bar{t}) = \prod_{i,j} (1+x_it_j)\ .$$

Identity 1.8 also is a simultaneous generalization of two classical Schur function identities due to Littlewood [8]. That is, by setting $k = 0$ and then $\ell = 0$ we get the following two identities.

1.13 $$\sum_{\lambda \text{ even}} S_\lambda(x_1,..,x_k) = \prod_{i\le j} \frac{1}{1-x_ix_j} = S_2 \otimes S_n(x_1,...,x_k)$$

1.14 $$\sum_{\lambda' \text{ even}} S_\lambda(y_1,..,y_\ell) = \prod_{i<j} \frac{1}{1-y_iy_j} = S_{1^2} \otimes S_n\ (x_1,...,x_k)$$

From the $\Lambda$-ring point of view, it is clear that all identities that hold for generalized Schur functions $S_\lambda(E-F)$ will hold for the $HS_\lambda(\bar{x},\bar{y})$'s. Thus the fact that identities 1.9 and 1.10 hold is no surprise from the $\Lambda$-ring point of view. However, we shall prove 1.9 combinatorially by extending Gessel's combinatorial proof of the Jacobi-Trudi identity [4]. Note that once again by specializing identity 1.9 by setting $k = 0$ or $\ell = 0$, we get both Jacobi-Trudi identities, i.e. the expansion of $S_{\lambda/\mu}(\bar{x})$ as a determinate in the homogeneous symmetric functions $h_n(\bar{x})$ and the expansion of $S_{\lambda/\mu}(\bar{y})$ as a determinate in the elementary symmetric functions $e_n(\bar{y})$. Having proven 1.9 combinatorially, we then can prove (1.10) combinatorially by simply proving combinatorially the analogue of Pieri's formula or Young's rule

1.15 $$HS_n(\bar{x};\bar{y})\ HS_\lambda(\bar{x};\bar{y}) = \sum_{\substack{\beta \\ |\beta/\lambda| = n \\ \beta/\lambda \text{ is a skew row}}} HS_\beta(\bar{x};\bar{y})$$

where a skew shape $\beta/\lambda$ is a skew row if no two squares of $\beta/\lambda$ lie in the same column.

As we shall see in the next few sections, identity 1.6 is given a bijective proof by using a variation of the Jue de Taquin game of Schützenberger [12]. Bijective proofs of identities 1.4 and 1.5 use one generalization of the Robinson-Schensted correspondence while bijective proofs of identities 1.7, 1.8, and 1.15 follow most easily by using a second generalization of the Robinson-Schensted correspondence.

## §2. Generalizations of the Robinson-Schensted Correspondence.

As mentioned in previous sections, we shall introduce two generalizations of the Robinson-Schensted (RS) correspondence which are most useful for constructing bijections to prove the various identities listed in section 1. We shall begin by defining two analogues of the Schensted bumping process for two sets of numbers. In all that follows, we shall assume that the primed numbers are strictly larger than the regular numbers, i.e., $1 < 2 < 3 < \ldots < 1' < 2' < 3' < \ldots$ . We shall also assume that the reader is familiar with the usual row insertion algorithm and column insertion algorithm as described in Knuth's book [6].

For our first generalization of the Schensted bumping algorithm, we bump the regular numbers from row to row as in the usual row insertion

algorithm and we bump the primed numbers from column to column as in the
usual column insertion algorithm.  This bumping process was first used by
Berle and Regev in [1].  Figure 4 gives two examples of inserting a number
z into a (k,ℓ)-semistandard tableau  T  via what we call RS 1  insertion
and denoted by  T ← z.  In figure 4, we indicate the sequence of elements
of  T  which are affected by this insertion or the so called <u>bumping path</u>
of  z  in  T  by circling the sequence of elements and then we give the
final result.

<u>Figure 4</u>    RS 1 insertion

$$
T = \begin{array}{llllll}
2' & 3' \\
1' & 2' & 3' \\
2 & 3 & 4 & 1' & 2' \\
1 & 1 & 2 & 2 & 1' & 2'
\end{array}
$$

T ← 1:

$$
\begin{array}{llllll}
2' & 3' \\
①{\to}②'{\to}③'{\to}\bigcirc \\
2 & ③ & 4 & 1' & 2' \\
1 & 1 & ② & 2 & 1' & 2' \\
& & \uparrow \\
& & 1
\end{array}
\qquad = \qquad
\begin{array}{llllll}
1' & 2' \\
3 & 1' & 2' & 3' \\
2 & 2 & 4 & 1' & 2' \\
1 & 1 & 1 & 2 & 1' & 2'
\end{array}
$$

T ← 1':

$$
\begin{array}{llllll}
1'{\to}②'{\to}③'{\to}\bigcirc \\
1' & 2' & 3' \\
2 & 3 & 4 & 1' & 2' \\
1 & 1 & 2 & 2 & 1' & 2'
\end{array}
\qquad = \qquad
\begin{array}{llllll}
1' & 2' & 3' \\
1' & 2' & 3' \\
2 & 3 & 4 & 1' & 2' \\
1 & 1 & 2 & 2 & 1' & 2'
\end{array}
$$

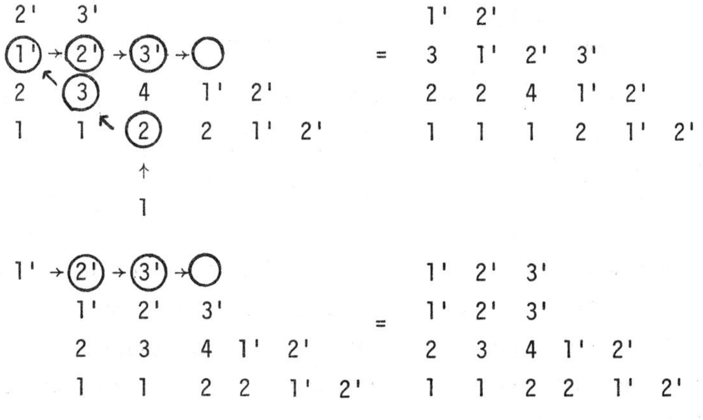

In our second generalization of the Schensted bumping algorithm, we
bump the regular numbers and the primed numbers from row to row with the
stipulation that we bump in such a way as to maintain the (k,ℓ)-semistandard
tableau conditions at each stage.  This bumping process was first used by
Berele and Remmel [2].  In figure 5, we give two examples of inserting a
number  z  into a (k,ℓ)-semistandard tableau  T  via this second bumping
algorithm which we call  RS 2  insertion and denote by  T⇐ z.  Once
again we shall circle the bumping paths for emphasis.

Figure 5:  RS 2 insertion

```
         2'  3'
         1'  2'  3'
T =      2   3   4   1'  2'
         1   1   2   2   1'  2'
```

```
           ◯                              2'
           ↑
          ②   3'                         1'  3'
           ↑
T ⟸ 1:    ①  2'  3'              =       3   2'  3'
          2   ③   4   1'  2'             2   2   4   1'  2'
          1   1   ②   2   1'  2'         1   1   1   2   1'  2'
                  ↑
                  1
```

```
           ◯                              2'
           ↑
          ②   3'                         1'  3'
           ↑
T ⟸ 1:    ①  2'  3'              =       1'  2'  3'
          2   3   4   ①  2'              2   3   4   1'  2'
          1   1   2   2   ①   2'         1   1   2   2   1'  2'
                      ↑
                      1'
```

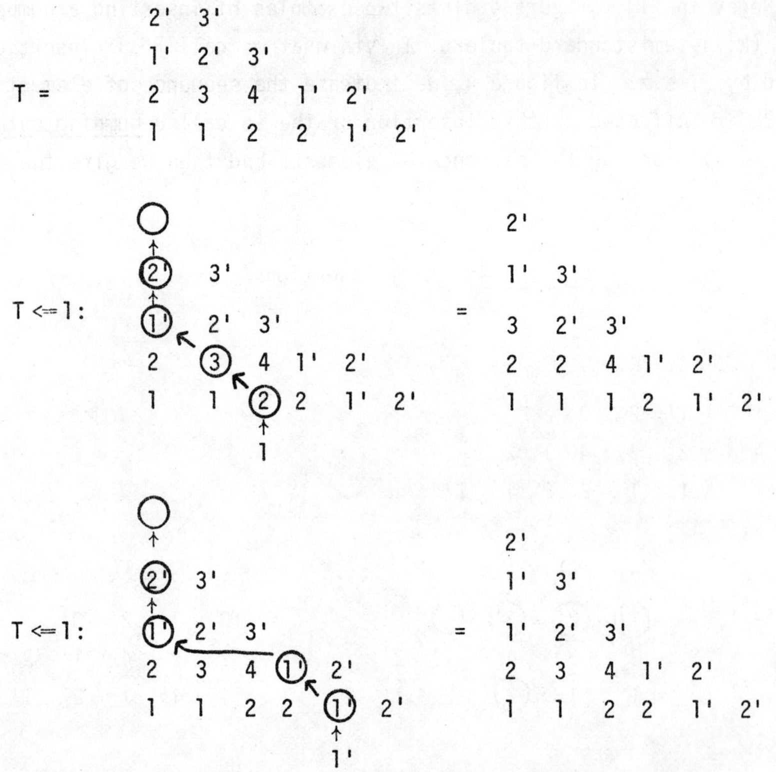

Note that the only difference between the RS 2 insertion procedure and the
usual row insertion procedure is that a primed number $z'$ bumps the least
element $x$ such that $z' \leq x$ when it is inserted into a row rather than
bumping the least number $x$ such that $z < x$ when a regular number is
inserted into a row.

Formal definitions of $T \leftarrow z$ and $T \Leftarrow z$ by recursion may be found
in [10] and [2] respectively. In both cases, it is routine to check that we
always end up with a $(k,\ell)$-semistandard tableau after the insertion process
and that the insertion process can be reversed if we know the new cell
that was created by the insertion process. Thus we can use either bumping
procedure to get an analogue of the Robinson-Schensted correspondence between
sequences $s = (s_0, s_1,\ldots,s_n)$ from $\{1,\ldots,k, 1',\ldots,\ell'\}$ and pairs $(P_i, Q_i)$
of tableaux of the same shape where $P_i$ is a $(k,\ell)$-semistandard tableau
and $Q_i$ is a standard tableau. That is, given such a sequence $s$, and a
semistandard tableau $T$, we define $T \leftarrow s_0,\ldots,s_n$ as equal to
$(\ldots((\emptyset \leftarrow s_0) \leftarrow s_1)\ldots) \leftarrow s_n$. ($T \Leftarrow s_0,\ldots,s_n$ is defined similarly). Then

Figure 6

Constructing the pair of tableaux for 2 1 2' 1' via RS 1 and RS 2

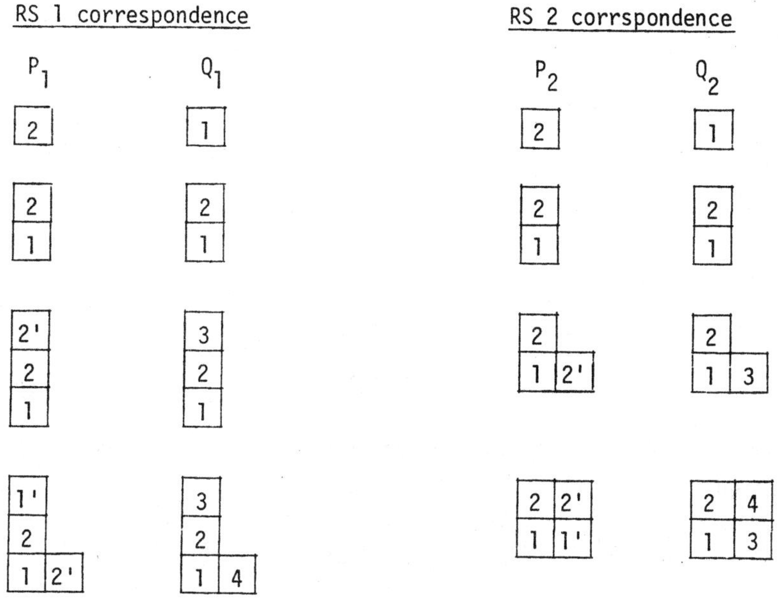


using RS 1 insertion, we define a map $s_0,\ldots,s_n \to (P_1,Q_1)$ where $P_1 = \phi \leftarrow s_0,\ldots,s_n$ and $Q_1$ is the standard tableau such that the number i is in cell $(a_i,b_i)$ if $(a_i,b_i)$ is the new cell created in going from $\phi \leftarrow s_0,\ldots,s_{i-1}$ to $\phi \leftarrow s_0,\ldots,s_i$. Similarly using RS 2 insertion, we can define a map $s_0,\ldots,s_n \to (P_2,Q_2)$ where $P_2 = \phi \Leftarrow s_0,\ldots,s_n$ and $Q_2$ records the growth of the $P_2$ tableau as above. In figure 6, we exhibit the steps of these two correspondences.

Actually from most points of view, our second analogue of the Schensted correspondence is the most natural extension of the usual Robinson-Schensted correspondence via row insertion. Indeed we can reduce RS 2 to standard row insertion. That is, suppose $\omega = \omega_1\ldots\omega_n$ is a word with $a_i$ i's for $i=1,\ldots,k$ and $b_j$ j''s for $j'=1',\ldots,\ell'$. We can associate a permutation $p(\omega) = p_1\ldots p_n$ with $\omega$ by labelling the 1's from left to right with $1,\ldots,a_1$, labelling the 2's from left to right with $a_1+1,\ldots,a_1+a_2$, etc. and then labelling 1''s from right to left with $1+\sum_{i=1}^{k} a_i,\ldots, b_1+\sum_{i=1}^{k} a_i$, labelling the 2''s from right to left with $1 + b_1 + \sum_{i=1}^{k} a_i$, $b_1 + b_2 + \sum_{i=1}^{k} a_i$, etc. For example, if

$\omega = 2\ 1\ 1'\ 2'\ 1\ 2\ 1'$, $p(\omega) = 3\ 1\ 6\ 7\ 2\ 4\ 5$. It is proven in [2] that if $p(\omega)$ is associated to $(P,Q)$ via the standard Robinson-Schensted correspondence, then $\omega$ is associated to $(P',Q)$ via RS 2 correspondence where $P'$ results from $P$ by replacing each $p_i$ by $\omega_i$ in $P$. For example,

$$3\ 1\ 6\ 7\ 2\ 4\ 5 \xrightarrow{RS} \begin{matrix} 3\ 6\ 7 \\ 1\ 2\ 4\ 5 \end{matrix},\ \begin{matrix} 2\ 5\ 6 \\ 1\ 3\ 4\ 7 \end{matrix}$$

and

$$2\ 1\ 1'\ 2'\ 1\ 2\ 1' \xrightarrow{RS\ 2} \begin{matrix} 2\ 1'\ 2' \\ 1\ 1\ 2\ 1' \end{matrix},\ \begin{matrix} 2\ 5\ 6 \\ 1\ 3\ 4\ 7 \end{matrix}\ .$$

However, no such reduction is possible for our first analogue of the Robinson-Schensted correspondence. For example

$\omega = 3'\ 2'\ 1\ 2\ 3 \xrightarrow{RS\ 1} \left(\begin{array}{|c|c|c|c|c|}\hline 1&2&3&2'&3'\\\hline\end{array}\ ,\ \begin{array}{|c|c|c|c|c|}\hline 1&2&3&4&5\\\hline\end{array}\right)$ which would force

the corresponding permutation $p(\omega)$ to be $p(\omega) = 1\ 2\ 3\ 4\ 5$. But then

$p(\omega) \xrightarrow{RS} \left(\begin{array}{|c|c|c|c|c|}\hline 1&2&3&4&5\\\hline\end{array}\ ,\ \begin{array}{|c|c|c|c|c|}\hline 1&2&3&4&5\\\hline\end{array}\right)$ and substituting $\omega_i$ by $p_i$

in the left tableau results in $\begin{array}{|c|c|c|c|c|}\hline 3'&2'&1&2&3\\\hline\end{array}$ not $\begin{array}{|c|c|c|c|c|}\hline 1&2&3&2'&3'\\\hline\end{array}$ .

We end this section by stating the key properties of the RS 1 and RS 2 correspondence that are at the heart of most of the bijective proofs using these correspondences. Essentially what is needed is two lemmas concerning the result of inserting a sequence of "increasing" or a sequence of "decreasing" elements into a $(k,\ell)$-semistandard tableau $T$. The situation is relatively straightforward for RS 2. We define a sequence $s = s_1,\ldots,s_n$ with each $s_i \in \{1,\ldots,k,1',\ldots,\ell'\}$ to be <u>$(k,\ell)$-increasing of type 2</u> if $s$ consists of a weakly increasing sequence of regular numbers followed by a strictly increasing sequence of primed numbers. For example, $1\ 1\ 2\ 3\ 1'\ 2'\ 4'$ is a $(3,4)$-increasing sequence of type 2. Similarly, we say $s$ is a <u>$(k,\ell)$-decreasing sequence of type 2</u> if $s$ consists of a weakly decreasing sequence of primed numbers followed by a strictly decreasing sequence of regular numbers. For example, $4'\ 4'\ 3'\ 2'\ 3\ 2\ 1$ is a $(3,4)$-decreasing sequence of type 2. The following two lemmas are proven in [2]. The lemmas can either be proven by reducing them to the corresponding properties for the usual Robinson-Schensted correspondence or directly by induction.

Lemma 2.1. (RS 2). Suppose $s = s_1, s_2, \ldots, s_n$ is a sequence from $\{1, \ldots, k, 1', \ldots, \ell'\}$. Let $T$ be a $(k,\ell)$-semistandard tableau and $T_i$ for $i = 0, \ldots, n$ be defined inductively by $T_0 = T$ and $T_{i+1} = T_i \Leftarrow s_{i+1}$. Let $(a_i, b_i)$ denote the new cell created in going from $T_{i-1}$ to $T_i$ for $i = 1, \ldots, n$. Then $s$ is a $(k,\ell)$-increasing sequence of type 2 iff for each $i$, $(a_{i+1}, b_{i+1})$ lies strictly to the right and weakly below $(a_i, b_i)$, i.e., $a_{i+1} > a_i$ and $b_{i+1} \leq b_i$.

Lemma 2.2. (RS 2). Let $s = s_1, s_2, \ldots, s_n$, $T_i$, and $(a_i, b_i)$ be as in Lemma 2.1. Then $s$ is a $(k,\ell)$-decreasing sequence of type 2 iff for each $i$, $(a_{i+1}, b_{i+1})$ lies weakly to the left and strictly above $(a_i, b_i)$, i.e. $a_{i+1} \leq a_i$ and $b_{i+1} > b_i$.

Thus for example, if $s = s_1 \ldots s_n$ is a $(k,\ell)$-increasing sequence of type 2 and $T' = T \Leftarrow s_1 \ldots s_n$, then Lemma 2.1 tells us that the difference between the shapes of $T'$ and $T$, $\lambda(T')/\lambda(T)$, is a skew row and the order in which the new cells were created in the insertion process is from left to right and top to bottom. Moreover, Lemma 2.1 tells us that if we reverse the RS 2 correspondence by removing a sequence of cells in a skew row from right to left and bottom to top, we will produce a $(k,\ell)$-increasing sequence of type 2. Similarly, Lemma 2.1 tells us that if $s$ is a $(k,\ell)$-decreasing sequence of type 2, then $\lambda(T')/\lambda(T)$ is a skew column, i.e., no two cells lie in the same row, and that the order in which the new cells were created is from right to left and from bottom to top.

The situation for the RS 1 correspondence is only a bit more complicated. Given a sequence $s = s_1, \ldots, s_n$ where each $s_i \in \{1, \ldots, k, 1', \ldots, \ell'\}$, we say that $s$ is a (k,ℓ)-decreasing sequence of type 1 if $s$ consists of a strictly decreasing sequence of regular numbers followed by a weakly increasing sequence of primed numbers. For example, 3 2 1 1' 1' 2' 4' is a (3,4)-decreasing sequence. Similarly, we say $s$ is a (k,ℓ)-increasing sequence of type 1 if $s$ consists of a strictly decreasing sequence of primed numbers followed by a weakly increasing sequence of regular numbers. For example, 4' 3' 2' 1 1 2 3 is a (3,4)-increasing sequence of type 1. Then the exact analogues of lemma 2.1 and lemma 2.2 hold with the RS 1 correspondence replacing the RS 2 correspondence, $(k,\ell)$-increasing sequences of type 1 replacing $(k,\ell)$-increasing sequences of type 2, and $(k,\ell)$-decreasing sequences of type 1 replacing $(k,\ell)$-decreasing sequences of type 2. The proofs of such lemmas may be found in [10].

We should also note that we can give analogues of Schensted's Theorems [11] for the interpretation of the lengths of the first row and first column of the P-tableaux. That is, the following is proven in [2] for the RS 2 correspondence.

Theorem 2.3. Let $s_1,\ldots,s_n$ be a sequence from $\{1,\ldots,k, 1',\ldots,\ell'\}$ and $P = \phi \Leftarrow s_1,\ldots,s_n$. Then the length of the first row of $P$ is the length of the longest $(k,\ell)$-increasing sequence of type 2 which is a subsequence of $s$ and the length of the first column of $P$ is the length of the longest $(k,\ell)$-decreasing sequence of type 2 which is a subsequence of $s$.

Moreover, the exact analogue of Theorem 2.3 holds for the RS 1 correspondence if we replace type 2 increasing and decreasing sequences by type 1 increasing and decreasing sequences.

## §3. Previous applications of the RS 1 and RS 2 correspondences.

In this section we shall briefly describe the bijections given in [2] and [10] which prove the identities 1.4, 1.5, 1.6, and 1.7. We shall start with the bijective proof of 1.6 since the bijection which proves 1.6 will be part a several other bijections. The first step in proving

3.0. $$HS_\lambda(x_1,\ldots,x_k;\, y_1,\ldots,y_\ell) = HS_{\lambda'}(y_1,\ldots,y_\ell;\, x_1,\ldots,x_k)$$

is to give an appropriate combinatorial interpretation of the right hand side. We let $\overline{\mathfrak{T}}_{\ell,k}(\lambda)$ denote the set of all fillings $T$ of $F_\lambda$ with elements from $\{1',\ldots,\ell', 1,\ldots,k\}$ so that the primed part of $T$ is a row strict tableau of some shape $\beta \subseteq \lambda$ and the regular part of $T$ is a column strict tableau of shape $\lambda/\beta$. Observe that if $T \in \overline{\mathfrak{T}}_{(\ell,k)}$, then $T'$, the conjugate tableau of $T$ (i.e., the tableau that results by flipping $T$ about the diagonal), is a tableau of shape $\lambda'$ where the primed part of $T'$ is a column strict tableau of some shape $\beta'$ and the regular part of $T'$ is a row strict tableau of shape $\lambda'/\beta'$. Given this observation, it is then clear that

3.1. $$HS_{\lambda'}(y_1,\ldots,y_\ell;\, x_1,\ldots,x_k) = \sum_{T \in \overline{\mathfrak{T}}_{(k,\ell)}(\lambda)} \omega(T).$$

and that we can prove 3.0 by constructing a weight preserving bijection $R:\mathfrak{T}_{(k,\ell)}(\lambda) \to \overline{\mathfrak{T}}_{(\ell,k)}(\lambda)$.

We call $R$ the reversing bijection since it will simply take a $T \in \mathfrak{T}_{(k,\ell)}(\lambda)$ and reverse the relative positions of the regular and primed

numbers in $T$. That is, we shall transform $T$ to a tableau
$R(T) \in \overline{\mathfrak{J}}_{(\ell,k)}(\lambda)$ by bringing the primed numbers of $T$ from the outside
to the inside and the regular numbers of $T$ from the inside to the
outside. The reversing bijection $R$ is constructed by playing a variant
of the Jue de Taquin game of Schützenberger [12] on $T$. At the start of the
game, we declare all the regular numbers to be active. We then take largest
regular number $x$ in $T$ which is in the rightmost cell among all those
cells in $T$ which contain $x$ and move it to the outside, i.e., past all
the primed numbers, via the set of moves pictured in Figure 7. We call the
regular number which is being moved the _special_ number and we move the
special number as long as it has a primed number either directly above it or
directly to its right. Once a special number is no longer bordered by a
primed number to its right or directly above, we stop moving it and declare
it to be inactive. Then we take the rightmost largest active regular number
that remains, make it the next special number, and move it past all the
primed numbers using the moves of figure 7. We continue in this way until
we have processed all the regular numbers in $T$ and hence we will produce a
new filling $R(T)$ of $F_\lambda$. For example, Figure 8a pictures the first step
of the reversing bijection and figure 8b pictures the final tableau
$R(T)$ for a tableau $T \in \mathfrak{J}_{(3,4)}(3,4,6,6)$.

Figure 7: The basic moves of the reversing bijection. (s denotes the
special number).

Case 1. There are primed numbers both above and to the right of $s$. Then

$$\begin{array}{c}\boxed{y'}\\ \boxed{s}\boxed{x'}\end{array} \longrightarrow \begin{array}{c}\boxed{y'}\\ \boxed{x'}\boxed{s}\end{array} \qquad \text{if } x' \leq y' \text{ and}$$

$$\begin{array}{c}\boxed{y'}\\ \boxed{s}\boxed{x'}\end{array} \longrightarrow \begin{array}{c}\boxed{s}\\ \boxed{y'}\boxed{x'}\end{array} \qquad \text{if } y' < x' \; .$$

Case 2. There is a primed number above but not to the right of $s$. Then

$$\begin{array}{c}\boxed{y'}\\ \boxed{s}\end{array} \longrightarrow \begin{array}{c}\boxed{s}\\ \boxed{y'}\end{array} \; .$$

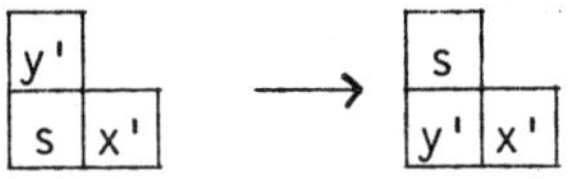

Case 3.  There is a primed number to the right of but not above s.  Then

$$\boxed{\begin{array}{c|c} s & x' \end{array}} \longrightarrow \boxed{\begin{array}{c|c} x' & s \end{array}} \quad .$$

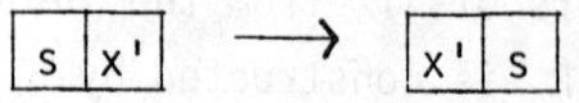

Figure 8.  An example of the reversing bijection.

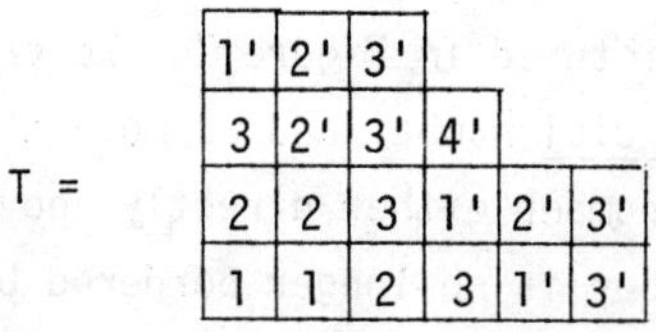

$$T = \begin{array}{|c|c|c|c|c|c|}
\hline 1' & 2' & 3' \\
\hline 3 & 2' & 3' & 4' \\
\hline 2 & 2 & 3 & 1' & 2' & 3' \\
\hline 1 & 1 & 2 & 3 & 1' & 3' \\
\hline
\end{array}$$

8a)

| 1' | 2' | 3' |
| 3 | 2' | 3' | 4' |
| 2 | 2 | 3 | 1' | 2' | 3' |
| 1 | 1 | 2 | ③ | 1' | 3' |

$\longrightarrow$

| 1' | 2' | 3' |
| 3 | 2' | 3' | 4' |
| 2 | 2 | 3 | 1' | 2' | 3' |
| 1 | 1 | 2 | 1' | ③ | 3' |

| 1' | 2' | 3' |
| 3 | 2' | 3' | 4' |
| 2 | 2 | 3 | 1' | ③ | 3' |
| 1 | 1 | 2 | 1' | 2' | 3' |

$\longrightarrow$

| 1' | 2' | 3' |
| 3 | 2' | 3' | 4' |
| 2 | 2 | 3 | 1' | 3' | ③ |
| 1 | 1 | 2 | 1' | 2' | 3' |

8b)

$$R(T) = \begin{array}{|c|c|c|c|c|c|}
\hline 2' & 3' & 3 \\
\hline 1' & 3' & 2 & 2 \\
\hline 1' & 2' & 3' & 1 & 3 & 3 \\
\hline 1' & 2' & 3' & 4' & 1 & 2 \\
\hline
\end{array}$$   $\varepsilon \ \overline{\mathfrak{J}}_{(4,3)} \ (3,4,6,6).$

It is clear that  R  is weight preserving.  Moreover, it is easy to
see that the moves of figure 7 are designed to ensure that the row
strictness conditions among the primed numbers are maintained.  The facts
that the regular numbers in  R(T)  do form a column strict skew tableau and
that the steps of the reversing bijection are reversible follow from the
particular order in which the regular numbers are processed.  The reader is

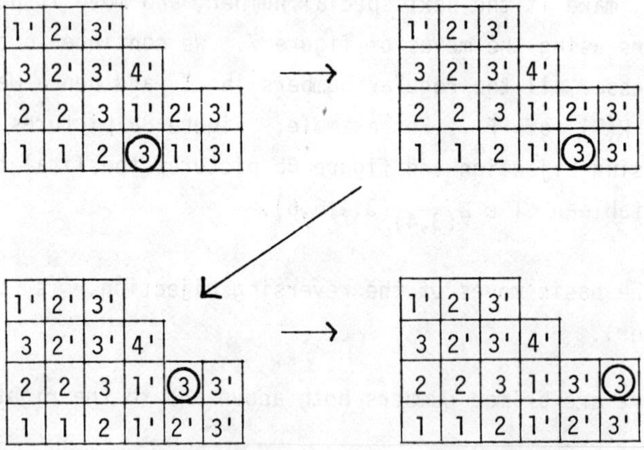

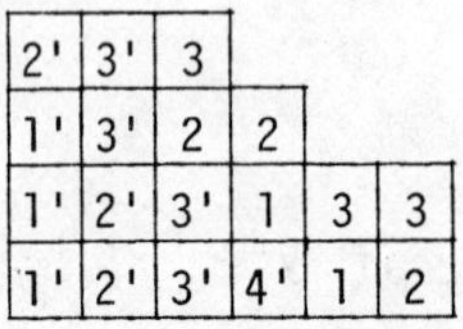

refered to [10] for detailed proofs.

Next we consider the bijections to prove 1.4 and 1.5 which use the RS 1 correspondence. First consider the right hand side of 1.4, $\prod_{j=1}^{\ell} \prod_{i=1}^{k} (x_i + y_j)$. There are clearly $2^{k\ell}$ monomials in the expansion of this product. Now for fixed $j$, a typical monomial in the expansion of $\prod_{i=1}^{k} (x_i + y_j)$ is of the form $x_{i_1} \ldots x_{i_p} y_j^{k-p}$ where $0 \leq p \leq k$ and $k \geq i_p > \ldots > i_1 \geq 1$. To such a monomial, we associate a word $\alpha_j = i_p \ldots i_1 (j')^{k-p}$ where $(j')^n$ denotes the word consisting of $n$ $j'$'s. We call such a word a <u>$j^{th}$-block word</u> and let the weight of $\alpha_j$ be given by $\omega(\alpha_j) = x_{i_1} \ldots x_{i_p} y_j^{k-p}$. Then if $A(k,\ell)$ is the class of all words $\alpha = \alpha_1 \ldots \alpha_\ell$ where each $\alpha_j$ is a $j^{th}$-block word and we let $\omega(\alpha) = \prod_{j=1}^{\ell} \omega(\alpha_j)$, then it is easy to see that

$$3.2 \qquad \sum_{\alpha \in A(k,\ell)} \omega(\alpha) = \prod_{j=1}^{\ell} \prod_{i=1}^{k} (x_i + y_j)$$

Thus to prove 1.4 bijectively, we need only give a weight preserving bijection $\Gamma : A(k,\ell) \to \mathfrak{I}_{(\ell)^k}(k,\ell)$. We define $\Gamma$ as follows. Suppose $\alpha = \alpha_1 \ldots \alpha_\ell$ is a word in $A(k,\ell)$ where each $\alpha_j$ is a $j^{th}$-block word. We construct a tableau $T = \Gamma(\alpha) \in \mathfrak{I}_{(\ell)^k}(k,\ell)$ in $\ell$ steps. At each step $j$, we will construct a $(k,j)$-semistandard tableau of shape $(j)^k$, i.e., a tableau with entries from $\{1,\ldots,k, 1',\ldots,j'\}$ in the $k \times j$ rectangle. Then at the end, $T_\ell = \Gamma(\alpha)$.

<u>Step 1</u>. If $\alpha_1 = i_p \ldots i_1 (1')^{p-k}$, then we first form $S_1 = \phi \leftarrow i_p \ldots i_1$ via the RS 1 correspondence. Then we form $T_1$ by placing $p-k$ $1'$'s on top of $S_1$.

<u>Step j+1</u>. Having constructed $T_j$, a $(k,j)$-semistandard tableau of shape $(j)^k$, suppose $\alpha_{j+1} = i_p \ldots i_1 ((j+1)')^{k-p}$. Let $S_{j+1} = T_j \leftarrow i_p \ldots i_1$. Then we claim the shape of $S_{j+1}$ is that of the $k \times j$ rectangle with a column of height $p$ attached on its right, i.e., $\lambda(S_{j+1}) = (j)^{k-p} (j+1)^p$. Thus we complete $T_{j+1}$ by adding the $p$ $(j+1)'$'s on top of the $j+1^{th}$ column of $S_{j+1}$.

Figure 9 gives an example of the $\Gamma$ bijection.

Figure 9

The $\Gamma$ bijection for $k=5$ and $\ell=4$ .

Let $\alpha = 5\ 4\ 1\ 1'\ 1'\ 5\ 3\ 2'\ 2'\ 2'\ 2\ 1\ 3'\ 3'\ 3'\ 5\ 3\ 1\ 4'\ 4'$

Step 1.

$$\alpha_1 = 5\ 4\ 1\ 1'\ 1',\ \phi \leftarrow 5\ 4\ 1 = \begin{array}{c} 5 \\ 4 \\ 1 \end{array},\quad T_1 = \begin{array}{c} 1' \\ 1' \\ 5 \\ 4 \\ 1 \end{array}$$

Step 2.

$$\alpha_2 = 5\ 3\ 2'\ 2'\ 2',\ T_1 \leftarrow 5\ 3 = \begin{array}{ll} 1' & \\ 1' & \\ 5 & \\ 4 & 5 \\ 1 & 3 \end{array},\quad T_2 = \begin{array}{ll} 1' & 2' \\ 1' & 2' \\ 5 & 2' \\ 4 & 5 \\ 1 & 3 \end{array}$$

Step 3.

$$\alpha_3 = 2\ 1\ 3'\ 3'\ 3',\ T_2 \leftarrow 2\ 1 = \begin{array}{lll} 5 & 2' & \\ 4 & 1' & \\ 3 & 1' & \\ 2 & 5 & 2' \\ 1 & 1 & 2' \end{array},\quad T_3 = \begin{array}{lll} 5 & 2' & 3' \\ 4 & 1' & 3' \\ 3 & 1' & 3' \\ 2 & 5 & 2' \\ 1 & 1 & 2' \end{array}$$

Step 4.

$$\alpha_4 = 5\ 3\ 1\ 4'\ 4',\ T_3 \leftarrow 5\ 3\ 1 = \begin{array}{llll} 5 & 2' & 3' & \\ 4 & 1' & 3' & \\ 3 & 5 & 1' & 3' \\ 2 & 3 & 5 & 2' \\ 1 & 1 & 1 & 2' \end{array},\quad T_4 = \begin{array}{llll} 5 & 2' & 3' & 4' \\ 4 & 1' & 3' & 4' \\ 3 & 5 & 1' & 3' \\ 2 & 3 & 5 & 2' \\ 1 & 1 & 1 & 2' \end{array}$$

Again it is clear that $\Gamma$ is weight preserving. However the crucial fact that one must prove is that indeed each $S_j$ has the shape claimed. The fact that when one RS 1 inserts a strictly decreasing sequence of $p$ regular numbers into a $(k,j)$-semistandard tableau $T_j$; one produces a tableau $S_{j+1}$ of shape $(j)^{k-p}\ (j+1)^p$ follows immediately from the RS 1 analogue of lemma 2.2

and the easily proved fact that the RS 1 bumping path of a regular number $\leq k$
lies entirely in the first $k$ rows of the tableau. However, much more
is true. To describe the general situation, we need some notation. First
fix $k$ and $\ell$ and let $\alpha = (\alpha_1 \leq \ldots \leq \alpha_p)$ and $\beta = (\beta_1 \leq \ldots \leq \beta_q)$ be
partitions with $p \leq k$ and $\beta_q \leq \ell$. Then let $[\alpha,\beta]$ denote the partition
pictured below.

$[\beta,\alpha] =$

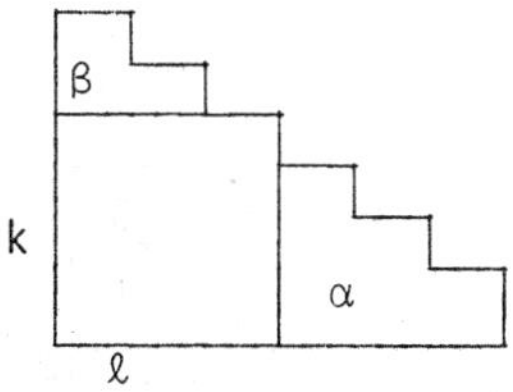

Thus $[0,0]$ is just the partition $(\ell)^k$. Next given a $(k,0)$-semistandard
tableau $T$, let the <u>word of</u> $T$, $\alpha(T)$, be the word which results by reading
down the columns from left to right. For example, if $T_1$ is the tableau
of Figure 1, $\alpha(T_1) =$ 5 3 2 1 5 4 2 1 5 4 2 3. Then the following is proven
by the author in [10].

<u>Theorem 3.1.</u> (Remmel [10]). Suppose $T$ is a $(k,\ell)$-semistandard tableau of
shape $[\beta,\phi]$ for some $\beta$. Then if $P$ is a $(k,0)$-semistandard tableau of
shape $\alpha$, $T^* = T \leftarrow \alpha(P)$ is a $(k,\ell)$-semistandard tableau of shape $[\beta,\alpha]$.
Vice versa, if $T^{**} = T \leftarrow \gamma$ for some sequence $\gamma$ from
$\{1,\ldots,k, 1',\ldots,\ell'\}$ and $\lambda(T^{**}) = [\beta,\alpha]$, then $\gamma = \alpha(Q)$ for some
$(k,0)$-semistandard tableau $Q$ of shape $\alpha$.

The special case of Theorem 3.1 where $\alpha$ is a shape $1^p$ is all that is
needed to prove that $\Gamma$ is well defined and that each step of the $\Gamma$
bijection is reversible. Also we should note that such a special case of
Theorem 3.1 is considerably easier to prove than the general case. However
Theorem 3.1 is precisely what is needed to prove the factorization theorem
given by 1.5. That is, to prove 1.5 bijectively one must construct a
bijection $\theta$ which takes triples of tableaux $(S,P,Q)$ where

$S \in \mathfrak{J}_{(k,\ell)} ([\phi,\phi])$, $P \in \mathfrak{J}_{(k,0)} (\alpha)$, and $Q \in \mathfrak{J}_{(0,\ell)} (\beta)$ to a tableau

$T \in \mathfrak{J}_{k,\ell} ([\beta,\alpha])$ such that $\omega(S)\ \omega(P)\ \omega(Q) = \omega(T)$. The $\theta$ bijection
consists of 4 steps: (1) Let $T_1 = S \leftarrow \alpha(P)$ so that by Theorem 3.1, $T_1$ is of
shape $[\phi,\alpha]$; (2) Let $T_2 = R(T_1)$ so the conjugate of $T_2$, $T_2'$, is
essentially a $(\ell,k)$-semistandard tableau of shape $[\alpha',0]$ except that the roles

of the primed numbers and regular numbers are reversed; (3) Next simply reverse the roles of the primed and regular numbers in RS 1 insertion to insert $\alpha(Q')$ into $T_2'$ to get $T_3 = T_2' \leftarrow \alpha(Q')$ which is essentially a $(\ell,k)$-semistandard tableau of shape $[\alpha',\beta']$; (4) Finally let $T_4 = R^{-1}(T_3')$ so that $T_4$ is a $(k,\ell)$-semistandard tableau of shape $[\alpha,\beta]$.

An example of the $\theta$ bijection is given in Figure 10.

Figure 10: The $\theta$ bijection, $k = 2$, $\ell = 3$.

$$S = \begin{array}{ccc} 2 & 1' & 3' \\ 1 & 1 & 2' \end{array}, \quad P = \begin{array}{cc} 2 & \\ 1 & 2 \end{array}, \text{ and } Q = \begin{array}{cc} 3' & \\ 1' & \\ 1' & 2' \end{array}.$$

Step 1.

$\alpha(P) = 2\ 1\ 2$. $T_1 = S \leftarrow \alpha(P) = \begin{array}{ccccc} 2 & 2 & 1' & 3' & \\ 1 & 1 & 1 & 2 & 2' \end{array}$.

Step 2.

$$T_2 = R(T_1) = \begin{array}{ccccc} 3' & 1 & 2 & 2 & \\ 1' & 2' & 1 & 1 & 2 \end{array}, \quad T_2' = \begin{array}{cc} 2 & \\ 1 & 2 \\ 1 & 2 \\ 2' & 1 \\ 1' & 3' \end{array}$$

Step 3.

$$Q' = \begin{array}{ccc} 2' & & \\ 1' & 1' & 3' \end{array}, \quad \alpha(Q') = 2'\ 1'\ 1'\ 3',$$

$$T_3 = T_2' \xleftarrow{*} \alpha(Q') = \begin{array}{ccccc} 2 & & & & \\ 1 & 2 & & & \\ 3' & 1 & & & \\ 2' & 2' & 2 & & \\ 1' & 1' & 1' & 3' & 1 \end{array}$$

(*) Reverse the roles of primed and regular numbers in the RS 1 insertion.

Step 4.

$T_4 = R^{-1}(T_3') =$

| | | | | |
|---|---|---|---|---|
| 3' | | | | |
| 1' | | | | |
| 1' | 2' | | | |
| 2 | 2 | 1' | 3' | |
| 1 | 1 | 1 | 2 | 2' |

Once again it is clear that $\theta$ is weight preserving. Repeated use of Theorem 3.1 then establishes that $\theta$ is well defined, one-one, and onto. Details of the bijections $R$, $\Gamma$, and $\theta$ may be found in [10].

To prove 1.7, Berele and Remmel [2] generalized the Knuth correspondence between nonnegative integer valued matrices and pairs of column strict tableaux via the RS 2 correspondence. Berele and Remmel considered classes of matrices $\mathcal{M}^{\vec{r},\vec{c}}(k_1, \ell_1, k_2, \ell_2)$ where $M$ is defined to be in $\mathcal{M}^{\vec{r},\vec{c}}(k_1, \ell_1, k_2, \ell_2)$ if $M$ has row sums $\vec{r} = (r_1,\ldots,r_{k_2}; r_{k_2+1},\ldots,r_{k_2+\ell_2})$ and columns sums $\vec{c} = (c_1,\ldots,c_{k_1}; c_{k_1+1},\ldots,c_{k_1+\ell_1})$ and satisfies the conditions picture in Figure 11. We let

$$\mathcal{M}(k_1, \ell_1, k_2, \ell_2) = \bigcup_{\vec{r},\vec{c}} \mathcal{M}^{\vec{r},\vec{c}}(k_1, \ell_1, k_2, \ell_2).$$

Figure 11. Condition for $\mathcal{M}(k_1, \ell_1, k_2, \ell_2)$ matrices.

| | $k_1$ | $\ell_1$ |
|---|---|---|
| $k_2$ | nonnegative integers | 0 or 1 |
| $\ell_2$ | 0 or 1 | nonnegative integers |

Given a matrix $M \in \mathcal{M}(k_1, \ell_1, k_2, \ell_2)$, we associate a word $\alpha(M)$ of biletters with $M$ as follows. First we label the columns of $M$ from left to right with $1,\ldots,k_1, 1',\ldots,\ell_1'$ and the rows of $\mathcal{M}$ from top to bottom with $1,\ldots,k_2, 1',\ldots,\ell_2'$. Then for each entry $M_{i,j}$ of $M$, $\alpha(M)$ will contain $M_{i,j}$ biletters of the form $\binom{p}{q}$ where $p$ is the label of the $i^{th}$ row of $M$ and $q$ is the label of the $i^{th}$ column. Finally the biletters of $\alpha(M)$ are arranged so that the top halves increase; for those biletters of the form $\binom{i}{z}$ with $i$ a regular number, the bottom halves form a

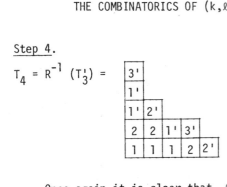

$(k_1,\ell_1)$-increasing sequence of type 2, and for those biletters of the form $\binom{i'}{z}$ with $i'$ a primed number, the bottom halves form a $(k_1,\ell_1)$-decreasing sequence of type 2. It is easy to check that the conditions on $M$ ensure that the biletters of $\alpha(M)$ can always be arranged in such a manner. For example, figure 12 gives
$\alpha(M)$ for $M \in \mathcal{M}^{\vec{r},\vec{c}}(3,2,2,2)$ where $\vec{r} = (3,4,4,3)$ and $\vec{c} = (3,3,2,2,4)$.

Figure 12.

$$
\begin{array}{cc}
 & \begin{array}{ccccc} 1 & 2 & 3 & 1' & 2' \end{array} \quad \leftarrow \text{row labels} \\
M = & \left|\begin{array}{ccccc|c} 1 & 1 & 1 & 0 & 0 & 1 \\ 2 & 0 & 0 & 1 & 1 & 2 \\ 0 & 1 & 0 & 1 & 2 & 1' \\ 0 & 1 & 1 & 0 & 1 & 2' \end{array}\right. \\
 & \qquad\qquad\qquad \uparrow \text{ column labels}
\end{array}
$$

$$\alpha(M) = \binom{1}{1}\binom{1}{2}\binom{1}{3}\binom{2}{1}\binom{2}{1}\binom{2}{1'}\binom{2}{2'}\binom{1'}{2'}\binom{1'}{2'}\binom{1'}{1'}\binom{1'}{2}\binom{2'}{2'}\binom{2'}{3}\binom{2'}{2}$$

$$\omega(M) = x_1^3\, x_2^3\, x_3^2\, s_1^2\, s_2^4\, y_1^3\, y_2^4\, t_1^4\, t_2^3 \; .$$

Now if $\alpha(M) = \binom{a_1}{b_1}\cdots\binom{a_n}{b_n}$, then we define the weight of $M$ by $\omega(M) = \prod_{i=1}^{n} \omega\binom{a_i}{b_i}$ where $\omega\binom{a}{b} = x_b\, y_a$ if both $a$ and $b$ are regular, $\omega\binom{a}{b} = s_b\, t_a$ if both $a$ and $b$ are primed, $\omega\binom{a}{b} = x_b\, t_a$ if $b$ is regular and $a$ is primed, and finally $\omega\binom{a}{b} = s_b\, y_a$ if $b$ is primed and $a$ is regular. Then it is not difficult to see that

3.3. $$\sum_{M \in \mathcal{M}(k_1,\, \ell_1,\, k_2,\, \ell_2)} \omega(M)$$

$$= \prod_{i,j=1}^{k_1,k_2} \frac{1}{1-x_i\, y_j} \prod_{i,j=1}^{\ell_1,\ell_2} \frac{1}{1-s_i\, t_j} \prod_{i,j=1}^{k_1,\ell_2} (1 + x_i\, t_j) \prod_{i,j=1}^{\ell_1,k_2} (1 + s_i\, y_j)$$

Since 3.3 gives a combinatorial interpretation to the right hand side of 1.7, it immediately follows that a bijective proof of 1.7 is given by the following theorem.

Theorem 3.2. (Berele and Remmel [ 2 ]). For any $k_1$, $k_2$, $\ell_1$, $\ell_2$, $\vec{r}$, and $\vec{c}$, there is a weight preserving bijection $\phi$ between $\mathcal{M}^{\vec{r},\vec{c}}(k_1,\ell_1,k_2,\ell_2)$ and $\{(P,Q) \mid \exists\, \lambda (P \in \mathcal{J}_\lambda(k_1,\ell_1)$ and $\tau(P) = \vec{r}$, $Q \in \mathcal{J}_\lambda(k_2,\ell_2)$ and $\tau(Q) = \vec{c})\}$.

The bijection $\phi$ is an immediate application of the RS 2 correspondence. That is, if $\alpha(M) = \binom{a_1}{b_1}\cdots\binom{a_n}{b_n}$, then $\phi(M) = (P,Q)$ where $P = \phi \Leftarrow b_1 \ldots b_n$ and $Q$ is the tableau such that $a_i$ is in cell $(e_i,f_i)$ if $(e_i,f_i)$ is the new cell created in going from $\phi \Leftarrow b_1\ldots b_{i-1}$ to $\phi \Leftarrow b_1 \ldots b_i$. For example, for the $M$ of figure 12,

$$\phi(M) = \left( \begin{array}{|c|c|c|c|c|} \hline 2' \\ \cline{1-1} 2' \\ \cline{1-1} 1' \\ \cline{1-2} 1' & 2' \\ \cline{1-4} 2 & 3 & 3 & 2' \\ \hline 1 & 1 & 1 & 2 & 2 \\ \hline \end{array} \;,\quad \begin{array}{|c|c|c|c|c|} \hline 2' \\ \cline{1-1} 1' \\ \cline{1-1} 1' \\ \cline{1-2} 1' & 2' \\ \cline{1-4} 2 & 2 & 1' & 2' \\ \hline 1 & 1 & 1 & 2 & 2 \\ \hline \end{array} \right)$$

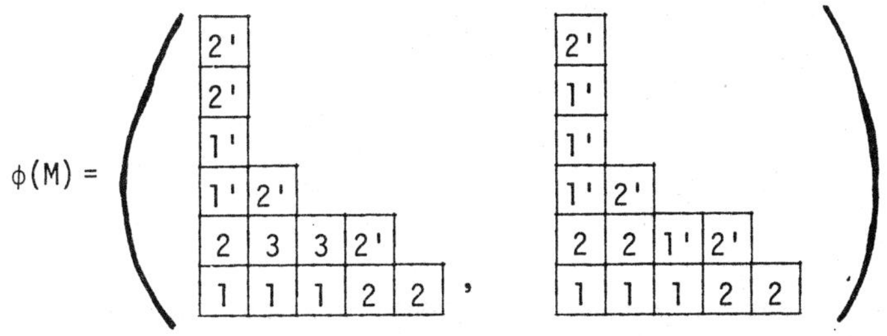

The fact that P has the desired properties is immediate from the basic properties of RS 2 insertion and the facts that $Q$ satisfies the $(k_2,\ell_2)$-semistandard conditions and that the steps of the bijection may be reversed follows lemmas 2.1 and 2.2. We refer the reader to [2] for details. Finally we should note that by applying the reversing bijection or by constructing a bijection similar to $\phi$, as is done in [2], between matrices $M \in \overline{\mathcal{M}}^{\vec{r},\vec{c}}(k_1,\ell_1,k_2,\ell_2)$, where $\overline{\mathcal{M}}^{\vec{r},\vec{c}}(k_1,\ell_1,k_2,\ell_2)$ has the same definition as $\mathcal{M}^{\vec{r},\vec{c}}(k_1,\ell_1,k_2,\ell_2)$ except that the conditions in Figure 11 are changed so that the entries of the diagonal matrices are 0 or 1 and the entries of the off-diagonal matrices are nonnegative integers, and pairs of semistandard tableaux with conjugate shapes one can also prove

$$3.4. \quad \sum_\lambda HS_\lambda(\bar{x};\bar{s})\, HS_{\lambda'}(\bar{y};\bar{t}) = \prod_{i,j}(1+x_iy_j) \prod_{i,j}(1+s_it_j) \prod_{i,j}\left(\frac{1}{1-x_it_j}\right) \prod_{i,j}\left(\frac{1}{1-y_is_j}\right).$$

Finally, we should note that by using similar methods, we can also generalize a new identity of Stanley [13] and prove the following for any $\beta$.

$$3.5. \quad \sum_{\lambda \supseteq \beta} HS_{\lambda/\beta}(\overline{x};\overline{s})\, HS_{\lambda/\beta}(\overline{y};\vec{t})$$

$$= \prod_{i,j} \frac{1}{1-x_iy_j} \prod_{i,j} \frac{1}{1-s_it_j} \prod_{i,j} (1+x_it_j) \prod_{i,j} (1+y_is_j) \sum_{\gamma \subseteq \beta} HS_{\beta/\gamma}(\overline{x},\overline{s})HS_{\beta/\gamma}(\overline{y},\overline{s}).$$

§4. Two more applications of the RS 2 correspondence.

In this section, we shall use the RS 2 correspondence to give bijective proofs of the two identities 1.8 and 1.15. We shall start by giving a bijective proof of 1.15. Recall that 1.15 simply states that the multiplication of a $(k,\ell)$-hook Schur function of a single row times an arbitrary $(k,\ell)$-hook Schur function behaves in precisely the same way as the multiplication of regular Schur functions. That is,

$$4.1. \quad HS_n(x_1,\dots,x_k;\, y_1,\dots,y_\ell)\, HS_\lambda(x_1,\dots,x_k;\, y_1,\dots,y_\ell)$$

$$= \sum_{\beta \in P_{n\times\lambda}} HS_\beta(x_1,\dots,x_k;\, y_1,\dots,y_\ell).$$

where $P_{n\times\lambda}$ denotes the set of all partitions $\beta$ such that $|\beta/\lambda| = n$ and $\beta/\lambda$ is a skew row.

To prove 4.1, we need only give a weight preserving bijection from $\mathcal{J}_{(k,\ell)}(n) \times \mathcal{J}_{(k,\ell)}(\lambda)$ onto $\bigcup_{\beta \in P_{n\times\lambda}} \mathcal{J}_{(k,\ell)}(\beta)$. In this case, the bijection is given directly by the RS 2 correspondence. That is, a tableau $S$ of shape $n$ is nothing but a $(k,\ell)$-increasing sequence of type 2, $s_1\dots s_n$. Then given a $(k,\ell)$-semistandard tableau $T \in \mathcal{J}_{(k,\ell)}(\lambda)$, Lemma 2.1 shows that $U = T \Leftarrow s_1\dots s_n$ is a $(k,\ell)$-semistandard tableau of some shape $\beta \in P_{n\times\lambda}$. Vice versa, if we start with a $(k,\ell)$-semistandard tableau $U$ of shape $\beta \in P_{n\times\lambda}$ and reverse the RS 2 correspondence by eliminating the cells in $\beta/\lambda$ from right to left, Lemma 2.1 shows that we produce a $(k,\ell)$-increasing sequence $s_1\dots s_n$ and a $(k,\ell)$-semistandard tableau $T$ of shape $\lambda$ such that $U = T \quad s_1\dots s_n$.

Next we turn our attention to 1.8. In this case, we shall use the RS 2 correspondence to give a modification of bijections given by Burge [3] and Lascoux [7] to prove the Schur function identity 1.13 which is a special case of 1.8. First we must give a combinatorial interpretation to the right hand side of 1.8,

$$\prod_{i\le j} \frac{1}{1-x_ix_j} \prod_{i<j} \frac{1}{1-y_iy_j} \prod_{i,j}(1+x_iy_j) .$$

We interpret the right hand side of 1.8 as the generating function by weight of a certain class of words $G(k,\ell)$ based an alphabet $A$ consisting of all ordered pairs $(i,j)$ where $i$ and $j$ are in $\{1,\ldots,k,\ 1',\ldots,\ell'\}$ and either (a) $j$ is regular and $i \le j$, b) $j$ is primed and $i$ is regular, or (i) $j$ is primed, $i$ is primed, and $i < j$. The weight of a pair $(i,j)$ in $A$ is given by $\omega(i,j) = x_i x_j$ if $i$ and $j$ regular, $\omega(i,j) = x_i y_j$ if $i$ is regular and $j$ is primed, and $\omega(i,j) = y_i y_j$ if both $i$ and $j$ are primed. Let $A^*$ denote the set of all words $\alpha = \alpha_1 \ldots \alpha_n$ with $\alpha_i \in A$ and define the weight of such an $\alpha$ by $\omega(\alpha) = \prod_{i=1}^{n} \omega(\alpha_i)$. Then let $G(k,\ell)$ denote the set of all words $\alpha = \alpha_1 \ldots \alpha_n$ in $A^*$ which satisfy the following two conditions:

1) The second elements in the pairs in $\alpha$ are weakly increasing and
2) If we let the s-block of $\alpha$ denote all those pairs $\alpha_p \ldots \alpha_q$ in with second element $s$, then the first elements of $\alpha_p \ldots \alpha_q$ form a $(k,\ell)$-increasing sequence of type 2 if $s$ is a regular number and the first elements of $\alpha_p \ldots \alpha_q$ form a $(k,\ell)$-decreasing of type 2 if $s$ is primed number.

For example, (1,1) (1,1) (1,2) (2,2) (2,2) (2,1') (1,1') (1',2') (1',2') (1',2') (1,2') is a word in $G(2,2)$. It is then easy to see that our definitions of $(k,\ell)$-increasing and $(k,\ell)$-decreasing sequences of type 2 imply that

$$4.2. \quad \sum_{\alpha \in G(k,\ell)} \omega(\alpha) = \prod_{1 \le i < j \le k} \frac{1}{1-x_i x_j} \prod_{1 \le i < j \le \ell} \frac{1}{1-y_i y_j} \prod_{i=1}^{k} \prod_{j=1}^{\ell} (1+x_i y_j).$$

Thus to prove 1.8 bijectively, we must give a weight preserving bijection $\Psi : G(k,\ell) \to \bigcup_{\lambda \text{ even}} \mathcal{J}_{(k,\ell)}(\lambda)$. The $\Psi$ bijection consists of two parts. In the first part we process the s-blocks in order for $s$ regular. If after $s$ steps in the first part, we have produced a tableau $T_s$ with all rows of even length, then at step $s+1$ in the first part, we let the head of the $(s+1)$-block consist of those pairs $(i, s+1)$ with $i < s+1$. Then if $i_1 \ldots i_m$ is the $(k,\ell)$-increasing sequence of type 2 that comes from the first elements of the head, we let $S_{s+1} = T_s \leftarrow i_1 \ldots i_m$. We then code the difference between the shapes $\lambda(S_{s+1})$ and $\lambda(T_s)$ with the $s+1$'s from the second elements of the head of the $(s+1)$-block by placing $j_t$ $s+1$'s in row $t+1$ if $j_t$ = the length of row $t$ of $S_{s+1}$ - the length of row $t$ of $T_s$ is even and by placing $j_t - 1$ $s+1$'s in row $t+1$ and one $s+1$ in row $t$ if $j_s$ odd. Finally we place all pairs $(s+1, s+1)$ in the $(s+1)$-block at the end of the first row to form a new tableau $T_{s+1}$ with all rows of even length. We illustrate this first part of $\Psi$ with an example.

<u>Part I of $\Psi$ bijection for $\alpha \in G(3,3)$</u>

<u>Step 1.</u> Suppose the 1-block of $\alpha$ is $\alpha_1 = (1,1)$. Then all 1's are placed in the first row that that $T_1 = \begin{array}{|c|c|}\hline 1 & 1 \\ \hline\end{array}$ .

<u>Step 2.</u> Suppose the 2-block of $\alpha$ is $\alpha_2\ \alpha_3\ \alpha_4\ \alpha_5 = (1,2)\ (1,2)\ (1,2)\ (2,2)$, then

$$S_2 = T_1 \Leftarrow 1\ 1\ 1 = \begin{array}{|c|c|c|c|c|}\hline 1 & 1 & 1 & 1 & 1 \\ \hline\end{array}$$

and

$$T_2 = \begin{array}{|c|c|c|c|c|c|c|c|}\cline{1-2} 2 & 2 \\ \hline 1 & 1 & 1 & 1 & 1 & 2 & 2 & 2 \\ \hline\end{array}$$

<u>Step 3.</u> Suppose the 3-block of $\alpha$ is $\alpha_6\ \alpha_7\ \alpha_8 = (1,3)\ (1,3)\ (2,3)$ so that

$$S_3 = T_2 \Leftarrow 1\ 1\ 2 = \begin{array}{|c|c|c|c|c|c|c|c|c|}\cline{1-4} 2 & 2 & 2 & 2 \\ \hline 1 & 1 & 1 & 1 & 1 & 1 & 1 & 2 & 2 \\ \hline\end{array}$$

and

$$T_3 = \begin{array}{|c|c|c|c|c|c|c|c|c|c|}\cline{1-2} 3 & 3 \\ \cline{1-4} 2 & 2 & 2 & 2 \\ \hline 1 & 1 & 1 & 1 & 1 & 1 & 1 & 2 & 2 & 3 \\ \hline\end{array}$$

The key fact to note is that by lemma 2.1, when we insert a $(k,\ell)$-increasing sequence of type 2, the new cells created form a skew row and the order in which the cells were created is from left to right. Thus, it is not difficult to see that our placement of the $s+1$'s at step $s+1$ allows us to recover $\lambda(S_{s+1})/\lambda(T_s)$ and hence we can recover the first elements of the head of the $(s+1)$-block by using the reverse of the RS 2 correspondence to eliminate the cells in $\lambda(S_{s+1})/\lambda(T_s)$ from right to left.

In part II of the $\Psi$ bijection, we process the $s'$-blocks for $s'$ a primed number. Again we assume that at step $s'$, we have produced a $(k,s)$-semistandard tableau $T_{s'}$ with all rows of even length. Then we take the

$(k,\ell)$-decreasing sequence of type 2, $j_1,..,j_n$, which results by taking the first elements in the $(s+1)'$-block and let $S_{(s+1)'} = T_{s'} \Leftarrow j_1 \cdots j_n$. Then by lemma 2.2, we know that $\lambda(S_{(s+1)'})/\lambda(T_{s'})$ is a skew column, i.e., no cells of $\lambda(S_{(s+1)'})/\lambda(T_{s'})$ are in the same row. Since all the rows of $T_{s'}$ are even, it follows that if we place an $(s+1)'$ to the left of each new cell created in $S_{(s+1)'}$ to get a new tableau $T_{(s+1)'}$, then $T_{(s+1)'}$ will be a $(k, s+1)$-semistandard tableau with all rows even. Note the placements of the $(s+1)'$'s cannot cause problems because our conditions on $G(k,\ell)$ ensure that $j_1,\ldots,j_n < (s+1)'$ so that $(s+1)'$ is strictly larger than any entry in $S_{(s+1)'}$. We shall illustrate part II of the $\Psi$ bijection by building on our example for part I.

Part II of the $\psi$-bijection.

Step 1'. Suppose the 1'-block of $\alpha$ is $\alpha_9 \alpha_{10} = (3,1')(2,1')$. Then

$S_{1'} = T_3 \Leftarrow 3\ 2 =$

| | | | | | | | | | | |
|---|---|---|---|---|---|---|---|---|---|---|
| 3 | 3 | | | | | | | | | |
| 2 | 2 | 2 | 2 | 3 | | | | | | |
| 1 | 1 | 1 | 1 | 1 | 1 | 1 | 2 | 2 | 2 | 3 |

and

$T_{1'} =$

| | | | | | | | | | | | |
|---|---|---|---|---|---|---|---|---|---|---|---|
| 3 | 3 | | | | | | | | | | |
| 2 | 2 | 2 | 2 | 3 | 1' | | | | | | |
| 1 | 1 | 1 | 1 | 1 | 1 | 1 | 2 | 2 | 2 | 3 | 1' |

.

Step 2'. Suppose the 2'-block of $\alpha$ is $\alpha_{11}\alpha_{12}\alpha_{13}\alpha_{14} = (1',2')(1',2')(3,2')(2,2')$, then

$S_{2'} = T_{1'} \Leftarrow 1'\ 1'\ 3\ 2 =$

| | | | | | | | | | | | |
|---|---|---|---|---|---|---|---|---|---|---|---|
| 1' | | | | | | | | | | | |
| 1' | | | | | | | | | | | |
| 1' | | | | | | | | | | | |
| 3 | 3 | 1' | | | | | | | | | |
| 2 | 2 | 2 | 2 | 3 | 3 | | | | | | |
| 1 | 1 | 1 | 1 | 1 | 1 | 1 | 2 | 2 | 2 | 2 | 3 |

$T_{2'} =$

| | | | | | | | | | | | |
|---|---|---|---|---|---|---|---|---|---|---|---|
| 1' | 2' | | | | | | | | | | |
| 1' | 2' | | | | | | | | | | |
| 1' | 2' | | | | | | | | | | |
| 3 | 3 | 1' | 2' | | | | | | | | |
| 2 | 2 | 2 | 2 | 3 | 3 | | | | | | |
| 1 | 1 | 1 | 1 | 1 | 1 | 1 | 2 | 2 | 2 | 2 | 3 |

Step 3'. Suppose the 3'-block is $\alpha_{15}\ \alpha_{16} = (2',3')(2',3')$, then

$S_{3'} = T_{2'} \Leftarrow 2'\ 2' =$

| | | | | | | | | | | | | |
|---|---|---|---|---|---|---|---|---|---|---|---|---|
| 1' | 2' | | | | | | | | | | | |
| 1' | 2' | | | | | | | | | | | |
| 1' | 2' | | | | | | | | | | | |
| 3 | 3 | 1' | 2' | | | | | | | | | |
| 2 | 2 | 2 | 2 | 3 | 3 | 2' | | | | | | |
| 1 | 1 | 1 | 1 | 1 | 1 | 1 | 2 | 2 | 2 | 2 | 3 | 2' |

and $T_{3'} =$

| | | | | | | | | | | | | | |
|---|---|---|---|---|---|---|---|---|---|---|---|---|---|
| 1' | 2' | | | | | | | | | | | | |
| 1' | 2' | | | | | | | | | | | | |
| 1' | 2' | | | | | | | | | | | | |
| 3 | 3 | 1' | 2' | | | | | | | | | | |
| 2 | 2 | 2 | 2 | 3 | 3 | 2' | 3' | | | | | | |
| 1 | 1 | 1 | 1 | 1 | 1 | 1 | 2 | 2 | 2 | 2 | 3 | 2' | 3' |

In general, we set $\Psi(\alpha)$ for $\alpha \in G(k,\ell)$ equal to the last tableau $T_{\ell'}$. Thus, in our specific example $\Psi(\alpha_1 \ldots \alpha_{16}) = T_{3'}$. By our remarks given in the description of parts I and II of the $\Psi$ bijection, it is easy to see that $\Psi(\alpha)$ is a $(k,\ell)$-semistandard tableau with only even rows. Moreover, each step of the process is reversible so that $\Psi$ is one to one. Indeed, it is easy to check that we can reverse the steps of $\Psi$ bijection starting with any $(k,\ell)$-semistandard tableau $T$ with only even rows. The fact that starting from such a $T$ we always produce an $\alpha \in G(k,\ell)$ follows from the fact that Lemma 2.2 guarantees that when we reverse the RS 2 correspondence by eliminating cells in a skew column from top to bottom in a step in Part II, we always produce a $(k,\ell)$-decreasing sequence of type 2 and the fact that Lemma 2.1 guarantees that when we reverse the RS 2 correspondence by eliminating cells in a skew row from right to left in step in Part I, we always produce a

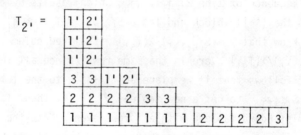

$(k,\ell)$-increasing sequence of type 2. Thus we have the following.

Theorem 4.1. The $\Psi$ map is a weight preserving bijection between $G(k,\ell)$ and $\bigcup_{\beta \text{ even}} \mathcal{T}_{(k,\ell)}(\beta)$ which proves

$$\sum_{\beta \text{ even}} HS_\beta(\bar{x};\bar{y}) = \prod_{i \leq j} \frac{1}{1-x_i x_j} \prod_{i<j} \frac{1}{1-y_i y_j} \prod_{i,j} (1+x_i y_j) .$$

Finally if we follow the $\Psi$ bijection with the reversing bijection $R$ of section 3, then it is easy to see that we get a bijective proof of the following identity.

4.3 $$\sum_{\beta' \text{ even}} HS_\beta(\bar{x};\bar{y}) = \prod_{i<j} \frac{1}{1-x_i x_j} \prod_{i \leq j} \frac{1}{1-y_i y_j} \prod_{i,j} (1+x_i y_j) .$$

## §5. The multiplication of $(k,\ell)$-hook Schur functions

The main purpose of this section is to show combinatorially that the multiplication of $(k,\ell)$-hook Schur functions behaves precisely the same as the multiplication of ordinary Schur functions. That is,

5.1. if $S_\nu(\bar{x})\ S_\mu(\bar{x}) = \sum_{\nu} g^\lambda_{\nu,\mu}\ S_\lambda(\bar{x})$, then $HS_\nu(\bar{x};\bar{y})\ H_\mu(\bar{x};\bar{y})$

$$= \sum_{\nu} g^\lambda_{\nu,\mu}\ HS_\lambda(\bar{x};\bar{y}).$$

Recall that in section 4, we used the RS 2 correspondence to give a combinatorial proof of 5.1 in the special case where $\nu$ has only one part. Thus one way to complete a combinatorial proof of 5.1 for arbitrary $\nu$ and $\mu$ is to give a combinatorial proof of the $(k,\ell)$-hook Schur function analogue of the Jacobi-Trudi identity,

5.2 $$HS_{\lambda/\mu}(x_1,\ldots,x_k;\ y_1,\ldots,y_\ell) = \det \| HS_{\hat{\lambda}_i-\hat{\mu}_j}(x_1,\ldots,x_k;\ y_1,\ldots,y_\ell) \| .$$

Gessel gave an elegant lattice path proof of the classical Jacobi-Trudi identity in [4]. We shall prove 5.2 combinatorially by extending Gessel's proof to the setting of $(k,\ell)$-hook Schur functions.

We shall consider lattice paths in the basic grid pictured in Figure 13. We shall call the rows labeled by $0,\ldots,k$, the regular part of the grid and the rows labeled by $1',\ldots,\ell'$, the primed part of the grid. We let $(s,t)$ denote the point in the $s^{th}$ row and $t^{th}$ column of the grid.

Figure 13: <u>The basic grid</u>

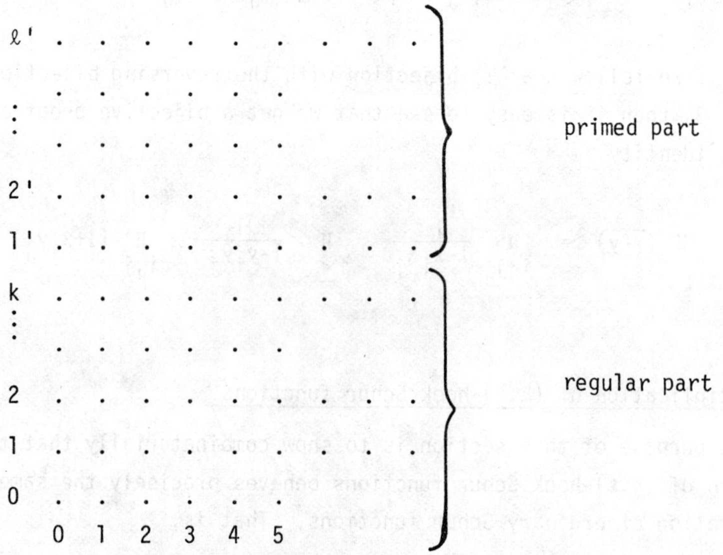


We let $P_i^j$ denote the collection of lattice paths connecting $(0,i)$ to $(\ell',j)$ consisting of directed segments which are either vertical ↑ or horizontal → if the end points of the segments are in the regular part of the grid and which are either vertical ↑ or diagonal ↗ if their end points are in the primed part of the grid. Note that this implies $P_i^j$ is empty unless $j \geq i$. Given a path $p$ in $P_i^j$, we define the weight of $p$, $\omega(p)$, to be the product of the weights of the individual segments of $p$ where the weight of any vertical segments is 1, the weight of a horizontal segment ending at $(s,t)$ is $x_s$, and the weight of a diagonal segment ending at $(s',t)$ is $y_{s'}$. For example, when $k=3$ and $\ell=4$, the path $p$ pictured in Figure 14 is an element in $P_1^7$ and has weight $x_1x_1x_1x_3y_1y_2$.

It is then clear from our definitions that

$$5.3. \qquad HS_{j-i}(x_1,\ldots,x_k;y_1,\ldots,y_\ell) = \sum_{p \in P_i^j} \omega(p)$$

where $HS_{j-i}(\overline{x},\overline{y}) = 0$ if $j-i$ is negative and $HS_0(\overline{x},\overline{y}) = 1$. We let

Figure 14

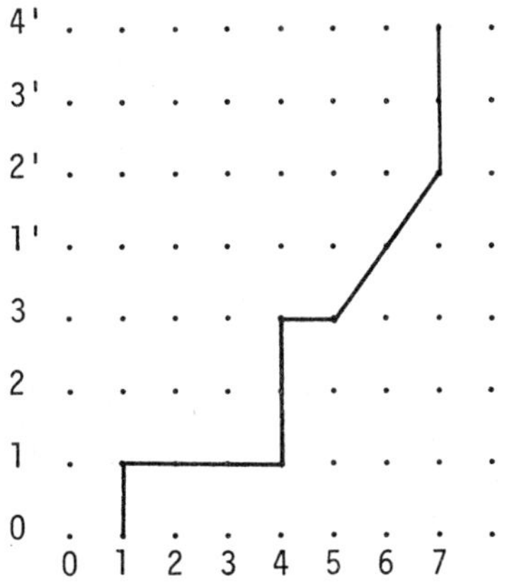


$$P_{i_1\cdots i_n}^{j_1\cdots j_n} = \underset{k=1}{\overset{n}{\times}} P_{i_k}^{j_k}$$

and define the weight of $(p_1,\ldots,p_n) \in P_{i_1\cdots i_n}^{j_1\cdots j_n}$ by $\omega(p_1,\ldots p_n) = \prod_{k=1}^{n} \omega(p_k)$.

With these conventions, it is also easy to see that if $\lambda \geq \mu$, $\lambda = (\lambda_1,\ldots,\lambda_n)$, and $\mu = (\mu_1,\ldots,\mu_n)$ where $0 \leq \lambda_1 \leq \cdots \leq \lambda_n$ and $0 \leq \mu_1 \leq \cdots \leq \mu_n$, then

5.4. $$\det \| HS_{\hat{\lambda}_i - \hat{\mu}_j} \| = \sum_{\sigma \in \mathcal{S}_n} \operatorname{sgn}(\sigma) \sum_{(p_1,\ldots,p_n) \in P_{\hat{\mu}_1 \cdots \hat{\mu}_n}^{\hat{\lambda}_{\sigma(1)} \cdots \hat{\lambda}_{\sigma(n)}}} \omega(p_1,\ldots,p_n)$$

where $\mathcal{S}_n$ denotes the symmetric group of all permutations of $\{1,\ldots,n\}$ and $\operatorname{sgn}(\sigma)$ is 1 for even permutation and -1 for odd permutations.

Next we let $P_\mu^\lambda = \bigcup_{\sigma \in \mathcal{S}_n} P_{\hat{\mu}_1 \cdots \hat{\mu}_n}^{\hat{\lambda}_{\sigma(1)} \cdots \hat{\lambda}_{\sigma(n)}}$. We say a n-tuple of paths $(p_1,\ldots,p_n)$ in $P_\mu^\lambda$ is intersecting if two of the paths go through the same point and is nonintersecting otherwise. Let $NP_\mu^\lambda$ denote the set of nonintersecting paths in $P_\mu^\lambda$. The proof of 5.2 now comes down to proving two lemmas. The first will state that the right hand side of 5.4 reduces to the sum of the weights of the nonintersecting paths and the second will state that there is a weight preserving bijection between $\mathcal{T}_{(k,\ell)}(\lambda/\mu)$ and $NP_\mu^\lambda$.

Lemma 5.1.

$$\sum_{(p_1,\ldots,p_n)\in NP_\mu^\lambda} \omega(p_1,\ldots,p_n) = \sum_{\sigma\in S_n} \operatorname{sgn}(\sigma) \sum_{(p_1,\ldots,p_n)\in P^{\hat\lambda_{\sigma(1)}\cdots\hat\lambda_{\sigma(n)}}_{\hat\mu_1\cdots\hat\mu_n}} \omega(p_1\ldots p_n)$$

Proof. The prove Lemma 5.1, we simply give an involution I on $P_\mu^\lambda$ whose fixed points are $NP_\mu^\lambda$ and which pairs off any intersecting n-tuple of paths $(p_1,\ldots,p_n)$ with another intersecting n-tuple of paths $I(p_1,\ldots,p_n)$ with the same weight but opposite sign. The involution I is very simple. For any given intersecting n-tuple of paths $(p_1,\ldots,p_n)$, we first locate the point $(s,t)$ in the grid which is farthest to the right and then the highest which has two paths which go through $(s,t)$. Then we locate the pair of paths $p_i$ and $p_j$ with the largest indices which go through $(s,t)$ and switch the tails of $p_i$ and $p_j$. That is, $I(p_1,\ldots,p_n) = (p_1',\ldots,p_n')$ where $p_k' = p_k$ if $k \notin \{i,j\}$, $p_i'$ is the path which follows $p_i$ to $(s,t)$ and then follows $p_j$, and $p_j'$ is the path that follows $p_j$ to $(s,t)$ and then follows $p_i$. Figure 15 gives an example of the I map.

Figure 15

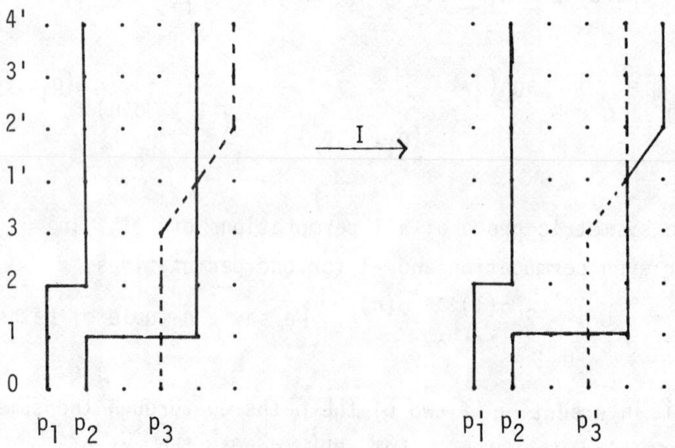


It is easy to see that $I^2$ equals the identity and that I is weight preserving. Moreover, if $(p_1,\ldots,p_n)\in P^{\hat\lambda_{\sigma(1)}\cdots\hat\lambda_{\sigma(n)}}_{\hat\mu_1\cdots\hat\mu_n}$, then

$I(p_1,\ldots,p_n)\in P^{\hat\lambda_{\tau(1)}\cdots\hat\lambda_{\tau(n)}}_{\hat\mu_1\cdots\hat\mu_n}$ where $\sigma$ and $\tau$ simply differ by a transposition $(i,j) \in \mathfrak{S}_n$ . Thus $(p_1,\ldots,p_n)$ and $I(p_1,\ldots,p_n)$ have opposite signs and hence $I$ proves lemma 5.1. □

Finally to complete the proof of 5.2., we need only prove the following.

Lemma 5.3.
$$\sum_{T \in \mathfrak{T}_{(k,\ell)}(\lambda/\mu)} \omega(T) = \sum_{(p_1,\ldots,p_n) \in NP_\mu^\lambda} \omega(p_1,\ldots,p_n)$$

Proof: We shall give a weight preserving bijection $F$ between $\mathfrak{T}_{(k,\ell)}(\lambda/\mu)$ and $NP_\mu^\lambda$ . Suppose $\lambda \geq \mu$, $\lambda = (\lambda_1,\ldots,\lambda_n)$, and $\mu = (\mu_1,\ldots,\mu_n)$ where $0 < \lambda_1 \leq \cdots \leq \lambda_n$ and $0 \leq \mu_1 \leq \cdots \leq \mu_n$. It is easy to see that if $(p_1,\ldots,p_n)$ is nonintersecting, then $(p_1,\ldots,p_n) \in P_{\hat{\mu}_1\ldots\hat{\mu}_n}^{\hat{\lambda}_1\ldots\hat{\lambda}_n}$ . Now if $T \in \mathfrak{T}_{(k,\ell)}(\lambda/\mu)$, we shall code each row of $T$ by a path. That is, the $k^{th}$ row of $T$ from the top will have length $\lambda_k - \mu_k$ and we will code it by a path $p_k$ in $P_{\hat{\mu}_k}^{\hat{\lambda}_k}$ . Thus if the $k^{th}$ row of $T$ is $s_1 \leq \cdots \leq s_p \leq t_1' \leq \cdots \leq t_q'$ reading from left to right, we code this by the path $p_k$ whose horizontal segments from bottom to top respectively end at $(s_1, \hat{\mu}_k+1),\ldots,(s_p, \hat{\mu}_k+p)$ and whose diagonal segments from bottom to top respectively end at $(t_1', \hat{\mu}_k + p+1),\ldots,(t_q', \hat{\mu}_k + p+q)$. The path $p_k$ is then completed by adding vertical segments which start at $(0, \hat{\mu}_k)$ and connect all the horizontal segments and diagonal segments above so that the path ends at $(\ell', \hat{\lambda}_k)$. Then $F(T) = (p_1,\ldots,p_n)$. Figure 16 gives an example of the $F$ bijection.

Figure 16

$\lambda = (3,5,5,7)$, $\hat{\lambda} = (3,6,7,10)$, $\mu = (0,1,3,3)$, and $\hat{\mu} = (0,2,5,6)$

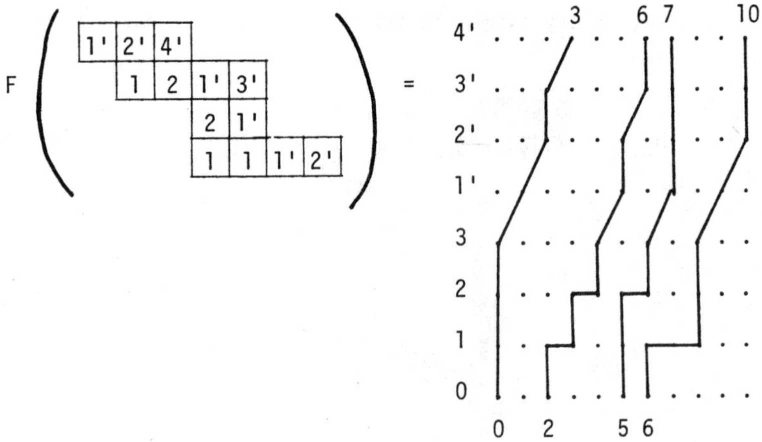

It is easy to see that $F$ is weight preserving and that we can reverse the process and start with any $(p_1,..,p_n) \in NP_\mu^\lambda$ and produce a filling $T$ of $F_\lambda$ such that the $k^{th}$ row of $T$ is coded by $p_k$ so that $F^{-1}$ is also defined. Our definitions ensure that if $T = F^{-1}(p_1,..,p_n)$, then within each row of $T$, there is a weakly increasing sequence of regular elements followed by a strictly increasing sequence of primed elements. Thus, the essential fact we need to check to prove that $F$ is a bijection is that the nonintersecting property of $(p_1,...,p_n)$ ensures that the columns of $T$ satisfy the conditions for a $(k,\ell)$-semistandard tableau and vice versa.

So fix a column $j$ of $T$, let $x$ and $y$ denote the elements of $T$ in the $j^{th}$ column which lie in the $i^{th}$ row and $(i+1)^{th}$ row of $T$ reading from top to bottom. Thus $x$ sits on top of $y$ in $T$ and $x$ is coded by a horizontal segment between $(j-1 + i-1, x)$ and $(j + i-1,x)$ if $x$ is regular and a diagonal segment between $(j-1 + i-1, x-1)$ and $(j + i-1, x)$ if $x$ is primed. Similarly $y$ is coded by a horizontal segment between $(j-1 + i, y)$ and $(j+i, y)$ if $y$ is regular and a diagonal segment between $(j-1 + i, y-1)$ and $(j-1, y)$ if $y$ is primed. By considering the picture, one can see that $(p_1,..,p_k)$ is nonintersecting iff the segment coded by $x$ lies strictly above the segment coded by $y$ for all such pairs $x$ and $y$. But then it is a routine case by case verification for the cases $x$ and $y$ both regular, $x$ regular and $y$ primed, etc. to see that the segment coded by $x$ lies strictly above the segment coded by $y$ iff $x$ and $y$ satisfy the appropriate conditions required for a $(k,\ell)$-semistandard tableau. □

Finally, we should note that by a entirely similar direct proof or by applying the reversing bijection of section 3, one can give a combinatorial proof of the analogue of the determinental expression for the Schur function in terms of the elementary symmetric function. That is, one can also prove combinatorially that

5.5 $$HS_{\lambda'/\mu'}(\overline{x};\overline{y}) = \det \| HS_{(1^{\hat{\lambda}_i-\hat{\mu}_j})}(\overline{x};\overline{y})\|$$

## BIBLIOGRAPHY

1. A. Berele and A. Regev, "Hook Young diagrams with applications to combinatorics and to representations of Lie superalgebras", preprint.

2. A. Berele and J. Remmel, "Hook Flag Characters and their combinatorics", preprint.

3. W.H. Burge, "Four correspondences between graphs and generalized Young tableaux", J. Combinatorial Theory (A) 17 (1974), 12-30.

4. I.M. Gessel, "Determinates and plane partitions", preprint.

5. D.E. Knuth, "Permutations, matrices and generalized Young tableaux", Pacific J. Math. 34 (1970), 709-727.

6. ________, The Art of Computer Programming, vol. 3, Addison-Wesley, Reading, Mass. (1973).

7. A. Lascoux, "Fonctions de Schur-Littlewood", Séminaire Pisot, Paris (1974)

8. D.E. Littlewood, The Theory of Group Characters, 2nd ed., Oxford University Press, London (1950).

9. I.G. Macdonald, Symmetric Functions and Hall Polynomials, Clarendon Press, Oxford, (1979).

10. J.B. Remmel, "A bijective proof of a factorization theorem for (k,ℓ)-hook Schur functions", preprint.

11. C. Schensted, "Longest increasing and decreasing subsequences", Canad. J. Math. 13 (1961), 179-191.

12. M.P. Schützenberger, "La correspondence de Robinson", Combinatoire et représentation du group symétrique (Actes Table Ronde C.N.R.S. Strasbourg 1976), Lecture Notes in Math., vol. 579, (1977), 59-113.

13. R. Stanley, "The generalized exponents of SL(n,ℂ) and the inner product of Schur functions", preprint.

DEPARTMENT OF MATHEMATICS
UNIVERSITY OF CALIFORNIA, SAN DIEGO
LA JOLLA, CALIFORNIA 92093

Current Address:

Department of Mathematics
Cornell University
Ithaca, New York 14853

Contemporary Mathematics
Volume **34**, 1984

# Multipartite P-Partitions and Inner Products of Skew Schur Functions

Ira M. Gessel[1]

ABSTRACT. A generalization of Richard Stanley's theory of P-partitions to multipartite partitions is developed, which is used to count r-tuples of permutations whose product is a given permutation according to their descent sets. The theory gives a combinatorial interpretation to numbers associated with inner products of certain skew Schur functions.

1. INTRODUCTION. Let $\pi$ be a permutation of $[n] = \{1,2,\cdots, n\}$. The descent set of $\pi$ is defined to be $D(\pi) = \{i \mid \pi(i) > \pi(i+1)\}$. A general problem of permutation enumeration is to find the number of permutations in a given set of permutations with a given descent set. We shall be concerned here primarily with the set of permutations which extend some poset to a total order. In section 2 we review some of Richard Stanley's theory of P-partitions [13] and show how a special case leads to formulas involving symmetric functions. In particular we derive a formula equivalent to that of Foulkes [3] for the number of permutations $\pi$ for which $D(\pi)$ and $D(\pi^{-1})$ have specified values.

In section 3 we consider the following problem: given a permutation $\pi$ in the symmetric group $S_n$ and subsets $A_1,\cdots, A_r$ of $[n-1]$, how many r-tuples $\sigma_1,\cdots, \sigma_r$ of permutations are there with $\sigma_1\sigma_2\cdots\sigma_r = \pi$ and $D(\sigma_i) = A_i$? The solution of this problem leads to a combinatorial interpretation for the scalar product of an arbitrary skew Schur function with an inner product of skew Schur functions of skew hook shape.

A more comprehensive account of the applications of P-partitions and multipartite partitions to permutation enumeration will be given elsewhere.

2. P-PARTITIONS. Let $P$ be an arbitrary partial order of $[n]$. We write $<_P$ for the partial order of $P$, and $<$ for the usual total order on $[n]$. Let $X$ be an infinite totally ordered set (for example, the positive integers). A P-partition is a function $f:[n] \to X$ satisfying

1) $i <_P j$ implies $f(i) \le f(j)$

---

1980 Mathematics Subject Classification. 05A05, 05A15, 05A17.

[1]Partially supported by NSF Grant MCS 8105 188.

2) $i <_P j$ and $i > j$ implies $f(i) < f(j)$.

We denote the set of P-partitions by $A(P)$. Thus, if $P$ is the poset

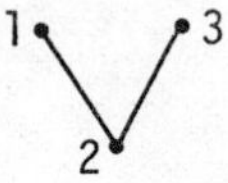


then $A(P)$ is the set of functions $f:[3] \to X$ satisfying $f(2) < f(1)$ and $f(2) \le f(3)$.

The definition of P-partitions is due to Stanley [13]; see also Knuth [7]. (Stanley defined P-partitions to be order-reversing rather than order-preserving.)

If $\pi$ is a permutation of $[n]$ we may identify $\pi$ with the total order

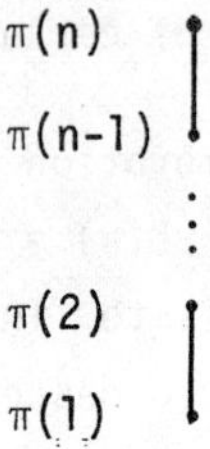


in which $\pi(i) <_\pi \pi(j)$ iff $i < j$. Then $A(\pi)$ is the set of functions $f\colon [n] \to X$ satisfying $f(\pi(1)) \sim_1 f(\pi(2)) \sim_2 \cdots \sim_{n-1} f(\pi(n))$, where $\sim_i$ is $\le$ if $\pi(i) < \pi(i+1)$ and $\sim_i$ is $<$ if $\pi(i) > \pi(i+1)$.

We now define $L(P)$ to be the set of permutations of $[n]$ which extend $P$ to a total order; that is, $\pi$ is in $L(P)$ iff $i <_P j$ implies $\pi^{-1}(i) < \pi^{-1}(j)$. The connection between $L(P)$ and $A(P)$ is given by the following theorem, which was proved in some special cases by MacMahon [11, Vol. 2, pp. 188-212], for "naturally labeled posets" by Knuth [7], and in general by Stanley [13]. All unions in this paper are disjoint.

THEOREM 1. $A(P) = \bigcup_{\pi \in L(P)} A(\pi)$.

PROOF. We proceed by induction on the number of incomparable pairs of elements in $P$. If there are none, then $P$ is a total order and the theorem is trivial. Now suppose that $i$ and $j$ are incomparable in $P$, with $i < j$. Let $P_{ij}$ be the poset obtained from $P$ by adding the relation $i <_P j$ and let $P_{ji}$ be defined similarly. It is easily seen that $L(P) = L(P_{ij}) \cup L(P_{ji})$ and $A(P) = A(P_{ij}) \cup A(P_{ji})$. Thus the theorem follows by induction.

To every $x$ in $X$ we assign a weight $w(x)$, which will always be a monomial in a power series ring. It is convenient to identify $x$ with its weight. In this section we assume that the weights of elements of $X$ are algebraically independent. We define the weight of a function $f\colon [n] \to X$ to be $w(f) = w(f(1))\, w(f(2)) \cdots w(f(n))$ and we define the generating function for $P$ to be

$$\Gamma(P) = \sum_{f \in A(P)} w(f).$$

It follows from Theorem 1 that

$$\Gamma(P) = \sum_{\pi \in L(P)} \Gamma(\pi). \tag{1}$$

Thus the power series of the form $\Gamma(\pi)$ will be important. To study them we introduce some notation. To any subset $A = \{a_1 < a_2 < \cdots < a_k\}$ of $[n-1]$ we may associate the composition $(a_1, a_2 - a_1, \cdots, a_k - a_{k-1}, n - a_k)$ of $n$. Then for any permutation $\pi$ in $S_n$ we define the descent composition $C(\pi)$ to be the composition associated to the descent set $D(\pi)$. It is convenient to transfer the partial order of inclusion of subsets to compositions (where it is reverse refinement).

It is clear that $\Gamma(\pi)$ depends only on $C(\pi)$. Thus we may define formal power series $F_L$, indexed by compositions, by $F_{C(\pi)} = \Gamma(\pi)$. For example,

$$F_{12} = \sum_{x_1 < x_2 \le x_3} x_1 x_2 x_3 .$$

$F_L$ is not symmetric in the elements of $X$ unless $L = 1^n$ or $L = n$. However $F_L$ does have the property that if $x_1 < \cdots < x_m$ and $y_1 < \cdots < y_m$ for $x_i, y_j \in X$ then the coefficient of $x_1^{k_1} \cdots x_m^{k_m}$ in $F_L$ is equal to the coefficient of $y_1^{k_1} \cdots y_m^{k_m}$. We call power series in $\mathbb{Z}[[X]]$ with this property quasisymmetric. We denote by Qsym the ring of quasisymmetric power series, and by $\mathrm{Qsym}_n$ the $\mathbb{Z}$-module of quasisymmetric power series homogeneous of degree $n$.

If $L = (L_1, \cdots, L_k)$ is a composition of $n$, we define $M_L$ by

$$M_L = \sum_{x_1 < \cdots < x_k} x_1^{L_1} \cdots x_k^{L_k} .$$

It is clear that the $M_L$ are a basis for $\mathrm{Qsym}_n$. Moreover, by the definition of $F_L$ and the partial order on compositions, we have

$$F_L = \sum_{K \ge L} M_K , \tag{2}$$

so by inclusion-exclusion,

$$M_L = \sum_{K \ge L} (-1)^{|K| - |L|} F_K , \tag{3}$$

where $|L|$ is the number of parts of $L$. Thus the $F_L$ form a basis for

$\mathrm{Qsym}_n$. It follows that $\Gamma(P)$ can be expressed uniquely as a linear combination of the $F_L$ and hence the number of $\pi$ in $L(P)$ with $C(\pi) = L$ is determined by $\Gamma(P)$. This number can be computed by inclusion-exclusion from the following:

COROLLARY 2. If $L = (L_1, \cdots, L_k)$ is a composition of $n$ and $x_1 < \cdots < x_n$ then the coefficient of $x_1^{L_1} \cdots x_k^{L_k}$ in $\Gamma(P)$ is the number of permutations $\pi$ in $L(P)$ with $C(\pi) \le L$.

PROOF. By Theorem 1, we need only consider the case where $P$ is a permutation. In this case the assertion follows from (2).

If $\Gamma(P)$ is symmetric, we can use the machinery of symmetric functions to express the number of permutations $\pi$ in $L(P)$ with $C(\pi) = L$ in a more concise form. We shall need some basic facts about symmetric functions. For more details, see Macdonald [10].

If $\lambda = (\lambda_1 \ge \cdots \ge \lambda_k > 0)$ is a partition, we define the monomial symmetric function $m_\lambda$ by

$$m_\lambda = \sum x_1^{\lambda_1} x_2^{\lambda_2} \cdots x_k^{\lambda_k} ,$$

where the sum is over distinct monomials with the $x_i$ distinct.
Note that

$$m_\lambda = \sum_L M_L \tag{4}$$

where the sum is over all rearrangements $L$ of the parts of $\lambda$. (Thus $m_{221} = M_{122} + M_{212} + M_{221}$.) The complete homogeneous symmetric functions are defined by

$$h_n = \sum_{x_1 \le \cdots \le x_n} x_1 x_2 \cdots x_n$$

and $h_\lambda = h_{\lambda_1} h_{\lambda_2} \cdots h_{\lambda_n}$. More generally, we set $h_L = h_{L_1} h_{L_2} \cdots h_{L_k}$ for any composition $L = (L_1, \cdots, L_k)$. Let $\mathrm{Sym}_n$ be the $\mathbb{Z}$-module of symmetric functions which are homogeneous of degree $n$ (with integer coefficients). Then the $m_\lambda$ and the $h_\lambda$, where $\lambda$ ranges over partition of $n$, are bases for $\mathrm{Sym}_n$. There is a symmetric scalar product on symmetric functions for which

$$\langle m_\lambda, h_\mu \rangle = \delta_{\lambda\mu} . \tag{5}$$

Thus if $g$ is a symmetric function, the coefficient of $x_1^{L_1} \cdots x_k^{L_k}$ in $g$ is $\langle g, h_L \rangle$.

Now let us define symmetric functions $S_L$, where $L$ is a composition, by

$$S_L = \sum_{K \le L} (-1)^{|L|-|K|} h_K .$$

Thus $S_{(2,1,3)} = h_{(2,1,3)} - h_{(3,3)} - h_{(2,4)} + h_{(6)}$. As we shall see later, these symmetric functions are skew Schur functions corresponding to skew hook shapes. They were first considered by MacMahon [11, Vol. 1, pp. 199-202], who wrote $h_L$ for our $S_L$. He showed that $S_L$ is the sum of all $x_1 x_2 \cdots x_n$ where, if $A$ is the subset of $[n-1]$ corresponding to $L$,

$$x_i \le x_{i+1} \quad \text{if} \quad i \notin A$$

$$x_i > x_{i+1} \quad \text{if} \quad i \in A .$$

Thus $S_L = \Gamma(P_L)$ where $P_L$ is the poset defined by

$$i <_P i+1 \quad \text{if} \quad i \notin A$$

$$i+1 <_P i \quad \text{if} \quad i \in A .$$

If $g$ is a symmetric function then $g$ is quasisymmetric and hence is a linear combination of the $F_L$.

THEOREM 3. Let $g$ be a symmetric power series in the elements of $X$. Then $g = \sum_L a_L F_L$, where $a_L = \langle g, S_L \rangle$ .

PROOF. Suppose that $g = \sum_\lambda b_\lambda m_\lambda$. Then by (5), $b_\lambda = \langle g, h_\lambda \rangle$ . Thus by (4), $g = \sum_K b_K M_K$, where $b_K = \langle g, h_K \rangle$ . Then by (3),

$$g = \sum_K b_K \sum_{L \ge K} (-1)^{|L|-|K|} F_L$$

$$= \sum_L F_L \sum_{K \le L} (-1)^{|L|-|K|} b_K = \sum_L F_L \langle g, \sum_{K \le L} (-1)^{|L|-|K|} h_K \rangle$$

$$= \sum_L F_L \langle g, S_L \rangle .$$

COROLLARY 4. Suppose that $\Gamma(P)$ is symmetric. Then the number of permutations $\pi$ in $L(P)$ with $C(\pi) = L$ is $\langle \Gamma(P), S_L \rangle$ .

We have found one class of posets with symmetric generating functions, the posets $P_L$ for which $\Gamma(P_L) = S_L$ . A permutation $\pi$ is in $L(P_L)$ if

$$\pi^{-1}(i) < \pi^{-1}(i+1) \quad \text{for} \quad i \notin A$$

$$\pi^{-1}(i+1) < \pi^{-1}(i) \quad \text{for} \quad i \in A ,$$

where $A$ is the subset of $[n-1]$ associated with $L$. In other words, $\pi \in \mathcal{L}(P_L)$ iff $C(\pi^{-1}) = L$. Thus we have:

THEOREM 5. The number of permutations $\pi$ of $[n]$ with $C(\pi^{-1}) = K$ and $C(\pi) = L$ is $\langle S_K, S_L \rangle$.

A formula closely related to Theorem 5 was found by Foulkes [3]. If we expand $S_K$ and $S_L$ in Schur functions and evaluate the scalar product using the orthogonality of Schur functions, we obtain Foulkes's formula.

If $L = (L_1, \cdots, L_k)$, we define $\bar{L}$ to be $(L_k, \cdots, L_1)$. Since $S_L = S_{\bar{L}}$, we have the following result of Foata and Schützenberger [2].

COROLLARY 6. The number of permutations $\pi$ with $C(\pi^{-1}) = K$ and $C(\pi) = L$ is equal to the number with $C(\pi^{-1}) = K$ and $C(\pi) = \bar{L}$.

The only known posets $P$ for which $\Gamma(P)$ is symmetric are those which correspond to column-strict skew plane partitions. Suppose that $\lambda = (\lambda_1, \lambda_2, \cdots, \lambda_k)$ and $\mu = (\mu_1, \mu_2, \cdots, \mu_\ell)$ are partitions, with $\ell \le k$, and that $\mu_i \le \lambda_i$ for each $i$, where we take $\mu_i$ to be zero for $i > \ell$. Then we define a column-strict plane partition of shape $\lambda/\mu$ to be an array $a_{ij}$ of positive integers defined for $1 \le i \le k$, $\mu_i < j \le \lambda_i$ satisfying

$$a_{ij} \le a_{i,j+1} \quad \text{and} \quad a_{ij} < a_{i+1,j}$$

whenever both sides of the inequality are defined. (What we have described is often called a column-strict reverse skew plane partition.) Thus, for example, a column-strict plane partition of shape 443/21 is (in the French notation)

$$\begin{array}{cccc} 3 & 3 & 4 & \\ \cdot & 2 & 3 & 3 \\ \cdot & \cdot & 1 & 2 \end{array}$$

It is clear that column-strict plane partitions of shape $\lambda/\mu$ are P-partitions for a certain poset $P = P_{\lambda/\mu}$. To construct $P_{\lambda/\mu}$ we first fill in the shape $\lambda/\mu$ with the numbers from $1$ to $n = \Sigma\lambda_i - \Sigma\mu_i$ as follows:

$$\begin{array}{cccc} 1 & 2 & 3 & \\ \cdot & 4 & 5 & 6 \\ \cdot & \cdot & 7 & 8 \end{array} \tag{6}$$

Then we obtain $P_{\lambda/\mu}$ by rotating this array 45° counterclockwise:

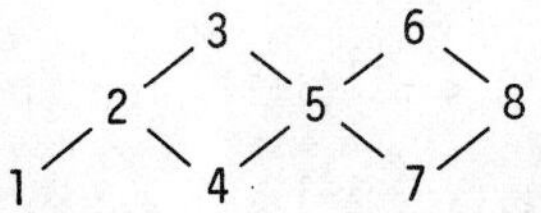

A <u>Young tableau</u> of shape $\lambda/\mu$ is a plane partition of shape $\lambda/\mu$ in which the parts are $1,2,\cdots,n$. The elements of $L(P_{\lambda/\mu})$ may be identified with Young tableaux of shape $\lambda/\mu$: given a permutation $\pi$ in $L(P_{\lambda/\mu})$ we replace $i$ in the array analogous to (6) by $\pi^{-1}(i)$, to obtain a Young tableaux $T(\pi)$. Thus in our example, if $\pi = 74581263$ then $\pi^{-1} = 56823714$ and $T(\pi)$ is

$$\begin{matrix} 5 & 6 & 8 & \\ \cdot & 2 & 3 & 7 \\ \cdot & \cdot & 1 & 4 \end{matrix}$$

The descent set of $\pi$ is the set of $i$ in $[n-1]$ for which $i+1$ appears in a higher row than $i$. Thus in our example; $D(\pi) = \{1,4,7\}$. We define $C(T(\pi))$ to be $C(\pi)$. It is convenient to say that a permutation $\pi$ is <u>compatible</u> with the shape $\lambda/\mu$ if $\pi$ is in $L(P_{\lambda/\mu})$.

It can be shown that that $\Gamma(P_{\lambda/\mu})$ is a symmetric function. (For a simple combinatorial proof, see Bender and Knuth [1].) It is called a <u>skew Schur function</u> and denoted $s_{\lambda/\mu}$. Note that $S_L$ is a skew Schur function corresponding to a special kind of shape called a <u>skew hook</u> (also called a <u>rim-hook</u> or <u>border strip</u>). Thus $S_{212} = s_{322/11}$, since $f$ satisfies $f(1) \leq f(2) > f(3) > f(4) \leq f(5)$ iff

$$\begin{matrix} f(1) & f(2) & \\ \cdot & f(3) & \\ \cdot & f(4) & f(5) \end{matrix}$$

is a column-strict plane partition of shape $322/11$.

Then from Corollary 4 we have:

THEOREM 7. $\langle s_{\lambda/\mu}, S_L\rangle$ is the number of skew Young tableaux $T$ of shape $\lambda/\mu$ with $C(T) = L$, or equivalently, the number of permutations $\pi$ compatible with the shape $\lambda/\mu$, with $C(\pi) = L$.

We note that Theorem 7 can be derived from a result of Zelevinsky [14] which gives a combinatorial interpretation for the general $\langle s_{\lambda/\mu}, s_{\alpha/\beta}\rangle$.

3. MULTIPARTITE P-PARTITIONS. Let $X_1, X_2,\cdots, X_r$ be totally ordered infinite sets. Then $X^{(r)} = X_1 \times X_2 \times\cdots\times X_r$ may be totally ordered lexicographically. (Thus if $X_1 = X_2 = \mathbb{N}$, we have $(0,0) < (0,1) < (1,0) < (1,1)$.) An <u>r-partite P-partition</u> is a P-partition with respect to the totally ordered set $X^{(r)}$.

If $f$ is a function from $[n]$ to $X^{(r)}$ we write $f(i) = (f_1(i), f_2(i),\cdots, f_r(i))$. We write $A_r(P)$ for the set of r-partite P-partitions.

As an example, take $r = 2$ and suppose that $P$ consists of the two incomparable points 1 and 2. By Theorem 1, $A_2(P) = A_2(12) \cup A_2(21)$. Now $A_2(12)$ consists of all $f: [2] \to X_1 \times X_2$ with $f(1) \le f(2)$; i.e., $(f_1(1), f_2(1)) \le (f_1(2), f_2(2))$ lexicographically. It is easily checked that $A_2(12)$ is the disjoint union of the sets

$$\{f: f_1(1) \le f_1(2) \text{ and } f_2(1) \le f_2(2)\}$$

and

$$\{f: f_1(1) < f_1(2) \text{ and } f_2(1) > f_2(2)\}.$$

A similar decomposition exists for $A_2(21)$.

We shall show that such a decomposition exists for any $P$ and any $r$. The basic idea of the decomposition is due to Gordon [5] and was further developed by Garsia and Gessel [4]. By Theorem 1,

$$A_r(P) = \bigcup_{\pi \in L(P)} A_r(\pi)$$

so we need only decompose $A_r(\pi)$. We shall see that the general case follows fairly easily from the case $r = 2$.

LEMMA 8. Let $g$ be a function from $[n]$ to the totally ordered set $X$. Then the following are equivalent:

(a) $g \in A(\sigma)$.

(b) If $i < j$ then $g(\sigma(i)) \le g(\sigma(j))$. If in addition $\sigma(i) > \sigma(j)$, then $g(\sigma(i)) < g(\sigma(j))$.

(c) If $g(i) < g(j)$, or if $i < j$ and $g(i) = g(j)$, then $\sigma^{-1}(i) < \sigma^{-1}(j)$.

PROOF. The defining condition for $A(\sigma)$ is obtained by replacing $i$ and $j$ in (b) by $\sigma^{-1}(i)$ and $\sigma^{-1}(j)$. Thus (a) and (b) are equivalent. Condition (c) is easily seen to be equivalent to the contrapositive of (b).

For any function $g$ defined on $[n]$ and any $\sigma$ in $S_n$, we define the function $g^\sigma$ by $g^\sigma(i) = g(\sigma i)$.

THEOREM 9. Let $f$ be a function from $[n]$ to $X_1 \times X_2$. Then $f$ is in $A_2(\pi)$ iff for some $\sigma$ in $S_n$, $f_2$ is in $A(\sigma)$ and $f_1^\sigma$ is in $A(\sigma^{-1}\pi)$. Moreover, $\sigma$ is unique.

PROOF. First suppose $f \in A_2(\pi)$. There is a unique $\sigma$ such that $f_2 \in A(\sigma)$. We must show that $f_1^\sigma \in A(\sigma^{-1}\pi)$, i.e., that

(i) $f_1(\pi(i)) \le f_1(\pi(i+1))$, and

(ii) If $f_1(\pi(i)) = f_1(\pi(i+1)$ then $\sigma^{-1}\pi(i) < \sigma^{-1}\pi(i+1)$.

Since $f \in A_2(\pi)$, we have (i) together with

(iii) If $f_1(\pi(i)) = f_1(\pi(i+1))$ then $f_2(\pi(i)) \le f_2(\pi(i+1))$.

(iv) If $f_1(\pi(i)) = f_1(\pi(i+1))$ and $\pi(i) > \pi(i+1)$ then $f_2(\pi(i)) < f_2(\pi(i+1))$.

Now suppose that $f_1(\pi(i)) = f_1(\pi(i+1))$. Then by (iii) and (iv), either $f_2(\pi(i)) < f_2(\pi(i+1))$ or $f_2(\pi(i)) = f_2(\pi(i+1))$ and $\pi(i) < \pi(i+1)$. Then since $f_2 \in A(\sigma)$, condition (c) of Lemma 8 gives $\sigma^{-1}\pi(i) < \sigma^{-1}\pi(i+1)$, and thus (ii) holds.

Conversely, suppose that $f_2 \in A(\sigma)$ and that $f_1^{\sigma} \in A(\sigma^{-1}\pi)$. We must show that (i), (iii), and (iv) hold. Since $f_1^{\sigma} \in A(\sigma^{-1}\pi)$, (i) is immediate. Now let us suppose that $f_1(\pi(i)) = f_1(\pi(i+1))$. Then $f_1^{\sigma}(\sigma^{-1}\pi(i)) = f_1^{\sigma}(\sigma^{-1}\pi(i+1))$, so since $f_1^{\sigma} \in A(\sigma^{-1}\pi)$, we have $\sigma^{-1}\pi(i) < \sigma^{-1}\pi(i+1)$. Thus since $f_2 \in A(\sigma)$, we have $f_2(\pi(i)) \le f_2(\pi(i+1))$, which is (iii), and by (b) of Lemma 8, if $\pi(i) < \pi(i+1)$ then $f_2(\pi(i)) < f_2(\pi(i+1))$, which is (iv).

Theorem 9 may be restated as follows:

COROLLARY 10. $A_2(\pi)$ is the disjoint union of the sets $A(\sigma_1,\sigma_2)$ over all $(\sigma_1,\sigma_2)$ such that $\sigma_2\sigma_1 = \pi$, where $A(\sigma_1,\sigma_2)$ is defined by

$$A(\sigma_1,\sigma_2) = \{f\colon [n] \to X_1 \times X_2 \mid f_1^{\sigma_2} \in A(\sigma_1) \text{ and } f_2 \in A(\sigma_2)\} .$$

We now discuss the consequences of Corollary 10 for generating functions. For simplicity we write $X$ and $Y$ for the totally ordered sets $X_1$ and $X_2$. We define the weight of $(x,y)$ in $X \times Y$ to be $w(x)w(y)$ where the weights of the elements of $X$ and $Y$ are independent indeterminates. As before, we identify $x$ with $w(x)$, $y$ with $w(y)$, and $(x,y)$ with $w(x,y) = xy$. (We assume that $X$ and $Y$ are disjoint.)

Then we may define $\Gamma_2(P)$ to be the sum of the weights of all bipartite P-partitions. If $q = q(X)$ is any quasisymmetric power series in the variables in $X$, then $q(XY)$ is well defined in the variables $(x,y) = xy$ for $x$ in $X$ and $y$ in $Y$. It is then clear that if $\Gamma(P) = \Gamma(P,X)$, then $\Gamma_2(P) = \Gamma(P,XY)$. Applying Corollary 10, we have:

THEOREM 11. $\Gamma_2(\pi) = \Gamma(\pi,XY) = \sum_{\tau\sigma=\pi} F_{C(\sigma)}(X)\, F_{C(\tau)}(Y)$.

It follows from Theorem 11 that the number of pairs $(\tau,\sigma)$ of permutations with $C(\tau) = K$, $C(\sigma) = L$, and $\tau\sigma = \pi$ depends only on $C(\pi)$. As a consequence, we have the following result of Solomon [12]:

COROLLARY 12. In the group algebra of $S_n$, let $g_L$ be the sum of all $\pi$ in $S_n$ with $C(\pi) = L$. Then the linear span of the $g_L$ over all compositions $L$ of $n$ is a subalgebra.

Solomon gave an analog of Corollary 12 for all Coxeter groups.

From Theorem 1 and Theorem 11 we obtain:

THEOREM 13. $\Gamma_2(P) = \Gamma(P,XY) = \sum_{\tau\sigma\in L(P)} F_{c(\sigma)}(X)\, F_{c(\tau)}(Y)$.

We now look at the special case of Theorem 13 in which $\Gamma(P)$ is symmetric. Here $\Gamma(P,XY)$ is symmetric in the terms $xy$, so the total ordering is irrelevant. Thus if $a(X)$ is any symmetric function, our definition of $a(XY)$ agrees with the standard "lambda-ring" definition. (See Knutson [8].)

If we expand $s_{\lambda/\mu}(XY)$ as a linear combination of the $F_K(X)\, F_L(Y)$, then the coefficient of $F_K(X)\, F_L(Y)$ is the number of pairs $(\pi,\sigma)$ of permutations such that $\sigma\pi$ is compatible with $\lambda/\mu$, $C(\pi) = K$, and $C(\sigma) = L$. Since $s_{\lambda/\mu}(XY)$ is symmetric in $X$ and $Y$, we could require instead that $\pi\sigma$ be compatible with $\lambda/\mu$.

We can extend the scalar product on symmetric functions to symmetric functions in two sets of variables by setting

$$\langle a(X)b(Y),\, c(X)d(Y)\rangle = \langle a(X),c(X)\rangle\, \langle b(X),d(X)\rangle ,$$

and extending by linearity. Then from Theorem 11 we obtain:

THEOREM 14. The number of pairs $(\pi,\sigma)$ of permutations such that $\pi\sigma$ is compatible with $\lambda/\mu$, $C(\pi) = K$, and $C(\sigma) = L$ is $\langle s_{\lambda/\mu}(XY)\, ,\, S_K(X)\, S_L(Y)\rangle$.

There is a multiplication $*$ defined on $\mathrm{Sym}_n$ called the <u>inner</u> or <u>internal</u> product which corresponds to the multiplication of characters of the symmetric group $S_n$. It may be characterized by the property that for any symmetric functions a, b, and c,

$$\langle a(XY),\, b(X)c(Y)\rangle = \langle a(X), b(X)*c(X)\rangle \ .$$

(To see that this characterization is equivalent to the usual definition, as given by Knutson [8], for example, one can verify it for power sum symmetric functions.) Then Theorem 14 may be restated as follows:

COROLLARY 15. $\langle s_{\lambda/\mu},\, S_K * S_L\rangle$ is the number of pairs $(\pi,\sigma)$ of permutations such that $\pi\sigma$ is compatible with $\lambda/\mu$, $C(\pi) = K$, and $C(\sigma) = L$.

A similar (but not obviously equivalent) interpretation to $\langle s_{\lambda/\mu},\, S_K * S_L\rangle$ was given by Lascoux [9] in the case in which $\mu = \phi$ and $K$ and $L$ are of the form $(1,\cdots, 1, j)$, so that $s_{\lambda/\mu}$, $S_K$, and $S_L$ are Schur functions.

We now generalize Corollary 10 to r-partite partitions.

THEOREM 16. $A_r(\pi)$ is the disjoint union of the sets $A(\sigma_1, \sigma_2, \cdots, \sigma_r)$ over all $(\sigma_1, \cdots, \sigma_r)$ such that $\sigma_r \sigma_{r-1} \cdots \sigma_1 = \pi$, where $A(\sigma_1, \cdots, \sigma_r)$ is the set of functions $f\colon [n] \to X_1 \times \cdots \times X_r$ satisfying

$$f_1^{\sigma_r \sigma_{r-1} \cdots \sigma_2} \in A(\sigma_1)$$

$$\cdots$$

$$f_{r-2}^{\sigma_r \sigma_{r-1}} \in A(\sigma_{r-2})$$

$$f_{r-1}^{\sigma_r} \in A(\sigma_{r-1})$$

$$f_r \in A(\sigma_r) \quad .$$

PROOF. We proceed by induction on $r$. We assume that $r \geq 2$. If $f$ is a function from $[n]$ to $X_1 \times \cdots \times X_r$, let us define $\overline{f}\colon [n] \to X_1 \times \cdots \times X_{r-1}$ by $\overline{f}(i) = (f_1(i), \cdots, f_{r-1}(i))$. Then since $X_1 \times \cdots \times X_r = (X_1 \times \cdots \times X_{r-1}) \times X_r$, by Corollary 10, $A_r(\pi)$ is the disjoint union of the sets $B(\sigma, \sigma_r)$ over all $(\sigma, \sigma_r)$ with $\sigma_r \sigma = \pi$, where

$$B(\sigma, \sigma_r) = \{f\colon [n] \to X_1 \times \cdots \times X_r \mid \overline{f}^{\sigma_r} \in A_{r-1}(\sigma) \text{ and } f_r \in A(\sigma_r)\} .$$

If $r \geq 3$, we apply induction to $A_{r-1}(\sigma)$.

In the same way that we derived Corollary 15 from Corollary 10, from Theorem 16 we derive the following:

THEOREM 17. $\langle s_{\lambda/\mu}, S_{L_1} * S_{L_2} * \cdots * S_{L_r} \rangle$ is the number of $r$-tuples $(\sigma_1, \sigma_2, \cdots, \sigma_r)$ of permutations such that $\sigma_1 \sigma_2 \cdots \sigma_r$ is compatible with $\lambda/\mu$ and $C(\sigma_i) = L_i$ for $1 \leq i \leq r$.

4. ADDITIONAL REMARKS. We have been concerned here with the descent sets of permutations. However, generating functions for permutations by their number of descents and greater index can easily be obtained from our results. Suppose that the elements of the totally ordered set $X$ are $x_0 < x_1 < x_2 < \cdots$. For $m > 0$ we define a homomorphism $\Lambda_m$ from formal power series in $X$ to formal power series in $q$ by setting $\Lambda_m(x_i) = q^i$ for $0 \leq i < m$ and $\Lambda_m(x_i) = 0$ for $i > m$. For any composition $L = (L_1, \cdots, L_k)$ of $n > 0$, we define $d(L) = k$ and $I(L) = L_2 + 2L_3 + \cdots + (k-1)L_k$. Thus if $L$ corresponds to the subset $A = \{a_1, \cdots, a_{k-1}\}$ of $[n-1]$ then $d(L) = |A| + 1$ and $\sum_{i=1}^{k-1} (n-a_i)$. So if $C(\pi) = L$, $d(L)$ is one more than the number of descents of $\pi$ and $I(L)$ is the "reversed" greater index of $\pi$. It is easily verified that if $L$ is a composition of $n$,

$$\sum_{m=1}^{\infty} t^m \Lambda_m(F_L) = t^{d(L)} q^{I(L)}/(1-t)(1-tq)\cdots(1-tq^n);$$

thus for any poset $P$ on $[n]$,

$$\sum_{m=1}^{\infty} t^m \Lambda_m(\Gamma(P)) = \sum_{\pi \in L(P)} t^{d(\pi)} q^{I(\pi)}/(1-t)(1-tq)\cdots(1-tq^n),$$

where $d(\pi) = d(C(\pi))$ and $I(\pi) = I(C(\pi))$, and similarly for multipartite P-partitions.

If $B$ is a set of permutations such that $\sum_{\pi \in B} F_{C(B)}$ is symmetric then a formula analogous to Theorem 17 counts r-tuples of permutations with product in B. It is not hard to show that the set of permutations with a given number of inversions has this property. It is also true, but harder to prove, that the set of permutations with a given cycle structure also has this property.

It seems likely that much of the theory developed here generalizes to Coxeter groups other than the symmetric groups. Instead of studying inequalities of the form $f(i) \leq f(j)$ we study inequalities determined by the reflecting hyperplanes of the Coxeter group. Thus for the hyperoctahedral groups we consider inequalities of the form $f(i) - f(j) \geq 0$, $f(i) + f(j) \geq 0$, and $f(i) \geq 0$.

Some of the results we have obtained can be expressed most elegantly in the language of coalgebras. (See [6] for definitions and basic properties.) We recall that a coalgebra is a vector space $V$ together with a comultiplication map $\Delta: V \to V \otimes V$ satisfying certain conditions. Any coalgebra $V$ has a dual algebra, which is the dual of $V$ as a vector space, with a multiplication defined by convolution.

First we define a coalgebra $S_n^*$ with basis $S_n$ and comultiplication given by

$$\Delta\pi = \sum_{\tau\sigma=\pi} \sigma \otimes \tau .$$

Next we define another coalgebra $\mathrm{Qsym}_n^*$ on the quasisymmetric power series of degree $n$, where if $F_L(XY) = \sum_{J,K} c_{J,K}^L F_J(X) F_K(Y)$, then

$$\Delta F_L = \sum_{J,K} c_{J,K}^L F_J \otimes F_K .$$

Theorem 11 asserts that the map $\pi \to F_{C}(\pi)$ extends by linearity to a (surjective) coalgebra homomorphism $\varphi^*: S_n^* \to \mathrm{Qsym}_n^*$. The dual of $S_n^*$ is just the group algebra of $S_n$ and the dual of $\mathrm{Qsym}_n^*$ may be identified with $\mathrm{Qsym}_n$ with a multiplication defined by

$$F_J * F_K = \sum_L C^L_{J,K} F_L \quad .$$

The dual of $\varphi^*$ is an injective homomorphism $\varphi$ from $Qsym_n$ to the group algebra of $S_n$ given by $\varphi(F_L) = \sum_{C(\pi)=L} \pi$ .

The comultiplication on $Qsym^*_n$ restricts to symmetric functions to give a coalgebra $Sym^*_n$ . The dual of $Sym^*_n$ is the algebra $Sym_n$ of symmetric functions with the inner product, which is isomorphic to the algebra of generalized characters, or class functions, on $S_n$. The dual of the injection $Sym^*_n \to Qsym^*_n$ is a surjection $\theta$: $Qsym_n \to Sym_n$ in which $\theta(F_L) = S_L$ . Some properties of $\theta$ have been given by Solomon [12].

## BIBLIOGRAPHY


1. E.A. Bender and D.E. Knuth, "Enumeration of plane partitions", J. Combin. Theory Ser. A, 13 (1972), 40-54.

2. D. Foata and M.-P. Schützenberger, "Major index and inversion number of permutations", Math. Nachrichten, 83 (1978), 143-159.

3. H.O. Foulkes, "Enumeration of permutations with prescribed up-down and inversion sequences", Discrete Math., 15 (1976), 235-252.

4. A.M. Garsia and I. Gessel, "Permutaion statistics and partitions", Advances in Math., 31 (1979), 288-305.

5. B. Gordon, "Two theorems on multipartite partitions", J. London Math. Soc., 38 (1963), 459-464.

6. S.A. Joni, G.-C. Rota, W. Nichols, and M. Sweedler, Umbral Calculus and Hopf Algebras, ed. R. Morris, Contemporary Mathematics, Vol. 6, American Mathematical Society, Providence, 1982.

7. D.E. Knuth, "A note on solid partitions", Math. Comp., 24 (1970), 955-962.

8. D. Knutson, λ-Rings and the Representation Theory of the Symmetric Group, Lecture Notes in Mathematics 308, Springer-Verlag, Berlin, 1973.

9. A. Lascoux, "Produit de Kronecker des répresentations du groupe symétrique", Séminaire d'Algèbra Paul Dubreil et Marie-Paule Malliavin, 32ème année (Paris, 1979), pp. 319-329, Lecture Notes in Mathematics 795, Springer-Verlag, Berlin, 1980.

10. I.G. Macdonald, Symmetric Functions and Hall Polynomials, Oxford University Press, 1979.

11. P.A. MacMahon, Combinatory Analysis, Chelsea, New York, 1960. Originally published in two volumes by Cambridge University Press, 1915, 1916.

12. L. Solomon, "A Mackey formula in the group ring of a Coxeter group", J. Algebra, 41 (1976), 255-268.

13. R.P. Stanley, Ordered Structures and Partitions, Mem. Amer. Math. Soc., 119 (1972)

14. A.V. Zelevinsky, "A generalization of the Littlewood-Richardson rule and the Robinson-Schensted-Knuth correspondence", J. Algebra, 69 (1981), 82-94.


DEPARTMENT OF MATHEMATICS
MASSACHUSETTS INSTITUTE OF TECHNOLOGY
CAMBRIDGE, MA 02139

Contemporary Mathematics
Volume 34, 1984

# PROBLEM SESSION

PROBLEM 1. (Sergey Yuzvinsky)

Let $B_n$ be the poset of all subsets of a set of $n$ elements (ordered by inclusion). For $x \in B_n$ denote by $\bar{x}$ the complement of $x$. Call $C \subset B_n$ a cut if (a) no pair $\{x, \bar{x}\}$ lies in a connected component of $B_n \setminus C$, (b) $C$ is minimal with respect to (a), i.e., deleting any element from $C$ violates (a).

Prove (or disprove) that $|C| \geq \binom{n}{\left[\frac{n}{2}\right]} = c_n$.

REMARK: There exist cuts of cardinality $> c_n$.

PROBLEM 2. (Sergey Yuzvinsky)

Let $(a_{ij})$ be an $m \times n$ matrix whose entries $a_{ij}$ are some $p$ objects (say integers 1, 2, ..., p). Let the two properties hold:
(a) $a_{ij} \neq a_{ik}$, if $j \neq k$, $i = 1, 2, \ldots, m$; $a_{rs} \neq a_{ts}$, if $r \neq t$, $s = 1, 2, \ldots, n$; (b) if $a_{ij} = a_{k\ell}$ then $a_{i\ell} = a_{kj}$.
Prove (or disprove) that such a matrix exists only if the triple (m, n, p) satisfies the Hopf condition:

$$\binom{p}{k} = 0 \pmod 2 \quad \text{for} \quad p - n < k < m$$

REMARK: The easiest way to construct such a matrix is to add subsets in a group of exponent 2 ($A + B = C$ where $|A| = m$, $|B| = n$, $|C| = p$). For these matrices the statement has been proved.

PROBLEM 3. (I.G. Macdonald)

Let $\lambda$ be a partition, $s_\lambda$ the S-function indexed by $\lambda$, and let $h(\lambda)$ denote the product of the hook-lengths of $\lambda$. Then $\tilde{s}_\lambda = h(\lambda)\, s_\lambda$ is a polynomial in the power-sums $p_1, p_2, \ldots$ with integral coefficients. Let $\ell$ be a positive integer and suppose that a rim hook of length $\ell$ can be removed from $\lambda$, leaving a partition $\mu$. Then I conjecture that

$$\tilde{s}_\lambda = \tilde{s}_{(\ell)}\, \tilde{s}_\mu \pmod{\ell},$$

both sides of this congruence being considered as polynomials in the p's. This conjecture, if true, would give a very simple proof of Nakayama's conjecture (if $\alpha$, $\beta$ are partitions of $n$, and $\ell$ is prime, then $\tilde{s}_\alpha \equiv \tilde{s}_\beta \pmod{\ell}$ iff $\alpha$, $\beta$ have the same $\ell$-core).

<u>PROBLEM 4</u>. (R.P. Stanley)

Let $x = (x_1, x_2, \ldots)$. Let $P_\lambda(x\,;\,t)$ be the Hall-Littlewood function, as defined on p. 104 of I.G. Macdonald, <u>Symmetric Functions and Hall Polynomials</u>. Let $\rho_k = (k, k-1, \ldots, 1, 0, 0, \ldots)$.

CONJECTURE:

$$P_{\rho_m + \rho_n}(x\,;\,-1) = s_{\rho_m}(x)\, s_{\rho_n}(x), \qquad (*)$$

where $s_\lambda(x)$ is a Schur function.

This conjecture has a purely combinatorial interpretation, given as follows. Recall that

$$s_\lambda(x) = \sum_T x^T ,$$

where $T$ ranges over all column-strict plane partitions of shape $\lambda$ and positive integer entries, and where $x^T = x_a x_b x_c \ldots$ if the entries of T are a, b, c, ... Moreover, it follows from (5.11') on p. 120 of Macdonald that if $\lambda$ has distinct parts, then

$$P_\lambda(x\,;\,-1) = \sum_R 2^{s(R)} x^R ,$$

where $R$ ranges over all shifted (= each row begins one space to the right of the row above) tableau of shape $\lambda$ and positive integer entries, satisfying:

(i) The rows and columns weakly decrease.

(ii) The diagonals from upper left to lower right strictly decrease.

(iii) $s(R)$ is the number of entries $j$ of $R$ such that: (a) $j$ is not the first element of its row, (b) the element immediately to the left is strictly larger, and (c) the element immediately below is strictly smaller or does not exist.

EXAMPLE: $\lambda = (5, 4, 2)$

```
5 5 ④ ② 1
  4 ③ 1 1      s(R) = 4
    2 ①
```

Thus the coefficient of a given monomial $x^\alpha$ in $s_{\rho_m}(x)s_{\rho_n}(y)$ and in $P_{\rho_m+\rho_n}(x\,;\,-1)$ can be given a purely combinatorial interpretation, so the conjecture could be proved by finding an appropriate bijection.

EVIDENCE: (1) It can be proved when $n = 0$, i.e., $P_{\rho_m}(x\,;\,-1) = s_{\rho_m}(x)$. A combinatorial proof would be desired.

(2) It has been verified in some additional small cases.

(3) The coefficients of the squarefree term $x_1 x_2 \ldots x_r$ $(r = \binom{m+1}{2} + \binom{n+1}{2})$ on both sides of (*) are equal.

(4) For any partition $\lambda$ into $\leq n + 1$ parts, it can be shown that $P_{\lambda + \rho_n}(x_1, \ldots, x_{n+1}; -1) = s_\lambda(x_1, \ldots, x_{n+1})\, s_{\rho_n}(x_1, \ldots, x_{n+1})$. Thus (*) is true if we set $x_{n+2} = x_{n+3} = \ldots = 0$.

NOTE: The symmetric function $Q_\lambda(x) = 2^{\ell(\lambda)} P_\lambda(x\,;\,-1)$ was used by Schur in his fundamental work on the projective (or spin) characters of the symmetric group. Is there some representation-theoretic significance of (*) ?

PROBLEM 5. (R.K. Gupta and P. Hanlon)

Fix $n$ and $\alpha$ a partition of $r$, with $\ell = \ell(\alpha) \leq n - r$. Define three things:

1. A polynomial $P_\alpha(q)$. This polynomial is

$$P_\alpha(q) = \prod_{(i,j)} (q^{(2i-1)} + q^{2j})$$

the product over all squares $(i,j)$ in $\alpha$.

2. The shape $\lambda_\alpha$. This is the shape obtained by adding a column of length $n - r$ to the left of $\alpha$.

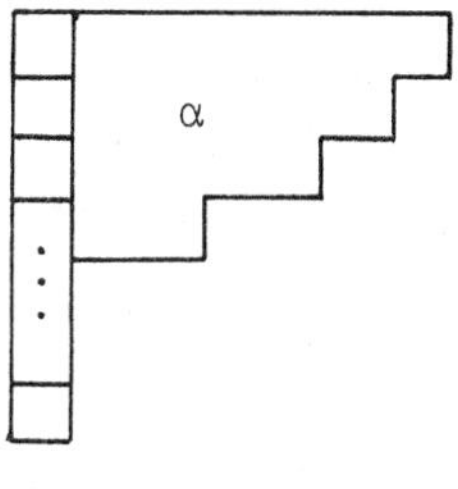


3. The mixed tensor representation $[\alpha, 1^r]$. This is the irreducible $gl_n(C)$ representation with highest weight

$(\alpha_1, \alpha_2, \ldots, \alpha_\ell, 0, 0, \ldots, 0, -1, -1, \ldots, -1)$.

The adjoint action of $gl_n(C)$ on itself lifts naturally to an action of $gl_n(C)$ on $\bigwedge gl_n(C)$.

Let $m_\alpha(p)$ denote the multiplicity of $[\alpha, 1^r]$ in the $p^{th}$ graded piece of $\bigwedge gl_n(C)$. Define $M_\alpha(q) = \sum_{p \geq 0} m_\alpha(p) q^p$.

CONJECTURE: $M_\alpha(q) = \left\{(1+q)(1+q^3)\ldots(1+q^{2(n-r)-1})\right\} P_\alpha(q)\left[\frac{n!}{\Pi\, h_{\lambda_\alpha}}\right]_{q^2}$

where $\left[\frac{n!}{\Pi\, h_{\lambda_\alpha}}\right]_{q^2}$ is the $q^2$ - analog of $n!$ divided by the product of the hook-lengths of $\lambda_\alpha$ (i.e., for each factor $j$ substitute $1 - q^{2j}$).

EVIDENCE: We have the following evidence to support this conjecture.

(A) By computer we have verified the conjecture for all possible $\alpha$, $n$ with $n \le 6$.

(B) It is known that the conjecture is correct for $q = 1$.

(C) It is easily seen that the smallest power of $q$ on the right hand side of our conjecture with nonzero coefficient is $p$, where $p$ is the sum of the numbers appearing in a box of $\alpha$ when $\alpha$ is superimposed on the grid

$$\begin{array}{llllll} 1 & 1 & 1 & 1 & 1 & \ldots \\ 2 & 3 & 3 & 3 & 3 & \ldots \\ 2 & 4 & 5 & 5 & 5 & \ldots \\ 2 & 4 & 6 & 7 & 7 & \ldots \\ \cdot & \cdot & \cdot & & & \\ \cdot & \cdot & \cdot & & & \end{array}$$

Moreover the coefficient of $q^p$ on the right hand side is 1. The same is known to be true for $M_\alpha(q)$.

(D) It is known that the coefficients of $M_\alpha(q)$ are symmetric around their middle term. More precisely,

$$q^{n^2} M_\alpha\left(\tfrac{1}{q}\right) = M_\alpha(q).$$

It can be shown that the same is true of the right hand side of our conjecture.

(E) The conjecture holds for $\alpha = \square$ and all values of $n$.

(F) For any $\alpha$, the conjecture holds in the limit on $n$ as $n$ goes to infinity.

TWO ALTERNATIVE DEFINITIONS OF $M_\alpha(q)$: The definition of $M_\alpha(q)$ given above can be replaced by either of the following two definitions.

(A) Let $F(x_1, x_2, \ldots, x_n, q)$ be the symmetric function given by

$$F(x_1, x_2, \ldots, x_n, q) = \prod_{i=1}^{n} \prod_{j=1}^{n} \left(1 + q\left(\frac{x_i}{x_j}\right)(x_1 \ldots x_n)\right).$$

Modulo the relation $x_1 \ldots x_n = 1$, $F$ can be uniquely written as

$$F = \sum f_\lambda(q)\, s_\lambda(x_1, \ldots, x_n)$$

where the sum is over partitions $\lambda$ with less than $n$ rows. One can define $M_\alpha(q)$ to be $f_{\lambda_\alpha}(q)$.

(B) Let $\eta$ be a skew shape and $\mu$ a proper shape. Let T be a semi-standard $\eta$-tableau of type $\mu$. Define seq(T) to be the sequence of integers obtained by reading the entries of T from right to left and then top to bottom. We say T is <u>proper</u> if at any point in seq(T) the number of previous occurrences of $\ell$ is greater than or equal to the number of previous occurrences of $\ell + 1$ for all $\ell$. Define $n(\eta, \mu)$ to be the number of proper $\eta$-tableau of shape $\mu$.

For $\mu$ a partition fitting in the $n \times n$ square $S_n$ let $\tilde{\mu}$ denote the partition given by taking the complement of $\mu$ in $S_n$ and then reflecting that shape across the line $y = x$. For example with $n = 3$ and $\mu$ = 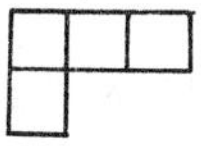 we have $\tilde{\mu}$ = 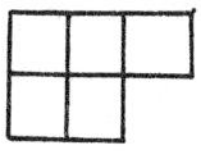

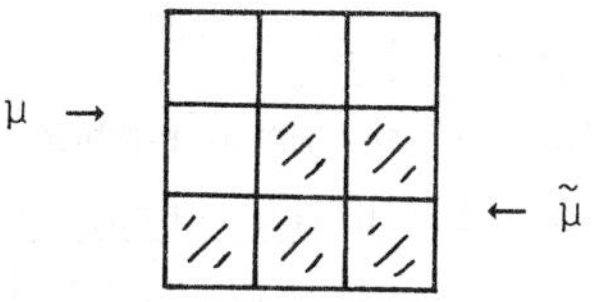


Lastly define $\hat{\alpha}$ to be the shape

$$\hat{\alpha}_i = \begin{cases} n + \alpha_i & \text{for } 1 \leq i \leq \ell . \\ n - 1 & \text{for } (n+1) - r \leq i \leq n. \\ n & \text{otherwise.} \end{cases}$$

$\hat{\alpha}$ = 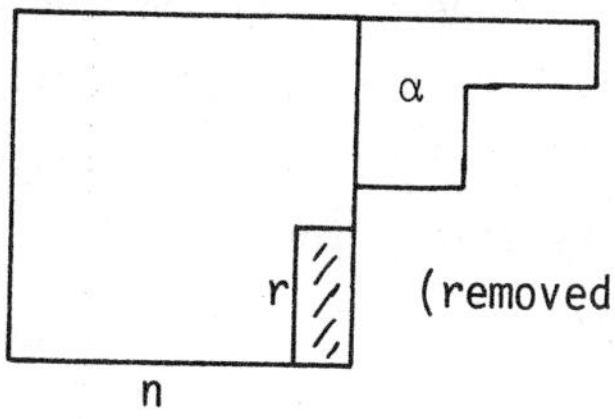


Then $$M_\alpha(q) = \sum_{\mu \subseteq S_n} n(\hat{\alpha}/\mu, \tilde{\mu})q^{|\mu|} .$$

<u>PROBLEM 6</u>. (L. Billera)

Let $\Delta$ be a finite simplicial complex whose vertices are "colored" by labels from $I = \{0, 1, \ldots, r\}$, i.e. no simplex has two vertices with the same label. (In particular, $\dim \Delta \leq r$). Let $\phi : 2^I \longrightarrow N$ be defined by

$\phi(S)$ = # of simplices in $\Delta$ with color set S.

PROBLEM: Characterize functions $\phi$ which arise in this fashion.

REFERENCE: R. Stanley, Balanced Cohen-Macaulay Complexes, Trans. AMS 249(1979), 139-157. Here it is shown that h-vectors of completely balanced Cohen-Macaulay complexes are such functions. (The converse is also true.) We seek an elementary characterization. Stanley gives necessary conditions related to the fact that $f_i(\Delta) = \sum_{|S| = i+1} \phi(S)$ is the f-vector of a simplicial complex. Similar conditions can be derived by looking at "color selected" subcomplexes. A further set of conditions is given by

$$(*) \qquad \phi(S \cup T) \leq \phi(S) \cdot \phi(T) \quad \text{whenever } S \cap T = \phi$$

For $r = 1$, (*) is necessary and sufficient ( $\Delta$ is just a bipartite graph with $\phi(0) + \phi(1)$ vertices).
For $r = 2$, the entire list of conditions is already not sufficient.

PROBLEM 7. (R. Gupta and R. Stanley)

Let $\alpha$ and $\beta$ be partitions of a common number $r$. Then one can form $(S_\alpha * S_\beta)(x_1, x_2, \ldots)$, the so called inner product of the Schur functions $S_\alpha(x_1, x_2, \ldots)$ and $S_\beta(x_1, x_2, \ldots)$. So $(S_\alpha * S_\beta)(x_1, x_2, \ldots)$ is a combinatorially defined formal symmetric function in an infinite set of variables $x_1, x_2, \ldots$ . For definitions and basic facts, see some of the articles in this volume or Macdonald's book Symmetric Functions and Hall Polynomials.

Interesting identities arise when one specializes the variables, in particular in the case $x_i = q^i$. Stanley has proven, verifying our earlier conjecture, that the power series $(S_\alpha * S_\beta)(q, q^2, \ldots)$, which we may denote by $(S_\alpha * S_\beta)(q^i)$, has a very simple form:

$$(S_\alpha * S_\beta)(q^i) = \frac{P_{\alpha,\beta}(q)}{\prod_{i=1}^{r} (1 - q^{h_i})}$$

where (1) $h_1, . . , h_r$ are the hook numbers associated to the Ferrers diagram of $\alpha$ , and (2) $P_{\alpha,\beta}(q)$ is an integral polynomial in $q$ whose value at $q = 1$ is exactly the number of standard Young tableaux of shape $\beta$ . Later, P. Hanlon found a completely different proof of this theorem. But we actually conjectured more originally.

CONJECTURE: The coefficients of the polynomial $P_{\alpha,\beta}(q)$ are all non-negative.

For example,

$$(S_{(2,1)} * S_{(2,1)})(q^i) = \frac{q^3 + q^5}{(1 - q)(1 - q)(1 - q^3)}$$

When $\alpha$ and $\beta$ are different, note that the theorem actually gives two distinct expressions for $(S_\alpha * S_\beta)(q^i)$.

Using a computer, Hanlon has found our conjecture is true for all cases, with $r \leq 9$. In the particular case where the shape of $\beta$ is a hook, i.e. where $\beta$ is of the form $(b, 1, \ldots, 1)$, A. Lascoux and M. Schützenberger have proven our conjecture.

We also note, that this is all well-known in the event $\alpha$ or $\beta$ is simply a row or a column. In fact, here one can determine the polynomials $P_{\alpha,\beta}(q)$ combinatorially by writing down standard Young tableaux $T$ of shape $\beta$, assigning a number $n_\alpha(T)$ to each tableau, and then summing the terms $q^{n_\alpha(T)}$. So we also ask:

QUESTION: What is the combinatorial significance of the degrees of the terms (counted with multiplicity) in $P_{\alpha,\beta}(q)$ ?

PROBLEM 8. (R. Gupta)

This is an extension of the last problem, and again has two parts: finding some purely formal equation involving functions of $q$, and then finding a combinatorial interpretation of the equation.

I now consider the truncated specialization $(S_\alpha * S_\beta)(q, q^2, \cdots, q^n)$ where $x_i$, for $i > n$, have been set to 0. Let $f(\beta)$ denote the number of standard Young tableaux of shape $\beta$, and let $h_i(\alpha)$ denote the hook numbers of the diagram of $\alpha$. Finally for any set of positive integers $x_1, x_2, \ldots, x_r, y_1, y_2, \ldots, y_r$, define a rational function of $q$ by setting

$$\begin{bmatrix} x_1, \ldots, x_r \\ y_1, \ldots, y_r \end{bmatrix} = \frac{(1 - q^{x_1}) \cdots (1 - q^{x_r})}{(1 - q^{y_1}) \cdots (1 - q^{y_r})}.$$

CONJECTURE:

$$(S_\alpha * S_\beta)(q, \ldots, q^n) = \sum_{i=1}^{f(\beta)} q^{d(i)} \begin{bmatrix} n - c(i,1), \ldots, n - c(i,r) \\ h_1(\alpha) \quad , \ldots, h_r(\alpha) \end{bmatrix}$$

For example, when $\alpha = \beta = (2,1)$, there are at least two ways of satisfying the conjecture.

$$(S_{(2,1)} * S_{(2,1)})(q, \ldots, q^n) = \begin{bmatrix} n, n, n+2 \\ 1, 1, 3 \end{bmatrix} + q^2 \begin{bmatrix} n-1, n-1, n \\ 1, 1, 3 \end{bmatrix}$$

$$= \begin{bmatrix} n, n+1, n+1 \\ 1, 1, 3 \end{bmatrix} + q^2 \begin{bmatrix} n-2, n, n \\ 1, 1, 3 \end{bmatrix}$$

QUESTION: Clearly choices of the numbers $c(i,j)$ and $d(i)$ depend on both $\alpha$ and $\beta$. How can they be determined combinatorially? Is there a preferred system of choices?

As before, the Conjecture is well-known when $\alpha$ or $\beta$ is a row or a column. Note also that the polynomial $\sum_{i=1}^{f(\beta)} q^{d(i)}$ must be the polynomial $P_{\alpha,\beta}(q)$ of the last problem.

PROBLEM 9 (R. Gupta)

Let $K_{\lambda,\mu}(q)$ denote the Kostka-Foulkes polynomial in $q$. These polynomials are certain natural "q-analogues" of the Kostka numbers $K_{\lambda,\mu}$ which count column-strict tableaux of shape $\lambda$ and weight $\mu$. See Macdonald's book for definitions and for the charge theorem of Lascoux and Schützenberger (answering Foulkes' problem).

I conjecture that various sorts of "stability theorems" hold for the $K_{\lambda,\mu}(q)$ in analogy to the special case considered in my talk. But first one must find out about "containment" of the polynomials. Given polynomials $f = a_0 + a_1 q + \ldots + a_n q^n$, $g = b_0 + b_1 q + \ldots + b_m q^m$, say <u>f is contained in g</u>, written $f \trianglelefteq g$, if $a_i \leq b_i$ for all $i$. Also write $K(\lambda, \mu)$ for $K_{\lambda,\mu}(q)$.

Next let $a \oplus \beta$ denote the column sum of partitions, so that the parts of $\alpha \oplus \beta$ are the parts of $\alpha$ and $\beta$, reordered as necessary. For example, $(3,3,1) \oplus (5,2,1) = (5,3,3,2,1,1)$. As usual, $(a^n)$ is the partition with $n$ parts, all equal to $a$. Then an example of the stability I have in mind is

CONJECTURE: Given partitions $\lambda$ and $\mu$ (of a common number) and some number $a$, then

$$K(\lambda \oplus (a^m), \mu + (a^m)) \trianglelefteq K(\lambda \oplus (a^{m+1}), \mu \oplus (a^{m+1}))$$

for $m = 0, 1, 2, \ldots$ .

Furthermore, if we fix a degree $p$, then the coefficient of $q^p$ in $K(\lambda \oplus (a^m), \mu \oplus (a^m))$ is constant for $m$ large enough. (Different values of $p$ will require different bounds on $m$.)

So the limit of the polynomials $K(\lambda \oplus (a^m), \mu \oplus (a^m))$ is some new power series we could call $K(\lambda,\mu ; a)$.

PROBLEM 10. (Ira Gessel)

By a permutation we mean a linear arrangement of distinct integers. Two permutations $\alpha$ and $\beta$ are called similar if for each $j$ the $j^{th}$ smallest elements of $\alpha$ and $\beta$ occur in the same position. (Thus 3514 and 4826 are similar.)

If $\alpha$ and $\beta$ have no elements in common, we define their shuffle product $S(\alpha, \beta)$ to be the set of all permutations obtained by interweaving $\alpha$ and $\beta$ .

A congruence is an equivalence relation $\equiv$ on permutations such that

1) if $\alpha$ and $\beta$ are similar they are equivalent (under $\equiv$).
2) if $\alpha_1 \equiv \alpha_2$ and $\beta_1 \equiv \beta_2$ then the sets $S(\alpha_1, \beta_1)$ and $S(\alpha_2, \beta_2)$ are elementwise equivalent.

PROBLEM: Classify all congruences (or say as much as possible about them).

A congruence determines an algebra spanned by equivalence classes $[\alpha]$:

Define $$[\alpha][\beta] = \sum_{\gamma \in S(\alpha,\beta)} [\gamma].$$

Examples are most easily described by giving a "congruence function" f such that $\alpha \equiv \beta$ iff $f(\alpha) = f(\beta)$. Some congruence functions are the descent set, the number d of descents, the greater index i (the sum of the descents), and the pair (d,i).

PROBLEM 11. (Russell Merris) [See Lin. Mult. Alg. 14 (1983), 21-35]

Suppose $\lambda, \mu \vdash n$. Confuse $\lambda$ with the character of $S_n$ to which it corresponds. Let $S_\mu = S_{\mu_1} \times S_{\mu_2} \times \dots \subset S_n$. The Kostka coefficient is given by

$$K_{\lambda,\mu} = (\lambda, 1_\mu)_{S_\mu}$$

EXAMPLE: n = 4

| | (4) | (3,1) | $(2^2)$ | $(2,1^2)$ | $(1^4)$ |
|---|---|---|---|---|---|
| (4) | 1 | 1 | 1 | 1 | 1 |
| (3,1) | 0 | 1 | 1 | 2 | 3 |
| $(2^2)$ | 0 | 0 | 1 | 1 | 2 |
| $(2,1^2)$ | 0 | 0 | 0 | 1 | 3 |
| $(1^4)$ | 0 | 0 | 0 | 0 | 1 |

With respect to reverse lexicographic order, $(K_{\lambda,\mu})$ is diagonal with 1's on the main diagonal. Say $\lambda$ majorizes $\mu$ (write $\lambda \succeq \mu$) if

$$\sum_{i=1}^{t} (\lambda_i - \mu_i) \geq 0, \qquad \text{for all } t.$$

THEOREM (c.f. Liebler-Vitale 1973): $K_{\lambda,\mu} \neq 0 \Longleftrightarrow \lambda \succeq \mu$.

THEOREM (c.f. Liebler-Vitale, also D.E. White) $\mu \succeq \nu \Longrightarrow K_{\lambda,\mu} \leq K_{\lambda,\nu}$.

Denote by $\lambda'$ the partition of n conjugate to $\lambda$.

(1) CONJECTURE: $\lambda \succeq \lambda' \Longrightarrow K_{\lambda,\mu} \geq K_{\lambda',\mu}$.

(2) QUESTION: If (1) is true, what is the case of equality?

Equality occurs, e.g., for any $\lambda$ if $\mu = (1^n)$ because $\lambda(\text{id}) = \lambda'(\text{id})$.

Equality also occurs, e.g., when

$$\lambda = (5, 4, 1^3) \quad ; \quad \mu = (2^6)$$
$$\lambda = (6, 3, 1^4) \quad ; \quad \mu = (2^6, 1)$$
$$\lambda = (7, 3, 1^5) \quad ; \quad \mu = (2^7, 1)$$
$$\lambda = (7, 4, 1^5) \quad ; \quad \mu = (2^8).$$

Denote by $\langle\mu\rangle$ the monomial symmetric polynomial, $\langle\mu\rangle(x_1,x_2,\ldots) = \sum x^\alpha$, the sum over all distinct permutations $\alpha$ of $(\mu_1, \mu_2, \ldots)$. Then the Schur polynomial is given by

$$\{\lambda\} = \sum_{\mu \vdash n} K_{\lambda,\mu} \langle\mu\rangle$$

Thus, (1) implies $\{\lambda\}(x) \geq \{\lambda'\}(x)$ for all x in the nonnegative orthant.

PROBLEM 12. (J. Griggs, A. Kustin, D. Ross)

Let Q be a finite ranked poset. Let $q \in Q$. $Q \propto q$ denotes the poset in which a twin of q has been added to Q, that is, there is a new element, call it q', which is below (resp. above) the same elements that q is below (resp. above) and which is unrelated to q.

We have shown that:

Q is Cohen-Macaulay (resp., CL-shellable) $\Longleftrightarrow$ $Q \propto q$ is Cohen-Macaulay (resp. CL-shellable). Further,

Q is shellable $\Longrightarrow$ $Q \propto q$ is shellable.

Is it true that $Q \propto q$ is shellable $\Longrightarrow$ Q is shellable?

To show this is false, one would need a poset Q such that:

Q is C-M, not shellable, (and not C-L shellable)

$Q \propto q$ is (C-M), shellable, and not C-L shellable.

PROBLEM 13. (Jacob Towber)

If U is the unipotent radical of a Borel subgroup of $G = GL(E) \approx GL_n(K)$ (where E is an n-dimensional vector-space over the field K) the coordinate-ring

$$\bigwedge^+ E = K(G/U)$$

of the flag mainifold G/U has the property of being isomorphic to the direct sum of one copy of each "shape-functor" (also called "Schur functor") $\bigwedge^\alpha E$ (where $\alpha$ runs over all partitions with greatest element $\leq n$); under the multiplication in this ring (the "Cartan product" if K is a field of characteristic 0), $\bigoplus_\alpha \bigwedge^\alpha E = \bigwedge^+ E$ becomes a commutative associative algebra graded by partitions $\alpha$, with the monoid operation defined by $\cup$ :

1) $\langle a_1, \ldots a_s\rangle \cup \langle b_1, \ldots b_t\rangle = \langle a_1, \ldots a_s, b_1, \ldots b_t\rangle$

Remarkably, there is a second natural associative multiplication defined on the direct sum of all shape-functors: it is graded anti-commutative, with the suitable monoid operation being defined by + :

2) $\alpha + \beta = \beta + \alpha = \langle a_1 + b_1, \ldots, a_t + b_t, a_{t+1}, \ldots a_s \rangle$

if $\alpha = \langle a_1, \ldots a_s \rangle$, $\beta = \langle b_1, \ldots b_t \rangle$, $a_1 \geq \ldots \geq a_s > 0$, $b_1 \geq \ldots \geq b_t > 0$, $s \geq t$.

Then denoting the first product by $\bullet$ and the second by $\wedge$ , we have

$$\bigwedge^{\alpha} \bullet \bigwedge^{\beta} \subseteq \bigwedge^{\alpha \cup \beta} , \quad \bigwedge^{\alpha} \wedge \bigwedge^{\beta} \subseteq \bigwedge^{\alpha + \beta} \quad \text{and}$$

$$\alpha + \beta = (\alpha' \cup \beta')' \quad (\alpha' = \text{conjugate partition to } \alpha)$$

The construction of this second product $\wedge$ may be found in the author's paper "Two new functors," J. of Alg. 47 (1977), 80-104; more details may be found in the paper "Young symmetry, the Flag Manifold, and Representations of GL(n)", J. of Alg. 61 (1979), 414-462.

QUESTION 1: Call an element of $\bigwedge^{+} E$ "homogeneous" if it lies in some $\bigwedge^{\alpha} E$. Are the two products discussed above the only G-equivariant associative algebra structures on $\bigoplus \bigwedge^{\alpha} E$ with the property that the product of two homogeneous elements is again homogeneous?

Note that, if we drop the homogeneity requirement, a counter-example is furnished by the well-known fact that every finite-dimensional irreducible polynomial representation of GL(n,K) (where K is a field of characteristic 0) occurs exactly once in

$$S(E \oplus \bigwedge^2 E)$$

(which is thus $\approx \bigwedge^{+} E$ as G-module, though not of course as an algebra).

QUESTION 2: What $\mathbb{Z}_2$ -graded algebra plays the role of the shape-algebra $\bigwedge^{+} E$ for the super-algebra gl(m/n)?

This of course assumes that some such "shape-algebra" exists. Analogous shape-algebras have been constructed for the classical groups, and recently for the spin group and for $G_2$ ; the connection with the preceding discussion is furnished by the following:

CONJECTURE: If $E = E_0 \oplus E_1$, dim $E_0 = m$, dim $E_1 = n$ then the "correct" shape-algebra for gl(m/n) $\approx$ gl$(E_0/E_1)$ is furnished by $\bigwedge^{+} E_0 \otimes \bigwedge^{+} E_1$, with commutative product given by

$$(\omega \otimes \omega') \bullet (\omega_1 \otimes \omega_1') = (\omega \bullet \omega_1) \otimes (\omega' \wedge \omega_1')$$

(where $\bullet$ , $\wedge$ on the right side are the two operations on shapes described earlier.

PROBLEM 14 (Henry Crapo)

Consider a finite set $X$ and a 2-scheme, a subset $F \subseteq \binom{X}{3}$ of triples of elements of X. A drawing $D$ of $F$ is any assignment $D: X \longrightarrow P$ of the elements of $X$ to points in the real projective plane $P_1$ such that each triple in $F$ is collinear.

Roughly speaking, if the projections of certain of the points onto a fixed line from a fixed centre of projection be selected in general position, the projection of the remaining points will invariantly determined. To make the observation precise, notice that the collinearity of each triple $t$ in the scheme $F$ is given by the vanishing of a quadratic polynomial $f_t$ in the coordinates of the points in any drawing $D$ of $F$, while the selected projection of any point $a$ is given by the vanishing of a linear equation $\pi_a$ in the coordinates of the point $a$. The projection of a point $a$ is invariantly given by the scheme $F$ and by selected projections of a subset $E$ of points if and only if the polynomial $\pi_a$ is algebraically dependent upon the set of polynomials

$$\{f_t : t \in F\} \cup \{\pi_e : e \in E\} .$$

In this way, the set $X$ becomes the set of points of a combinatorial geometry (matroid) $T(X, F)$, the Tay geometry, in which dependence is algebraic dependence of the $\pi_a$ modulo the $\{f_t : t \in F\}$. The nullity of the set $X$ modulo the scheme $F$, that is, the number of elements in the complement $X \setminus B$ of a basis $B$ for the geometry $T(X, F)$, is the Tay number $t(x, F)$ of the scheme

EXAMPLE 1. The Tay number of the scheme

$$F = \{ace, adf, bcf, bde\}$$

is equal to 1. We use the symbol $\binom{a\ c\ e}{b\ d\ f}$ to represent this scheme $F$. Drawings of $F$ locate the six points of intersection of four general lines in the plane; the six points on the line form a quadrilateral set. The location of any one of these points is given invariantly by the other five, and the six points are related by the vanishing of a polynomial of degree 3 in the Euclidean coordinates of the points on the real line, of degree 6 in their projective coordinates:

$$\Delta\binom{a\ c\ e}{b\ d\ f} = [af][cb][ed] - [ad][cf][eb],$$

where the bracket $[af]$ in Euclidean coordinates is simply the difference $a - f$.

EXAMPLE 2. The scheme $F = \{abr, acs, bct, der, dfs, eft\}$ has Tay number 0, but if we add the triple $rst$, the Tay number increases to 2. The resulting scheme $F'$ contains two quadrilateral sets $\binom{a\ b\ c}{t\ s\ r}$ and $\binom{d\ e\ f}{t\ s\ r}$. In the enlarged scheme F', the element f is dependent upon the set $E = \{a\ b\ c\ d\ e\ s\ t\}$, but is not dependent upon any proper subset of E. The set $E \cup \{f\}$ contains no quadrilateral set of six points, but we can see that the point r completes the quadrilateral set $\binom{a\ b\ c}{t\ s\ r}$, five points of of which are in E, then f completes the quadrilateral set $\binom{d\ e\ f}{t\ s\ r}$, five points of which are in $E \cup \{r\}$.

A subset $E \subseteq X$ is <u>closed</u> if and only if no element $e \in X \setminus E$ is dependent upon the set E.

CONJECTURE: A subset $E \subseteq X$ is closed in the geometry T(X, F) of a 2-scheme F if and only if it is closed with respect to completion of quadrilateral sets, that is, if and only if there is no quadrilateral set Q and element $e \in Q$ such that $Q \setminus e \subseteq E$, but $e \notin E$.

We suspect that if this holds in general, it is a simple application of one of the fundamental theorems of invariant theory.

<u>PROBLEM 15</u>. (C.Y. Chao)

It is known that for each dimension $n \geq 10$, there exist uncountably many nonisomorphic nilpotent Lie algebras over the real number field R with dimension n. (See "Uncountably many nonisomorphic nilpotent Lie Algebras," Proc. Amer. Math Soc. <u>13</u>(1962), 903-906, and "Infinitely many nonisomorphic nilpotent algebras," Proc. Amer. Math. Soc. <u>24</u>(1970), 126-133.) Is 10 the smallest dimension? That is, is there a dimension $m < 10$ such that there are uncountably many nonisomorphic nilpotent Lie algebras over R with dimension m?

<u>PROBLEM 16</u>. (C.Y. Chao)

Can one classify the family of symmetric graphs with a composite number of vertices? (A symmetric graph is a graph whose group of automorphisms is transitive on the vertices as well as on the edges. See "On the classification of symmetric graphs with a prime number of vertices," Trans. Amer. Math. Soc. <u>158</u>(1971), 247-256, "A class of vertex-transitive digraphs", J. Comb. Theory (B) <u>14</u>(1973), 246-255, and "Point-symmetric graphs and digraphs of prime order and transitive permutation groups of prime degree," J. Comb. Theory (B) <u>15</u>(1973), 12-17.)

<u>PROBLEM 17</u>. (D. Frank Hsu)

DEFINITION 1: A <u>$(K, \lambda)$ complete mapping</u>, where $K = \{k_1, k_2, \ldots, k_s\}$ and the $k_i$ are integers such that $\sum_{i=1}^{s} k_i = \lambda(|G|-1)$, is an arrangement of the

non-identity elements of G (each counted $\lambda$ times) into s cyclic sequences of lengths $k_1, k_2, \ldots, k_s$, say

$$(g_{11}g_{12} \cdots g_{1k_1})(g_{21}g_{22} \cdots g_{2k_2}) \cdots (g_{s1}g_{s2} \cdots g_{sk_s}),$$

such that the elements $g_{ij}^{-1}g_{i,j+1}$ (where i = 1, 2, ..., s; and the second suffix j is added modulo $k_i$) comprise the non-identity elements of G each counted $\lambda$ times.

A <u>(K, $\lambda$) near complete mapping</u>, where K = $\{h_1, h_2, \ldots, h_r; k_1, k_2, \ldots, k_s\}$ and the $h_i$ and $k_j$ are integers such that $\sum_{i=1}^{n} h_i + \sum_{j=1}^{s} k_i = \lambda|G|$, is an arrangement of the elements of G (each used $\lambda$ times) into r sequences with lengths $h_1, h_2, \ldots h_r$ and s cyclic sequences with lengths $k_1, k_2, \ldots, k_s$, say

$$[g'_{11}g'_{12} \cdots g'_{1h_1}] \cdots [g'_{r1}g'_{r2} \cdots g'_{rh_r}](g_{11}g_{12} \cdots g_{1k_1}) \cdots (g_{s1}g_{s2} \cdots g_{sk_s})$$

such that the elements $(g'_{ij})^{-1}g'_{i,j+1}$ and $g_{ij}^{-1}g_{i,j+1}$ together with the elements $g_{ik_i}^{-1}g_{i1}$ comprise the non-identity elements of G each counted $\lambda$ times. (We have $\sum(h_i-1) + \sum k_j = \lambda(|G| - 1)$. So it follows that $r = \lambda$).

A <u>generalized (K, $\lambda$) complete mapping</u> of a group G is either a (K, $\lambda$) complete mapping or a (K, $\lambda$) near complete mapping of that group. For other details, the reader is referred to [3] and [4]. The relation between this subject and a non-associative algebraic structure, called <u>neofield</u>, can be found in [1], [2] and [3]. The construction of designs using generalized complete mappings can be found in [4].

DEFINITION 2: A <u>(k, $\lambda$) complete mapping</u> is a (K, $\lambda$) complete mapping with $K = \{k, k, \ldots, k\}$. Similarly, a <u>(k, $\lambda$) near complete mapping</u> is a (K, $\lambda$) near complete mapping with $K = \{h, h, \ldots, h; k, k, \ldots, k\}$ and $k-h = 1$.

It is immediate from Definition 2 that $s = \lambda(|G| - 1)/k$ for a (k, $\lambda$) complete mapping, and $s + \lambda = \lambda(|G| + 1)/k$ for a (k, $\lambda$) near complete mapping. This shows that a necessary condition for the existence of a generalized complete mapping is $k \mid \lambda(|G| + i)$ where $i = -1$ for a (k, $\lambda$) complete mapping, and $i = +1$ for a (k, $\lambda$) near complete mapping. We would like to know if the stated necessary condition is also sufficient. However, since we still don't know much about generalized complete mappings in non-abelian groups, we restrict ourselves to abelian groups G.

The following theorem, which is proved in [3], should be helpful in formulating our problem:

THEOREM: An abelian group $(G, \bullet)$ has a $(K,1)$ complete mapping if and only if the product of all its elements is the identity element e. It has a $(K,1)$ near complete mapping if and only if it has a unique element of order 2, and then the product of all its elements is this unique element of order 2.

PROBLEM: Given a finite abelian group G, can we always find $(k, \lambda)$ complete mappings for any k such that $k \mid \lambda(|G| - 1)$ or $(k, \lambda)$ near complete mappings for any k such that $k \mid \lambda(|G| + 1)$?

The author does not have enough evidence to say anything about the answers to this general problem. However, we would like to conjecture that the answer is "yes" for a cyclic group G.

CONJECTURE: Let G be a cyclic group. If $|G|$ is odd, then there exists a $(k, \lambda)$ complete mapping for any k such that $k \mid \lambda(|G| - 1)$. If $|G|$ is even, then there exists a $(k, \lambda)$ near complete mapping for any k such that $k \mid \lambda(|G| + 1)$.

Note that in the case of $\lambda = 1$ and $|G|$ odd, it was also conjectured to be true [1] for a cyclic group G. Some other related problems can be found in [5].

REFERENCES:

[1] R. J. Friedlander, B. Gordon and P. Tannenbaum; Partitions of groups and complete mappings, Pacific J. of Math., 92 (1981), 282-293.

[2] D. F. Hsu; Cyclic neofields and combinatorial designs, Lecture Note in Math., #824, Springer-Verlag, 1980.

[3] D. F. Hsu and A. D. Keedwell, Generalized complete mappings, neofields, sequenceable groups and blocks designs, I, Pacific J. of Math. V.III, No. 2 (1984), 317-332.

[4] D. F. Hsu, and A. D. Keedwell, Generalized complete mappings, neofields sequenceable groups and block designs, II, to appear in Pacific J. of Math.

[5] A. D. Keedwell, Sequenceable groups, generalized complete mappings, neofields and block designs, Combinatorial Mathematics X (Adelaide, 1982) Lecture Notes in Math. #1036, Springer-Verlag, Berlin, New York (1983), 49-71.

PROBLEM 18. (Gian-Carlo Rota)

Find a good formula for evaluating $\binom{n}{k}$ mod $p^2$, mod $p^3$, etc., where n and k are positive integers and p is a prime.

## ADDENDUM

Several proposers have provided additional information concerning problems submitted to the problem session, including solutions obtained since the Boulder meeting. The following summarizes all information received by the editor as of press time (August 1984):

PROBLEM 1 (S. Yuzvinsky) was proposed independently by P. Frankl (Marseille 1981, Oberwolfach 1982), and is still open.

PROBLEM 4 (R. Stanley) was solved by D. Worley, and the solution appears in his Ph.D. thesis (M.I.T. 1984).

PROBLEM 5 (P. Hanlon, R. Gupta) was solved by J. Stembridge, and the solution will appear in his M.I.T. Ph.D. thesis.

PROBLEM 18 (G.C. Rota) was solved by D. Coppersmith, and the solution will appear in Advances in Mathematics.

ABCDEFGHIJ–AMS–8987654